Fundamentals of Voice-Quality Engineering in Wireless Networks

Network operators are faced with the challenge of maximizing the quality of voice transmissions in wireless communications without impairing speech or data transmission. This book provides a comprehensive survey of voice-quality algorithms and features, interactions, and trade-offs, at the device and system level. It does so from the practitioner rather than the algorithm-designer angle, and this approach makes it unique. The author covers the root cause of the problems and available solutions, as well as methodologies for measuring and quantifying voice quality before and after applying the remedies. The book concludes by discussing interactions between impairments and treatments, trade-offs between solutions and their side-effects, and short-term versus longer-term solutions, using case studies to exemplify the key issues. Avoiding complex mathematics, the book bases its approach on real and sizable field experience backed up by scientific and laboratory analysis. This title is suitable for practitioners in the wireless-communications industry and graduate students in electrical engineering. Further resources for this title, including a range of audio examples, are available on-line at www.cambridge.org/9780521855952.

DR. AVI PERRY, a key contributor to the ITU-T G.168 standard, is currently an independent consultant. He was Vice President at NMS Communications, where he was responsible for core technology, system engineering, competitive analysis, documentation, and technical support. He has also held positions at Lucent Technologies and Bell Laboratories and the Northwestern University Graduate School of Management.

Fundamentals of Voice-Quality Engineering in Wireless Networks

AVI PERRY, PH.D.

CAMBRIDGE
UNIVERSITY PRESS

CAMBRIDGE UNIVERSITY PRESS
Cambridge, New York, Melbourne, Madrid, Cape Town, Singapore, São Paulo

Cambridge University Press
The Edinburgh Building, Cambridge CB2 2RU, UK

Published in the United States of America by Cambridge University Press, New York

www.cambridge.org
Information on this title: www.cambridge.org/9780521855952

© Cambridge University Press 2007

First published 2007

Printed in the United Kingdom at the University Press, Cambridge

A catalog record for this publication is available from the British Library

ISBN-13 978-0-521-85595-2 hardback
ISBN-10 0-521-85595-0 hardback

Contents

Preface   *page* xi
List of abbreviations   xvi

Introduction   1
    Plan of the book   3

Part I   Voice-quality foundations   9

1 **An overview of voice-coding architectures in wireless communications**   11
  1.1   Introduction   11
  1.2   Pulse-code modulation   12
  1.3   Wireless codecs   13
  1.4   Summary   22
  1.5   Transmission media   25
  1.6   Audio illustrations   26

2 **Quantitative assessment of voice quality**   28
  2.1   Introduction   28
  2.2   Measuring and quantifying voice quality   29
  2.3   Relating voice-quality metrics to perceived worth   32
  2.4   Mean opinion scores of wireless codecs   32
  2.5   Standard computerized tools for speech-quality assessment   37

Part II   Applications   49

3 **Electrical echo and echo cancelation**   51
  3.1   Electrical echo   51
  3.2   Echo in wireless and long-distance networks 56
  3.3   Echo control   58
  3.4   Placement in the network   66

3.5   Convergence and adaptation processes   69
3.6   Tail delay and its impact on performance   72
3.7   Far-end echo cancelation   73
3.8   Background music – music on hold   78
3.9   Voice-response systems – barging   79
3.10   Hot signals   79
3.11   Voice-band data and fax   80
3.12   Signaling systems nos. 5, 6, and 7   81
3.13   Conclusions   81

4   **Acoustic echo and its control**   83
4.1   Introduction   83
4.2   Round-trip delay and echo dispersion   85
4.3   Reasons for perceiving acoustic echo in wireless
       communications   87
4.4   Controlling acoustic echo in digital-wireless networks – main
       considerations   87
4.5   Acoustic-echo activity detection   92
4.6   Residual acoustic-echo control in the frequency domain 93
4.7   The impact of comfort-noise matching on the performance
       of acoustic-echo control   96
4.8   Correcting impairments caused by some acoustic-echo control
       algorithms built into mobile handsets   106
4.9   Network topology of acoustic-echo control   108
4.10   Conclusions   111

5   **Noisy ambience, mobility, and noise reduction**   112
5.1   Noise in wireless networks   112
5.2   Introduction to noise reduction   114
5.3   The noise-reduction process   115
5.4   Noise reduction and GSM DTX   124
5.5   Noise reduction and CDMA EVRC and SMV codecs   126
5.6   Noise-reduction level compensation   126
5.7   Noise reduction and signals other than voice   128
5.8   Network topology of noise reduction   130

6   **Speech-level control**   133
6.1   Introduction   133
6.2   Basic signal-level measurements and definitions   133
6.3   Automatic level control (ALC)   135
6.4   Noise compensation (NC)   141
6.5   Combined NC and ALC   143
6.6   High-level compensation   144

Contents

Part III   Wireless architectures   145

7   **Mobile-to-mobile stand-alone VQS architectures and their
    implications for data communications**   147
  7.1   Introduction   147
  7.2   Wireless-network architectures   148
  7.3   Data and fax communications in wireless-digital networks   153
  7.4   TFO (a.k.a. vocoder bypass)   159

8   **The VQS evolution to 3G**   163
  8.1   Introduction   163
  8.2   The 2.5G network voice architecture   166
  8.3   The 3G network architecture   167
  8.4   Conclusions   172

**Part IV   A network operator's guide for selecting,
    appraising, and testing a VQS**   173

9   **A network operator's guide to testing and appraising voice-quality systems**   175
  9.1   Introduction   175
  9.2   Judging audio quality and performance of hybrid-echo cancelation   176
  9.3   Subjective testing   183
  9.4   Judging audio quality and performance of acoustic-echo control   187
  9.5   Judging audio quality and performance of acoustic-noise reduction   191
  9.6   Equipment set-up for lab testing   201
  9.7   Automating MOS improvement estimate due to NR   201
  9.8   Judging audio quality and performance of automatic and
     adaptive level control   203
  9.9   Testing for data compatibility on a fully featured system   206
  9.10   Subjective evaluation through a simulated network   217
  9.11   Live network evaluation   218
  9.12   Subjective listening with voice recordings   218

10   **Service provider's system, management, and delivery requirements**   221
  10.1   Introduction   221
  10.2   Management-systems overview   221
  10.3   Customer support   248
  10.4   Training   248

11   **Making economically sound investment decisions
    concerning voice-quality systems**   249
  11.1   Introduction   249
  11.2   State of the art performance and feature richness   249

11.3 Cost considerations   250
11.4 Network-growth considerations   254
11.5 Operations   254
11.6 Conclusions   254
11.7 Fully featured VQS versus echo-cancelation only   255

Part V   Managing the network   259

12 **Trouble shooting and case studies**   261
12.1 Introduction   261
12.2 List of troubles   261

Part VI   Afterthoughts and some fresh ideas   287

13 **Tracer probe**   289
13.1 Introduction   289
13.2 Case study   289
13.3 Description of tracer probe   291
13.4 Requirements and options   293
13.5 Conclusion   293

14 **My sound**   294
14.1 Introduction   294
14.2 Controlling sound flavor   295
14.3 Architecture   296

15 **A procedure for evaluating and contrasting new wireless codecs'
    performance with and without VQS**   299
15.1 Introduction   299
15.2 Procedure   299

16 **The theory of sleep**   301

Part VII   Recordings   303

17 **Audio illustrations**   305
17.1 Codec illustrations   305
17.2 Echo as a function of delay and ERL   307
17.3 Noise types   308
17.4 Acoustic-echo control and comfort-noise matching   308
17.5 Before and after VQS   309

17.6   Trouble shooting   310
17.7   Ring-back and DTMF signals: example test tools   311

Glossary of common voice-quality systems terminology   313
Brief summary of echo cancelation and VQS major standards   330
Brief summary of key voice-quality assessment standards   332
Bibliography   333
Index   351

Preface

Since the early 1990s and for the 13 years that followed I was in charge of specifying and managing the design and development of voice-quality enhancement algorithms and systems in Lucent Technologies, and later in NMS Communications. When taking on the task, I was astonished by the secrecy surrounding the make-up of the minute algorithm details that separated the exceptional from the second-rate performers; the ones that elevate the performance of the voice-quality algorithms to significant heights versus their counterparts, which adhere to standard textbook and public-domain prescriptions.

And although I found out that there was no lack of technical material addressing the subject of voice quality, I learned that the many books, articles, and papers devoted to the subject focused on the math, while steering clear of the practical heuristics that are the backbone of any successful implementation. Their analysis was mostly academic – deep, technically precise, but nonetheless narrow. It addressed a single aspect of voice quality, whether it was electrical or hybrid-echo cancelation, non-linear acoustic-echo control and suppression, adaptive and automatic gain control, or noise reduction, rather than an interactive blend. It was intended for the few subject experts and algorithm designers rather than the user, the product manager, the troubleshooter, the marketing and sales personnel, and – most importantly – those responsible for making decisions affecting quality of service and investment in voice-quality products.

It is a fact of life that many voice-quality troubles, by and large, comprise a sum of specific impairment categories or unanticipated interactions among multiple applications. For example: treatment of electrical or acoustic echo must take into account the potential presence and impact of accompanying noise and the presence of level control that may amplify imperfections. Tandem treatments of a single impairment may bring about unwanted side effects, which could develop into more of a problem when the impairments and their associated remedies embrace a blend of multiple categories. Noise on one end of the communications channel and echo on the other may interact in a non-textbook fashion, and may spit out deviant impairments. Replacing a traditional analog or a standard PCM PBX with a new high-tech VoIP type may result in a sudden burst of fresh complaints concerning echo that would not fade away. An upgrade to an existing voice-enhancement unit intended to improve particular voice quality features may trigger a rush of customer complaints concerning the customers' inability to communicate with their voice mailboxes, etc.

There are many more real-life examples differing from an academic exercise confined to a single impairment category or a single application. There are many more individuals, whose job responsibilities require an ability to diagnose rather than design, decide rather than develop, convince rather than fabricate.

I wrote this book to fill a void.

This book is unique in the sense that it integrates all aspects of voice quality into a single objective. Solutions to voice-quality impairments such as echo, acoustic echo, noise, and improper level are all related to each other. Practical algorithms must address interactions and mutual trade-offs before they can claim success. One may not treat acoustic echo without considering noise and noise reduction, comfort noise matching, and level optimization simultaneously. This book is focused on expediency while replacing the math with logical and practical reasoning and treating it only as a necessary means rather than an end. It bases its approach on real and sizable field experience backed up by scientific and laboratory analysis.

Furthermore, this book is unique in what it does not do. It does not drive deep into the math that makes the algorithms perform. It does not scare away those who view Greek letters, long equations, differential calculus, and Fourier transforms as programmers and designers' cuisine. It does, however, employ logical reasoning, intuition, and real-life ordinary examples to drive a point, and to provide a comprehensive understanding of how it all works, why it does not, and what spice can take it up a notch.

This book is aimed at the network administrator, the product manager, the trouble-shooter, the marketing and sales personnel, and those responsible for making decisions affecting investment in voice-quality products. It is intended to help them expand their knowledge, better their understanding of their own craft, and propose effective solutions without yielding to those math wizards who can decipher the magic Σ and convert it to a DSP code.

This book addresses the big picture of what it takes to communicate clearly at maximum fidelity without impairing speech or data transmission. It provides a comprehensive survey of voice-quality algorithms and features, interactions and trade-offs. It does so from the practitioner rather than the algorithm-designer angle, and this approach makes it unique.

The great difference between algorithm designers and practitioners is inherent in the definition of the problem to be solved. The algorithm designer starts the process by defining his or her scope in detail. The problem triggering the need for a solution is well defined. Its parameters are known and understood, and the focus is confined to a relatively narrow range. On the other hand, the practitioner is faced with voice-quality issues that may be stimulated by a combination of problems. His or her main issue is diagnosis and scope definition. The practitioner must resort to root-cause analysis that may infer a diagnosis consisting of interactions among a variety of impairments. The solution must, therefore, be more wide-ranging while keeping its focus on overall voice quality rather than one of its specific facets.

This book addresses common voice-quality problems in wireless communications, their root cause, available solutions, remedies, and methodologies for measuring and quantifying voice-quality scores, before and after applying these remedies. The concluding segments put it all together by addressing interactions among impairments and treatments, trade-offs between remedies and their corresponding side effects, and short-term versus longer-term solutions. The methodology employed for the closing analysis rides its course on top of trouble shooting and case studies. This approach provides a proper platform for scrutinizing symptoms that may be realized through combinations of impairments, improper application of remedies, or a blend of both.

The common voice-quality issues contended with in the book encompasses the phenomena of electrical and acoustic echo and echo cancelation, noisy ambience and noise reduction, and mismatched or improper signal levels and level optimization. The issues are analyzed as they relate to the specific codec environment they play in. CDMA and GSM wireless codecs possess different voice-quality features, their relative performance varies in different ways in the presence of noise, and similar treatments have dissimilar effects on the consequent voice quality. This book brings to light these particular parameters.

In fact, the literature is filled with excellent analytical and mathematically inclined texts. The majority of these books deal with a specific aspect or application such as noise reduction, non-linear acoustic echo, and VoIP. Unlike this work, which is intended for a relatively larger audience adopting a more encompassing application domain, the focus of most other texts is relatively narrow and comparatively deep. One more on that level may only address an audience which is already surrounded by plenty of stimulating material.

Books dealing with noise reduction are intended for the few subject experts and algorithm designers. They place heavy emphasis on the math, and they do a good job analyzing and dissecting the algorithms details.

Probably one of the best books on noise reduction is *Noise Reduction in Speech Applications* edited by Gillian Davis (managing director, Noise Cancelation Technologies Ltd.).[1] Before launching into the noise-reduction algorithm details, it provides a short introduction on digital-signal processing and adaptive filtering. It also includes descriptions of systems aspects; digital algorithms and implementations of single-channel speech enhancers, microphone arrays, and echo cancelers; and applications in the more specialized areas of speech recognition, internet telephony, and digital hearing aids.

Other books on the subject of noise reduction dig deep into the math and are aimed at the experts in the specific subject matter.[2]

Echo cancelation is treated in the common literature largely as a signal processing exercise. Electrical-echo cancelation comprises two processes – linear and non-linear. The linear process consists of an elegant mathematical procedure implemented via numerical iterative least-squares algorithms. The non-linear part is based on heuristics and tricks of the trade. In fact, some heuristics measures are used to improve the linear portion as well. The next two examples provide a representative sample for the literature in this area. Their analysis is mostly academic. They focus on the math underlying the linear portion of the algorithm, while steering clear of the practical heuristics that are the backbone of any successful implementation.

[1]   Gillian M. Davis (ed.), *Noise Reduction in Speech Applications*, CRC Press, (2002).
[2]   S. V. Vaseghi, *Advanced Signal Processing and Digital Noise Reduction*, John Wiley and Teubner, (1996).
       E. Beranek and L. Leo, *Noise Reduction*, McGraw Hill, (1960).

Electrical-echo cancelation is not addressed in most of the literature as a comprehensive application. One typical (and probably the best) example is the signal processing book *Digital Signal Processing: Principles, Algorithms and Applications* by John G. Proakis and Dimitris Manolakis.[3] It treats the subject as a side effect to sampling and reconstruction of signals. The analysis focuses on the linear part, and the approach is thorough yet theoretical. It does not address the practicality of canceling echo in real applications, although it does do a good job in providing the mathematical analysis of some of the application aspects.

The book *Least-Mean-Square Adaptive Filters (Adaptive and Learning Systems for Signal Processing, Communications and Control Series)* edited by Simon Haykin and B. Widrow[4] – as well as *Adaptive Filter Theory* by Simon Haykin,[5] which is a reference book for adaptive filters, a collection of papers by various authors – focus on the mathematical properties of the linear portion of the echo-cancelation algorithm. They do not treat the application as a realistic problem in search of a practical solution, but rather as a mathematical exercise seeking an academic illumination.

Acoustic-echo control in wireless communications is covered in *Advances in Network and Acoustic Echo Cancellation* by Jacob Benesty, Thomas Gansler, Denis R. Morgan, M. Mohan Sondhi, and Steven L. Gay.[6]

Once more, books on the subject of echo cancelation[7] provide a thorough mathematical analysis, and mathematical algorithms ranging from filtering via fast Fourier transform (FFT) in Schobben's book to a profound mathematical analysis in Borys' book concerning discrete Volterra series in the Z domain, non-linear filters etc. These mathematical analyses are interesting but they do not reflect the present issues and the latest implementations and treatments of acoustic echo in wireless networks. They are intended for a very small audience whose focus is merely academic.

Voice-quality enhancements and echo cancelation (EC) solutions have been evolving and have become essential elements in virtually every type of telecommunications network. This upward growth trajectory is expected to swell far beyond singular applications. Both voice-quality systems (VQS) and echo-cancelation equipment elements have penetrated all the blueprints of impending network infrastructures. The fastest growing segments of the industry, voice-over-packet (VOP) and 2G–3G wireless communications, continue to be key drivers posting fresh challenges and fostering innovative solutions that deliver high-performance voice communications.

One more aspect that makes this work unique is its accompanying audio illustrations. If a picture is worth a thousand words when describing a scene, then an audio illustration is

[3] John G. Proakis and Dimitris G. Manolakis, *Digital Signal Processing Principles, Algorithms, and Applications*, Prentice Hall, 1996.

[4] Simon Haykin and Bernard Widrow, *Least-Mean-Square Adaptive Filters*, John Wiley & Sons, (2003).

[5] Simon Haykin, *Adaptive Filter Theory*, 3rd ed., Prentice Hall, (1997).

[6] J. Benesty, T. Gansler, D. R. Morgan, M. M. Sondhi, and S. L. Gay, *Advances in Network and Acoustic Echo Cancellation*, Springer, (2001).

[7] Andrej Borys, *Nonlinear Aspects of Telecommunications: Discrete Volterra Series and Nonlinear Echo Cancellation*, CRC Press, (2000).

Daniel W. E. Schobben, *Real-Time Adaptive Concepts in Acoustics: Blind Signal Separation and Multichannel Echo Cancellation*, Kluwer Academic, (2001).

worth no less when giving details of sound effects. Voice quality issues and solutions are best illuminated by way of audio demonstrations. The book would not be complete without accompanying audio files (www.cambridge.org/9780521855952) presenting specific cases pointed out within the text. These examples clarify and give life to the dry analysis while driving in convincing lines of reasoning.

My dear friend and colleague, Dr. Cristian Hera, a bona fide expert in the field of voice quality, put most of the audio recordings together. He also reviewed the manuscript and his valuable suggestions, amendments, and contributions greatly enhanced its merit.

The experience and know-how expounded in the book were accrued during my engineering-management tenure at Lucent Technologies and during my role as vice president of voice-quality systems technology in NMS Communications. During that time, I shared analysis, research, field experience, laboratory evaluations, and simulations, with my kitchen-cabinet team of expert engineers, Andy Stenard, Cristian Hera, Joseph Papa, and Neil Dennis. The time we spent together enriched and expanded my grasp of the subject matter. I am also indebted to another team member, Ron Tegethoff, for letting me include descriptions of his tracer probe and adaptive far-end echo-cancelation algorithms (see Chapter 13 and Chapter 3, respectively).

During the writing of the book, I benefited greatly from discussions I had with Bob Reeves, Rapporteur of the ITU study group 16, Question 6, who has been leading the effort pertaining to the publication of the latest issues of G.168, G.169 and G.160. Furthermore, Bob provided me with the plots used in Figures 9.2–9.6, for which I am grateful.

And finally, it has been my dear wife, Shelly, without whose pressing I would never have brought this work to a close. She has been the reason the writing has reached the editor's and publisher's desks.

She won. I dedicate this work to her.

If you ever thought (for even just a moment) that EC and VQS were "out-of-style equipment," "passé technologies," or "old economy paraphernalia," then reading this book and listening to the recordings will probably make you do an about-face and set your perspectives on a telling expedition. Try it and see!

Abbreviations

1XMC	1X multi channel
2.5G	2nd $+ \frac{1}{2}$ generation (enhanced 2G wireless network)
2G	2nd generation (wireless network)
3G	3rd generation (wireless network)
3GPP	3rd generation (wireless standards) partnership project
3XMC	3X multi channel
8-PSK	8-(state)phase-shift keying (modulation)
A/D	analog-to-digital converter
AbS	analysis by synthesis
ACELP	algebraic code-excited linear prediction
ACR	absolute category rating
ADPCM	adaptive differential pulse-code modulation
AEC	acoustic-echo control
AGC	automatic gain control (same as ALC)
AIUR	air-interface user rate
a.k.a.	also known as
ALC	automatic level control
AMI	alternate mark inversion
AMR	adaptive multi-rate
ANSI	American National Standards Institute
ARPU	average rate per user
ATM	asynchronous transfer mode
ATME	automatic test and measurement equipment
BER	bit error rate
BIST	built-in self-test
BSC	base-station controller
BSS	base-station subsystem
BTS	base transceiver station
CC	continuity check
CCITT	International Telephone and Telegraph Consultative Committee
CCS	common-channel signaling
CDMA	code division multiple access
CED	called-station identification
CELP	code-excited linear prediction

C/I	carrier to interference
CM	configuration management
CN	core network
CO	central office
Codec	encoder/decoder
CPE	customer-premises equipment
CPU	central processing unit
CSD	circuit-switched data
CSI	called-subscriber identification
CSS	composite-source signal
D/A	digital-to-analogue converter
D-AMP	digital advanced mobile phone service
dB	decibel
dBov	dB to overload point
DCME	digital circuit multiplication equipment
DCS	digital command signal
DEC	digital echo canceler
DIS	digital identification signal
DMOS	degradation mean-opinion score
DP	delay processor
DS0	digital-signal level 0
DS1	digital-signal level 1
DSL	digital subscriber line
DSN	delta signal-to-noise
DSP	digital signal processor/processing
DTDT	double talk-detection threshold
DTE	data terminal equipment
DTMF	dual-tone multi-frequency
DTX	discontinuous transmission
EAD	echo-activity detection
EC	echo canceler, echo cancelation
EDGE	enhanced data rates for GSM evolution
EEC	electrical-echo cancelation
EFR	enhanced full rate
EMC	electromagnetic compatibility
EMI	electromagnetic interference
EMS	element-management system
ERL	echo-return loss
ERLE	echo-return loss enhancement
ERP	ear reference point
ES	errored seconds; echo suppression
ESD	electromagnetic compatibility
ESF	extended super frame

EVRC	enhanced variable-rate codec
eX-CELP	extended code-excited linear prediction
Fax	facsimile
FD	flat delay
FDM	frequency-division multiplexing
FFT	fast Fourier transform
FIR	finite impulse response
FLR	far-end level regulator
FM	fault management
FNUR	fixed network user rate
FR	full rate
GMSC	gateway MSC
GPRS	general packet-radio service
GSGSN	gateway SGSN
GSM	Global-systems mobile
GUI	graphical user interface
HATS	head and torso simulator
HDLC	high-level data-link control
HEC	hybrid-echo canceler
HLC	high-level compensation
HLR	home-location register
HR	half rate
HSCSD	high-speed circuit-switched data
Hz	hertz
IEC	International Electro-technical Commission
IFFT	inverse fast-Fourier transforms
IMT-20	international mobile telecommunications – 2000
IP	internet protocol
IRS	intermediate reference system
IS	international standard
ISC	international switching center
ISDN	integrated-services digital network
ISLP	inter-system link protocol
ITU	International Telecommunication Union
IWF	inter-working function
IXC	inter-exchange carrier
kbps	kilobits per second
LAN	local-area network
LBR	low bit rate
LCP	linear-convolution processor
LEC	local-exchange carrier
LFV	line-format violation
LHD	long-haul delay

LPC	linear prediction coefficients
LQ	listening quality
LTP	long-term prediction
MCC	mobile-control center
MG	media gateway (CDMA-2000)
MGW	media gateway (3GPP)
MIPS	million instructions per second
MJ	major alarm
MN	minor alarm
MNB	measuring normalizing blocks
Modem	modulator–demodulator
MOS	mean-opinion score
MOS-LQO	mean-opinion score – listening quality objective
MOS-LQS	mean-opinion score – listening quality subjective
ms	millisecond
MS	mobile station (2G)
MSC	mobile-switching center
MTBF	mean time between failures
MUX	multiplexer
NC	noise compensation
NEBS	network equipment-building system
NEST	near-end speech threshold
NHE	near-end hybrid equalizer
NLMS	normalized least mean square
NLP	non-linear processor; non-linear processing
NM	noise matching
NPLR	noise-power-level reduction
NR	noise reduction
NRLC	noise reduction with level compensation
NS	noise suppression
NSF	non-standard facilities
NSS	non-standard set-up
OAM&P	operation, administration, maintenance, and provisioning
OC-3	optical carrier signal level 3
OMC	operation and maintenance control
OS	operating system
OSS	operations support system
PAMS	perceptual-analysis measurement system
PBX	private branch exchange (a small to medium switch)
PCC	per-call control
PCM	pulse-code modulation
PCME	packet circuit multiplication equipment
PCS	personal communication service

PCU	packet control unit
PDC	personal digital cellular
PESQ	perceptual evaluation of speech quality
PMG	packet-media gateway
PNLMS	proportionate normalized least mean square
POTS	plain old telephone service
PSI-CELP	pitch synchronous innovation – code-excited linear prediction
PSQM	perceptual speech-quality measure
PSTN	public switched-telephone network
PVC	permanent virtual circuit
QoS	quality of service
RAI	remote-alarm indication
RAN	radio-access network
RAS	remote-access system
RCELP	relaxed code-excited linear prediction
R_{in}	receive in (port of the VQS)
RMS	root mean square: also remote-management system
RNC	radio-network controller
R_{out}	receive out (port of the VQS)
RPE	regular pulse excited
RT	remote terminal
SDH	synchronous digital hierarchy
SF	superframe
SGSN	service GPRS support node
S_{in}	send in (port of the VQS)
SMV	selectable-mode vocoder
SNR	signal-to-noise ratio
SNRI	signal-to-noise ratio improvement
SONET	synchronous optical network
S_{out}	send out (port of the VQS)
SPC	signal-processing control
SPL	sound-pressure level
SPLR	signal power-level reduction
SPT	signal-processing terminal
STM-1	synchronous transport mode level 1
STS-1	synchronous transport signal level
TBD	to be determined
TC	transcoder (3GPP)
TCE	transcoder equipment
TCH	traffic channel
TDM/P	time-division multiple interface
TDMA	time-division multiple access
TFO	tandem-free operation

TNLR	total noise-level reduction
TRAU	transcoder unit (2G/3G)
TrFO	transcoder-free operation
TSI	transmitting-subscriber identification: time-slot interchange
UMTS	universal mobile-telecommunication system
UTRAN	3G radio-access network
VAD	voice-activity detector
VED	voice-enhancement device
VLR	visitor-location register
VoATM	voice-over-ATM
VoIP	voice-over-IP
VOP	voice-over-packet (ATM or IP)
VOX	voice-operated transmission
VPA	voice-path assurance
VQ	voice quality
VQS	voice-quality system
VSC	voice-signal classifier
VSELP	vector-sum-excited linear prediction
WAN	wide-area network
W-CDMA	wideband-CDMA
Z-bal	balancing impedance

Introduction

In survey after survey potential and actual users of wireless communications indicated that voice quality topped their reasons for selecting a specific service provider. While providers have been well aware of this key component powering their offering, they have not always been certain as to the specific methodology, resolution elements, equipment type, architecture, trade-offs, and rate of return on their particular investment that elevate the perceived voice-quality performance in their network.

It is only natural that voice quality in wireless networks has become a key differentiator among the competing service vendors. Network operators, network infrastructure planners, sales representatives of equipment vendors, their technical and sales support staff, and students of telecommunications seek information and knowledge continually that may help them understand the components of high-fidelity communicated sound.

Throughout the 1990s applications involving voice-quality enhancements, and specifically echo cancelation, have induced fresh inventions, new technology, and startling innovations in the area of enhanced voice performance. The initial echo canceler (EC) product implementations existed for about a decade before a diverse array of voice-quality enhancement realizations emerged to meet the evolving needs of digital wireless communications applications.

Early EC implementations were limited to very long distance (e.g., international) circuit-switched voice and fax applications where echo was perceived (in voice conversations) due to delays associated with signal propagation. The EC application soon expanded beyond strictly very-long-distance applications as further signal processing and dynamic routing along the communications path added delay to end-to-end voice transport. Consequently, EC equipment became a necessity for all long-distance calls (rather than just very-long-distance).

In the late 1980s, AT&T promoted a voice-transmission quality plan called the "zero-mile policy."[1] AT&T installed EC equipment next to each one of their toll switches. As a result, every trunk in the AT&T backbone network was equipped with EC coverage regardless of distance. The "zero-mile policy" made sense because of the dynamic routing architecture that AT&T put into practice, a treatment that essentially removed the correlation between geographic distance and physical wire span. Furthermore, innovations such

---

[1] AT&T introduced "dynamic routing" in the late 1980s. The procedure called for the use of lightly utilized links regardless of distance when routing calls along busy corridors. For example, morning calls between New York and Boston were routed via the West Coast, where traffic was still on its third shift. Late-night calls along the West Coast were routed through the East, taking advantage of the Late Night Show.

as call forwarding and call mobility have rendered geographic distance a weak predictor of signal delay. Consequently, there was a wider deployment of ECs by service providers, who have effectively duplicated the AT&T's experience.

Throughout the 1980s and until the mid nineties, AT&T with its leading and prominent Bell Laboratories, a body packed with Nobel-Prize scientists, continued to be the most innovative leader in the voice-quality arena. In 1991, AT&T revolutionized the notion of what voice-quality enhancement was about when it introduced TrueVoice. This new and innovative application was first to incorporate into their echo-canceler equipment a graphic-equalizer-type manipulation of speech levels by varying the extent of signal amplification across the various frequency bands of the speech impulse. Although the new technology improved voice quality, it did not interact well with certain modes of voice-band data. The new technology produced operational headaches, a fact that led to a slow withdrawal and an eventual suspension of TrueVoice in the mid to late nineties.

In the mid nineties AT&T introduced a subscription-based service – TrueVoice2 – a noise-reduction feature designed to enhance communications over a small cluster of international circuits. TrueVoice2 represented a first, a pioneering effort, intended to enhance voice quality by way of reducing circuit noise.

In the 1990s, mobile telephony introduced a new variety of impairments and challenges affecting the quality of voice communications. The impairments comprised acoustic echo, noisy environments, unstable and unequal signal levels, voice-signal clipping, and issues related to echo cancelation of vocoder compressed voice. The technological focus moved to developing remedies that would handle these new parameters properly. Furthermore, the challenges brought about by the wireless era imposed more stringent performance requirements on electrical-echo cancelation.

The 1990s experienced a vast growth in digital wireless communications. During the first half of the decade, voice-quality considerations other than echo cancelation were confined to low-bit-rate codec performance while hybrid-echo cancelation was delegated to a select group of stand-alone systems providing 8 E1/T1 per shelf with four shelves per bay (or 32 E1/T1) at best. During the second half of the decade, it was gradually acknowledged that mobile handsets failed to follow standards, and acoustic echo was not taken care of at the source. Echo-canceler equipment vendors such as Lucent, Tellabs, and Ditech Communications later on seized the opportunity for added revenues. They started incorporating acoustic-echo control while enhancing the offering with noise reduction and level optimization with their stand-alone echo-canceler systems. The market reacted favorably and voice-quality systems (VQS) embarked on an expandable path for replacing plain echo cancelers.

The commencement of the new millennium witnessed a continuing, yet substantial, advancement in microprocessor technology and in signal-processing software and algorithmic design. These developments spawned a fresh trend in the implementation and delivery of VQS. They allowed for high-density, lower cost per line, innovative products, and systems that could easily be integrated inside larger systems like a mobile switching center or a base-station controller. More and more VQS began selling as switch or base-station features and the stand-alone-system market was reduced to a secondary segment. Ericsson came up with their own integrated echo-canceler version. Nokia purchased echo-canceler technology from Tellabs, only to integrate it in its MSC,[2] Siemens created their version of application-specific integrated circuit (ASIC) echo canceler, and Alcatel acquired echo-canceler software technology from Ditech Communications. Lucent created a voice-quality (VQ) software version and integrated it with their codec software on their mobile-switching center, and Nortel purchased echo-canceler modules from Tellabs, which were inserted and integrated in their switch.

Stand-alone systems continued to hold a quality edge over their integrated cousins, which focused on hybrid-echo cancelation exclusive of acoustic-echo control, noise reduction, and level optimization. These voice-quality applications and a higher-performance echo cancelation were still part of a classy club of customers (or others whose mobile-switching center did not contain any voice-quality features) who preferred the higher voice quality and the operations inertia that history had bestowed upon them. Verizon Wireless, Nextel, T-Mobile USA, and T-Mobile Germany (as of 2004), Japan NTT DoCoMo and KDDI, most Korean and Chinese service providers, as well as numerous others continued to insist on stand-alone voice-quality systems rather than switch integrated voice-quality features.

Voice-quality systems and, more specifically, echo cancelation, have also become a crucial component of VoIP and ATM voice platforms. Still, the implementation is persistently integrated within the packet gateway, and stand-alone voice-quality systems have not been able to take a significant hold in this market segment.

This book regards a voice-quality system as a functional system. Most of the analysis and the descriptions are independent of whether the system is implemented inside a codec, a switch, a base-station controller, or as a stand-alone. The generic nature of the applications may be delivered via any of these, and most of the analysis except for a few sections, which declare themselves as particular to specific implementation, is not implementation specific.

Plan of the book

The book is divided into six major parts.

Part I – *Voice-quality foundations* – opens the discussion in Chapter 1 with an overview of voice-coding architectures in digital wireless networks. It provides an overview of the GSM, TDMA, and CDMA codecs from the early 1990s through the first half of the next decade. It provides a high level analysis of the architecture principles that make low bit-rate codecs effective in dictating the transformation of natural speech to an electronic signal and back into speech. And it articulates in plain language how this transformation alone produces changes in the quality of transmitted voice.

Chapter 1 reviews the major voice codecs, their history, and their relative perceived quality. Voice-coding architectures are the building blocks of transmitted voice. They are the core that shapes the characteristics and quality of transmitted speech. Nevertheless, they are treated in the book only as background to the main subject, which deals with

---

[2]   Nokia added a home-made voice-quality suite on their base-station controller as an optional module later on.

impairments due to transmission architecture and environment, and corresponding remedies that immunize and repair any potential or actual spoil. Since the effectiveness of the various remedies depends on that underlying coding, it is essential that these designs be understood so that remedies can be fine tuned and customized to suit the particular characteristics of the underlying voice architecture.

Quantitative assessment of the perceived voice quality as it relates to a particular codec is postponed to the next chapter. Instead, the presentation conveys a sense of relative performance standing, and how voice quality has been improving over time.

Chapter 2 – *Quantitative assessment of voice quality* – kicks off the presentation with an overview of the standard metrics and methodologies followed by a description of specialized tools employed for obtaining subjective voice quality scores through genuine opinion surveys and via computer modeling emulating people's perceptive evaluation of speech quality. It then relates voice-quality scores obtained from surveys or computer evaluations to the perception of worth. It elaborates on the relationships between opinion scores and the potential return on investment in voice-quality technology. It examines the results of voice-quality studies with reference to the three popular GSM codecs – full rate (FR), enhanced full rate (EFR) and half rate (HR). The presentation includes a discussion of the effect of noise and transmission errors on the relative performance of these codecs.

Part II – *Applications* – opens the discussion in Chapter 3 with an overview of echo in telecommunications networks, its root causes, and its parameters. It follows the presentation with the methods used for controlling electrical echo, including network loss, echo suppression, linear convolution, non-linear processing, and comfort noise injection. The chapter covers the application of echo cancelation in wireless communications. And in view of the fact that today's wireless networks include long-distance circuit-switched VoIP and VoATM infrastructures (specifically as part of third-generation architectures), the chapter covers echo cancelation in long distance and voice-over-packet applications as well.

Chapter 4 – *Acoustic echo and its control* – examines the sources and the reasons for the existence of acoustic echo in wireless networks. It explains how acoustic echo is different from hybrid or electrical echo, and how it can be diagnosed away from its hybrid relative. The chapter follows the description of the impairment by examining the present methods for properly controlling acoustic echo in wireless communications. It also gives details of how background noise makes it more difficult to control acoustic echo properly. It describes those particular impairments that may be set off by some acoustic-echo control algorithms, specifically those built into mobile handsets, and it describes how they can be remedied by proper treatment brought about by means of voice-quality systems (VQS) in the network.

Both electrical echo cancelation and acoustic-echo control require a comfort noise injection feature. Discussion of acoustic-echo control must include methods and algorithms designed to generate suitable comfort noise. Although comfort noise injection is not a stand-alone application resembling the treatment of electrical or acoustic echo, it is, nonetheless, an essential component supporting these applications, and it can make an enormous difference in the perception of how good the voice quality is. It may employ an algorithm chosen from a range of uncomplicated to very sophisticated and, hence, it is essential that the book allocates space to this critical feature.

Chapter 5 is devoted to the subject of noise reduction. Noise reduction is the most complicated feature among the voice-quality assurance class of applications. It also requires a higher-level understanding of mathematics. The discussion, however, substitutes numerous mathematical expressions for intuition, ordinary analogies, and logical reasoning, supplemented by graphical and audio illustrations.

The analysis gets underway with the definition of noise, a definition consistent with the principles and characterization employed by a typical noise-reduction algorithm. It then introduces and explains the mathematical concept of time and frequency domains and the transformation process between the two. Once the reader is armed with the understanding of time and frequency domain representations, the analysis proceeds to a discussion of the noise-estimation process. The presentation then moves ahead to examine the suppression algorithm, which employs the noise-estimation results in its frequency-band attenuation procedures. The next segment contains a presentation covering the final algorithmic steps, which involve scaling and inverse transformation from frequency to time domains.

The next section in Chapter 5 reflects on key potential side effects associated with noise-reduction algorithms including treatment of non-voice signals. It points to key trade-offs and adverse-feature interactions that may occur in various GSM and CDMA networks – a subject that is covered much more thoroughly in Chapter 12 – *Trouble shooting and case studies*. The final section offers an examination of the network topology and placement of the noise-reduction application within it.

Chapter 6 is dedicated to the subject of level-control optimization. The presentation is divided into three parts and an introduction. The presentation starts the ball rolling in the introduction by defining standard methodologies for measuring and quantifying signal levels. The first part deals with automatic level control (ALC), how it works, and its placement within the network. The second part describes the adaptive level control, a.k.a. noise compensation (NC), how it works under different codecs, and where it is placed in the network. The third part describes the high-level compensation procedure along the same outline.

Part III – *Wireless architectures* – is essential to the understanding of the VQS contributions, its relevance to the delivery of high-performance mobile voice-communications service, its compatibility with data services, and its place within the 3G network architecture.

Part III commences with Chapter 7: Mobile-to-mobile stand-alone voice-quality system architectures and their impact on data communications. The chapter reviews the optional placements of the voice-quality system functions relative to the mobile-switching center and the base-station controller, since placement impacts voice performance, applications, deployment cost, and data-detection algorithms. The second part of the chapter presents an analysis of the techniques employed by a voice-quality system when coping with data communications without interfering or blocking its error-free transmission. The analysis includes descriptions of data-detection algorithms based on bit-pattern recognitions. The scope encompasses circuit-switched and high-speed circuit-switched data (CSD and HSCSD respectively) services as well as tandem-free operations (TFO).

Chapter 8 – *The VQS evolution to 3G* – portrays the 2G- and 3G-network topologies and their impact on VQA feasibility and architecture. It provides an evolutionary examination of the process leading to 3G from the popular 2G wireless architecture. It presents a parallel progression of placement and applicability of the voice-quality system that supports the evolving infrastructure.

Part IV offers a practical guide for the service provider. It guides product managers who are faced with the dilemma of whether or not they should invest in voice-quality systems and, if so, what performance tests they ought to implement, and what system capabilities they must require from their supplier.

Chapter 9 – *A network operator guide to testing and appraising voice-quality systems* – describes test and evaluation procedures of performance associated with the various voice-quality applications. The telecommunications-equipment marketplace is filled with a variety of echo-canceler (EC) and voice-quality systems (VQS) promoted by different vendors. Noticeably, the characteristics and performance of these products are not identical. In addition, the non-uniformity and arbitrary judgment that is often introduced into the EC and VQS product-selection process makes the network operator's final decision both risky and error prone. Chapter 9 describes the criteria and standards that are available to facilitate methods for objectively analyzing the benefits of EC and VQA technology when confronted with multiple EC and VQS choices. The scope includes procedures for evaluating performance of electrical (hybrid), acoustic-echo control, noise reduction, and level optimization via objective, subjective, laboratory, and field-testing.

Chapter 10 – *Service-provider's system, management, and delivery requirements* – presents a basic template that may be used by service providers as part of their request for information from vendors. The chapter elaborates on the various elements beyond voice performance that make the voice-quality system easy to manage, and easy to integrate within the operation of the network. The information is rather dry, but highly useful as a reference. Readers of this book who are not interested in the system-engineering requirements may skip this chapter in their pursuit for understanding of the magic that makes voice-quality systems enhance speech communications.

Chapter 11 – *Making economically sound investment decisions concerning voice-quality systems* – discusses key differences between stand-alone and integrated systems. It points to the pros and cons of investing in a full VQS suite versus a minimal set containing a hybrid-echo canceler only, and it closes the chapter with a simple model providing guidelines for assessing return on investment.

Part V – *Managing the network* – presents an extensive account of conditions that must be accommodated for healthy voice communications to come about. The presentation is carried out as if the reader is in charge of running a mobile-switching center where all equipment, including voice-quality assurance gear, has been operating satisfactorily up to that moment in time when the specific trouble at issue has been reported. The specific troubles are analyzed for root cause and remedial actions.

Part VI – *Afterthoughts and some fresh ideas* – concludes the book with a discussion of application ideas and inventions that may be incorporated into forthcoming voice-quality systems.

Chapter 13 presents an application concept referred to as *Tracer probe*. This application may be highly useful in promptly detecting and isolating network troubles without the tedious and laborious effort of the current methods and procedures.

Chapter 14 presents an application concept, *My sound*, that assigns voice-processing control to subscribers of a preferred class, where they may be able to play games with sound effects.

Chapter 15 presents procedures for evaluating voice quality of new codecs and the impact of voice-quality systems on their overall voice quality. The main reason for including this chapter in this part is the fact that even as of today, after many studies and standard works that have been published, there are still skeptics. And they want to verify new codec introduction in their own network, and justify the notion that noise reduction and optimal level control do in fact make a difference in the perception of voice quality. For those who do not take published studies as truth, we outline a way to come up with their own studies.

Chapter 16 presents a concept I named the *Theory of sleep*. The discussion challenges the concept of present system default settings where voice applications are turned on and only disabled when detecting non-voice transmission. It presents a complementary concept, whereas certain voice applications are permanently disabled unless the system detects voice transmission. The writing does not recommend this approach. It simply challenges existing paradigms.

Part VII – *Recordings* – provides a summary of the recordings on the accompanying website, www.cambridge.org/9780521855952. These recordings are intended to highlight and clarify particular aspects of sound and voice communications.

A bibliography provides a list of references relevant to the material discussed throughout the book.

The glossary contains a compendium applied to echo cancelation and voice-quality enhancement technology. Uncertainty, confusion, and misinterpretation often result when acronyms or terminology that are specific to the field of echo cancelation and voice-quality enhancement are used freely. The glossary is a collection of commonly used acronyms and other terms, accompanied with brief descriptions. This material is intended to provide clarity and insight into the unique language of echo cancelation and voice-quality enhancement.

Part I

Voice-quality foundations

Part I reviews the major voice codecs, their history, and their relative perceived quality. Voice-coding architectures are the building blocks of transmitted voice. They are the core that shapes the characteristics and quality of transmitted speech. Nevertheless, they are treated in this book only as background to the main subject, which deals with impairments due to transmission architecture and environment, and their corresponding remedies that immunize and repair any potential or actual spoil. Since the effectiveness of the various remedies depends on that underlying coding, it is essential that these designs be understood so that remedies can be fine tuned and customized to suit the particular characteristics of the underlying voice architecture.

1 An overview of voice-coding architectures in wireless communications

1.1 Introduction

It must have happened to most of us. At some point through our lives we came across someone whom we deemed an "audio nut." (Some of us may have even impersonated that one special character.) That singular person would not listen to music unless it was played back on an exceedingly pricey hi-fi system. He or she actually did hear a titanic disparity in sound quality between what we would be taking pleasure in on a regular basis and what he or she regarded as a minimum acceptable threshold.

In all probability, we might have adopted the same mind-set had we been accustomed to the same high-quality sound system. It is a familiar human condition – once a person lives through luxury, it grows to be incredibly hard getting used to less. How quickly have we forgotten the pleasure we took in watching black-and-white TV, listening to the Beatles on vinyl records, Frank Sinatra on AM radio, etc. But hey, we have experienced better sound quality and, thus, we refuse to look back.

Wireless telecommunications is entering its third generation (3G). Infancy started with analog communications. It developed through childhood in the form of GSM and CDMA, and has crossed the threshold to puberty with cdma2000 and W-CDMA – its third generation. Voice quality in wireless telecommunications is still young and looking up to adolescence, but technology has advanced appreciably, and most of us have been content with its voice performance. We got mobility, connectivity, and voice performance that had never been better, except of course, on most wireline connections. But we were forgiving. We did not notice the relative degradation in voice quality. The little price paid was well worth the mobility.

Why then, why is voice quality over wireline networks – the older, almost boring, plain old telephone service (POTS) – so much better than the newer, cooler, wireless mobile-phone service? Why is there more distortion, more delay, more occasional echo, different noises that come and go, plenty of speech clipping, noise clipping, and voice fading, that we have been putting up with just to gain mobility? Why does it have to be this way? Or does it?

This chapter provides a partial rejoinder to the questions above. It provides an overview of the transmission plans and coding schemes that command the transformation of natural speech to an electronic signal and back into speech in wireless networks. It explains in simple terms how this transformation alone produces changes in the quality of transmitted voice. It focuses on the coding and decoding methods that

11

transform sound into a binary code and the network infrastructure that carries this digital signal to its final destination where it is transformed back into a close relative (rather than a perfect replica) of the original sound. It also addresses the transmission media that adds a variety of distortions to the original signal before it arrives at its final destination.

1.2   Pulse-code modulation

To date, pulse-code modulation (PCM)[1] and the PSTN G.711 ITU standard[2] have been the most common digital-coding scheme employed in wireline telecom networks worldwide. Voice quality associated with PCM is regarded as "toll quality" and is judged to be unsurpassed in narrowband telecommunications. Although PCM exhibits a degraded voice quality relative to the natural voice, it is perceived to be rather fitting for the application of speech transport over the wires.

Pulse-code modulation occupies a band of 0 to 4 kHz. Considering that the young human ear is sensitive to frequencies ranging from 16 Hz to 20 kHz, the bandwidth allocated to a telephone channel seems to be spectrally depriving. Nonetheless, since the principal body of speech signals (energy plus emotion) dwell in the $\sim$100 Hz to 4 kHz band of frequencies, the 4 kHz channel provides a dependable nominal speech channel. (Musicians may not welcome listening to a Beethoven symphony over that channel. Some of the musical instruments may sound heavily synthesized and somewhat distorted. Speech, on the other hand, is not as spectrally rich as a symphonic performance; neither does it have the benefit of a comparable dynamic range.)

When the band-limited nominal 4 kHz channel is sampled at a rate equal to twice the highest frequency, i.e., $(4000 \times 2 =)$ 8000 samples per second, then the sample takes account of the entire information contained in the sampled signal. Since each sample comprises 8 bits of information, the total bandwidth required for representing an analog signal in its PCM form is (8000 samples $\times$ 8 bits =) 64 kbps.[3]

Pulse-code modulation is a speech-coding method classified as waveform coding. This class of codecs employs extensive sampling and linear quantization representing the original analog signal in a digital form. The delay associated with PCM coding is negligible and is equal to its sampling rate, i.e., 0.125 milliseconds.[4] A round trip between two wireline (PSTN) phones involves one code/decode operation. Accordingly, the delay attributed to the PCM coding/decoding process is 0.25 ms. The significant portion of the delay experienced in wireline communications is, by and large, the result attributed to signal propagation over long distances.

[1]   K. W. Cattermole, *Principles of Pulse Code Modulation*, Elsevier, (1969).
[2]   ITU-T Recommendation G.711, *Pulse Code Modulation (PCM) of Voice Frequencies*, (1988).
[3]   As a reference point, it should be noted that a 22 kHz, 16-bit mono music sample requires a data rate of 44 100 samples per second.
[4]   8000 samples per second produce a single frame. The sampling frequency is 1/8000 seconds = 0.125 ms = 125 μs.

1.3 Wireless codecs

A rate of 64 kbps is suitable for wireline telecommunications where capacity is constrained in the short term only by the amount of wire or fiber that is buried under the surface. In contrast, wireless communications is accessed through the air, and the available spectrum is permanently unyielding, and has always been exceedingly limited. Allocating 64 kbps of expensive spectrum to each channel is without doubt uneconomical and impractical. Consequently, the technology afforded to supporting heavy traffic over a wireless access channel resorted to codec technology that traded off extensive sampling for sophisticated algorithms requiring computer power and improved digital signal-processing technology. Wireless codec technologies utilize perceptual irrelevancy in the speech signal by designing more efficient quantization algorithms and intelligent adaptive linear-prediction schemes. These take advantage of the high short-term correlation between consecutive speech frames consequent of vocal-tract anatomy, and of their long-term correlation – thanks to the prominent periodicity inherent in human speech. Perceptual irrelevancy is realized by the fact that the human threshold of audibility is relatively narrow, and is limited to a 16 kHz range. Furthermore, sensitivity to changes in frequency and intensity are dependent on frequency and duration. Consequently, irrelevancy is taken into account by employing quantization techniques that leave behind some valueless (in the sense that one may not benefit from) information about the signal, while at the same time, lack of sensitivity to small changes in frequency and intensity gives rise to prediction as a steadfast means for generating the next speech frame. Wireless codecs bring into play a class referred to as analysis by synthesis (AbS) codecs, which merges a linear prediction scheme that models properties of the vocal tract with an adaptive excitation signal chosen via an algorithm minimizing the error (using the least-squares algorithm) between the input speech and the reconstructed version. The linear-prediction part constructs the next speech sample by exercising linear combinations of the preceding speech samples. The mechanism involves splitting the input speech into small frames.[5] Each frame is then subjected to an excitation signal that is chosen by using vocal tract properties that are fine tuned by employing coefficients that provide the best fit to the original waveform. This approach is known as linear predictive coding (LPC).[6]

Linear predictive coding is a short-term predictive scheme that generates the next speech sample by resorting to a linear combination of the preceding 16 samples. An enhancement to this algorithm contains a long-term prediction (LTP)[7] scheme. Long-term prediction employs from between 20 and 120 samples apart from the predicted one. The predictive coefficients are chosen by the least squares method providing the greatest correlation to the actual speech sample.

The AbS approach utilizes a closed-loop algorithm whereas the signal driving the speech-reproduction engine is also employed in determining the best-fit parameter values, which drive the next reproduction run. The encoder analyzes the input speech by

---

[5] A typical frame is 20 ms long.
[6] Brian Douglas, *Voice Encoding Methods for Digital Wireless Communications Systems*, Southern Methodist University, EE6302, Section 324, (Fall 1997).
[7] Andreas Spanias, Speech coding: a tutorial review, *Proceedings of the IEEE*, **82**: 10 (1994), 1541–1582.

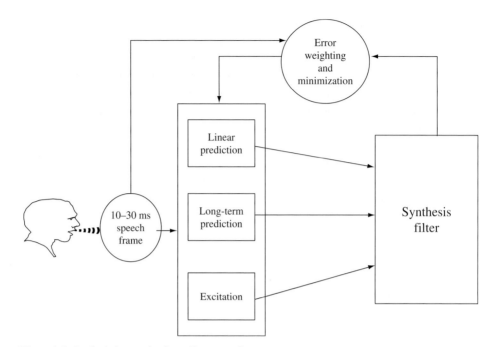

Figure 1.1 Analysis by synthesis coding procedure.

synthesizing many different approximations to it. The encoder output consists of information representing the synthesis-filter parameters and the excitation driving it. The decoder reconstructs the speech by passing the excitation through the synthesis filter. The encoding procedure is illustrated in Figure 1.1.

A key distinction among the various wireless codecs is the method used for the pulse excitation. The computational effort associated with passing each and every non-zero pulse for every excitation speech frame through the synthesis filter is considerably large. Accordingly, excitation procedures incorporate a variety of intelligent inferences in addition to a condensed number of pulses per millisecond.

One of the first to be introduced in GSM (phase I) was the regular pulse-excited (RPE) coder. The RPE uses uniform spacing between pulses. The uniformity eliminates the need for the encoder to locate the position of any pulse beyond the first one. The RPE distinguishes between voiced and unvoiced signals. When the signal is classified as unvoiced, the RPE ceases from generating periodic pulses. Its pulsing becomes random, and it corresponds to the functioning of the unvoiced signal.

The RPE-LTP is the basis for the GSM full-rate (FR) codec.[8] It was first implemented on digital-signal processors (DSPs) in the early 1990s. At that time, DSP technology

---

8   GSM 06.01, *Digital Cellular Telecommunications System (Phase 2+); Full Rate Speech; Processing Functions.*
    GSM 06.10, *Digital Cellular Telecommunications System (Phase 2+); Full Rate Speech; Transcoding.*
    GSM 06.11, *Digital Cellular Telecommunications System (Phase 2+); Full Rate Speech; Substitution and Muting of Lost Frames for Full Rate Speech Channels.*
    GSM 06.12, *Digital Cellular Telecommunications System (Phase 2+); Full Rate Speech; Comfort Noise Aspect for Full Rate Speech Traffic Channels.*

limited the implementation to a computationally efficient method bearing reasonable voice quality by means of practical computational effort. Still, the codec algorithm delay experienced by the RPE-LTP is about 40 ms, and the codec rate is 13 kbps.

The RPE-LTP approach is not suitable for codec rates below 10 kbps.

In the mid eighties Schroeder and Atal[9] proposed a coder concept that employs a codebook for generating the pulse excitation. The class of codec algorithms that resulted from the proposal has been referred to as code-excited linear prediction (CELP) coding. Using a codebook (look-up table) for constructing the finest match contributes to a reduction in algorithm complexity and the required bandwidth. In general, accuracy of speech reproduction for CELP-type algorithms improves with more extensive codebook size and codec rate.

In the mid 1990s, Qualcomm introduced an 8 kbps CELP (QCELP-8) codec[10] supporting CDMA communications. The transpired voice performance registered below acceptable levels, and three years later Qualcomm introduced its popular 13 kbps CELP (QCELP-13) codec,[11] which became a North American standard for CDMA at the expense of lower utilization of spectrum.

The QCELP-13 is a variable-rate codec. It supports a full rate (rate 1) at 13.3 kbps, half rate (rate $\frac{1}{2}$) at 6.2 kbps, a quarter rate (rate $\frac{1}{4}$) at 2.7 kbps, and rate $\frac{1}{8}$ at 1.0 kbps. The algorithm determines the rate on a frame-by-frame basis. The rate assessment and decision are based on the voice activity level. The process involves two phases. In the first phase, the frame is classified either as active speech, noise, or pauses. In the second phase the active speech is segmented into voiced speech (vowel sounds) or unvoiced speech (consonants). Rate 1 is used for transitional frames, or frames with poor periodicity, rate $\frac{1}{2}$ is used for stationary frames or frames with well-defined periodicity, rate $\frac{1}{4}$ is used for unvoiced speech, and rate $\frac{1}{8}$ is used for noise or pauses. Taken as a whole, the QCELP-13 average bit rate is slightly lower than 13.3 kbps, while its voice quality is almost equivalent to the GSM RPE-LTP full rate. The codec algorithmic delay is less than the RPE-LTP and it approaches 30 ms.

Although CELP speech coders deliver a reasonably fine voice quality at low bit rates, they required significant computational efforts. In 1990 Gerson and Jasiuk[12] proposed a modified version of the CELP they called vector-sum-excited linear prediction (VSELP) coder. This enhances the codebook search procedure by organizing frequently occurring speech combinations closer together. It employs three excitation sources, one of which is adaptive, and it generates a combined excitation sequence to drive the synthesis filter.

GSM 06.31, *Digital Cellular Telecommunications System (Phase 2+); Full Rate Speech; Discontinuous Transmission (DTX) for Full Rate Speech Traffic Channels*.
See www.etsi.org/.

[9] M. R. Schroeder and B. Atal, *Code-Excited Linear Prediction (CELP): High Quality Speech at Very Low Bit Rates*, Proceedings ICASSP-85, Tampa (1985), p. 937.

[10] Youngnam Han, Hang Gu Bahk, and Seungtaik Yang, CDMA mobile system overview: Introduction, background, and system concepts, *ETRI Journal*, **19**: 3 (1997), 83–97.

[11] Spanias, *Speech Coding*, p. 1541.

[12] I. Gerson and M. Jasiuk, Vector sum excited linear prediction (VSELP) speech coding at 8 kbits/s, Proceedings ICASSP-90, New Mexico (Apr. 1990), pp. 461–464.

Vector-sum-excited linear prediction is used in the GSM half rate,[13] North American D-AMPS full rate,[14] and the Japanese PDC full rate.[15] The GSM half rate utilizes a codec rate of 5.6 kbps. The D-AMPS TDMA TIA/EIA IS-85 full rate utilizes a codec rate of 8 kbps. The Japanese full rate PDC RCR-STD-27 utilizes a codec rate of 6.7 kbps. The VSELP codec algorithmic delay is approximately 30 ms.

Under quiet conditions, the transmission quality of the above VSELP codecs is almost equivalent to the GSM full rate. However, noisy environments tend to diminish the voice quality of the VSELP GSM half-rate codec more rapidly than any of the other GSM codecs.[16]

In 1991, NTT DoCoMo was looking to run a 3.45 kbps half-rate codec in a single digital signal processor while preserving the same voice quality of the 6.7 kbps full-rate VSELP defined in the same standard. They contributed to a Japanese standard RCR-STD-27 that, among other requirements, defined the pitch synchronous innovation – code-excited linear-prediction (PSI-CELP) half-rate codec.[17] PSI-CELP's main deviation from CELP is its non-conventional random-excitation vectors. It resorts to having the random-excitation vectors contain pitch periodicity by repeating stored random vectors and by using an adaptive codebook. During silent, unvoiced, and transient frames, the coder stops using the adaptive codebook and switches to fixed random codebooks.

As digital-signal processing power got better in the mid 1990s, codec algorithms could afford a more intensive processing muscle, and accordingly, in 1996 an algebraic CELP (ACELP)[18] procedure was adopted for the GSM enhanced full-rate (EFR) codec. The ACELP reduces the complexity of the codebook search and removes the need for fixed codebook storage at both the encoder and decoder. It trades this off by resorting to a more exhaustive series of nested loop procedures, which determines critical components of the excitation pattern algebraically.

Algebraic CELP procedures are widely used in a variety of applications. The codec rate varies considerably among these implementations. The EFR ACELP codec utilizes 12.2 kbps over a standard GSM full-rate channel. The EFR extra bits (the FR uses 13 kbps) are used for additional error checking of the most important codec parameters. These additional error-management capabilities provide for better quality signals in the presence of increased interference where lost parameters may be recovered. The quality

---

[13]   GSM 06.02, *Digital Cellular Telecommunications System (Phase 2+); Half Rate Speech; Half Rate Speech Processing Functions.*
       GSM 06.06, *Digital Cellular Telecommunications System (Phase 2+); Half Rate Speech; ANSI-C Code for the GSM Half Rate Speech Codec.*
       See www.etsi.org/.
[14]   TIA TR-45.5 subcommittee, *TDMA IS-136 (Time Division Multiple Access) Mobile Telephone Technology* (1994).
[15]   Jay E. Padgett, G. Gunter, and Takeshi Hattori, Overview of wireless personal communications, www.cs.bilkent.edu.tr/~korpe/courses/cs515-fall2002/papers/overview-wireless-pcs.pdf, *IEEE Communications Magazine* (January 1995).
[16]   Chapter 2 provides a more detailed discussion on the subject.
[17]   Gerson and Jasuik, VSELP speech coding.
[18]   GSM 06.51, *Digital Cellular Telecommunications System (Phase 2+); Enhanced Full Rate (EFR) Speech Coding Functions; General Description.*
       GSM 06.53, *Digital Cellular Telecommunications System (Phase 2+); ANSI-C Code for the GSM Enhanced Full Rate (EFR) Speech Codec.*
       See www.etsi.org/.

improvement of EFR over FR is not limited to an error-prone environment. The overall better accuracy afforded by the EFR ACELP approach over the RPE-LTP pays off under all-purpose transmission settings.

The EFR ACELP-codec algorithm has been adopted by the US PCS 1900, and a lower codec rate (7.4 kbps) was assumed by the D-AMPS enhanced-rate TIA/EIA 641 standard. The EFR codec delay is the same as the FR delay and is equal to 40 ms. The D-AMPS ACELP codec delay is about 25 ms.

New technology more optimized for variable-rate codecs was introduced in the late 1990s. The innovation was referred to as relaxed code-excited linear predictive coding (RCELP). The relaxation is delivered by replacing the original residual with a time-warped version of it, allowing only one pitch parameter per frame, thus utilizing a lesser amount of bandwidth on the pitch information. The relaxed algorithm provides equivalent voice quality to the QCELP-13 version at a much-reduced bit rate.

The first wireless RCELP codec was developed in Bell Labs. It was named enhanced variable-rate codec (EVRC), and was standardized under TIA IS-127.[19] The Korean CDMA carriers began deploying the new codec in 1998. US carriers started rolling it out in 1999.

The EVRC operates at rates of 8 kbps, 4 kbps, 1 kbps, and 0 kbps – rate 1, rate $\frac{1}{2}$, rate $\frac{1}{8}$, and blank, respectively. Reminiscent of QCELP-13 the EVRC applies rate 1 to voiced speech, rate $\frac{1}{2}$ to unvoiced speech, and rate $\frac{1}{8}$ or blank, to noise, pauses, or bit errors respectively. In addition to being spectrum efficient, the EVRC withstands a higher degree of data degradation due to its innovative treatment of bit errors. Instead of treating bit errors as odd frequency shifts in the caller's voice as done by QCELP-13, the EVRC inserts short breaks, which do not degrade the perceived audio when the bit errors are not too abundant.

EVRC is the first codec to include noise suppression as part of the codec. The EVRC codec algorithmic delay is about 30 ms.

With the emergence of 3G wireless standards and technology – W-CDMA and cdma2000 – in late 1999, wireless codecs have crossed a striking threshold by resorting to a more intelligent and dynamic allocation of spectrum resources. The first to be launched in 1999 as the latest GSM codec has been the adaptive multi-rate (AMR).

The AMR codec contains eight source codecs:

12.2 kbps (EFR)
10.2 kbps (FR)
7.95 kbps (FR +]HR)
7.40 kbps (ANSI-136 TDMA = D-AMPS = IS 641 EFR)
6.70 kbps (PDC EFR)
5.90 kbps (FR +]HR)
5.15 kbps (FR +]HR)
4.75 kbps (FR +]HR)

Each one of the above operates with a different level of error control. Since the GSM full rate channel contains 22.8 kbps, the leftover bandwidth – after the allocation for the codec – is allotted to error control. The same is true for the half-rate channel containing 11.4 kbps.

---

[19]   EIA/TIA Recommendation IS-127, *Enhanced Variable Rate Codec (EVRC)* (1998).

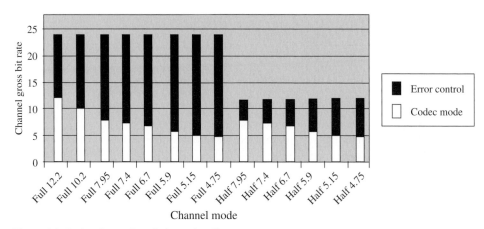

Figure 1.2 Ratio of speech and channel coding.

All of the eight AMR codecs are suitable for the full-rate channel when assigned, of which six are proper for the half-rate channel mode. Figure 1.2 presents the relationships between GSM-channel mode and the corresponding bandwidth allocation.

The AMR codec[20] employs the ACELP codebook procedure in all channel modes. The codec dynamically adapts the channel mode and the error-protection level to the channel error and traffic conditions. It uses a lower speech-coding bit rate and more error protection in channels where carrier to interference (C/I) and traffic conditions are worse. Consequently, the main advantages of AMR over its EFR predecessor are exposed during worsening channel and traffic conditions. At the same time, the AMR voice-quality performance is equivalent to the EFR under ideal channel conditions.

Figure 1.3 depicts the way AMR optimizes voice-quality performance for a given channel quality. Once channel conditions are determined or established by the wireless system, AMR picks the best codec with the highest speech-quality performance. Channel assignments of half rate are not included in the figure. However, those are assigned when traffic conditions become congested and more subscribers try to access the network. The additional air capacity is obtained by trading off speech quality under a lower C/I, by resorting to a lower error-control rate. The figure does not include half-rate codec performance under worsening channel-interference conditions. Since codec rates below 10.2 kbps may be coded either within a full-rate or a half-rate channel, the difference in speech quality performance between the full- and the half-rate channels is due to the magnitude of error control only. Accordingly, speech running inside a full-rate channel outperforms the half rate channel only when channel quality is reduced.

The CDMA response to AMR came in 2001 when the selectable-mode vocoder (SMV) – IS-893[21] for wideband CDMA was introduced. Upon its introduction, it was deemed the

20   W. Xu, S. Heinen, M. Adrat, *et al.*, An adaptive multirate (AMR) codec proposed for GSM speech transmission, *Int. Journal of Electronics and Communications (AEÜ)*, **54**: 6 (Dec. 2000).
       ITU-T Recommendation G.722.2, *Wideband Coding of Speech at around 16 kbit/s Using Adaptive Multi-Rate Wideband (AMR-WB)*, Geneva (January 2002).
21   3GPP2 C.S0030-0 Ver 2.0, *Selectable Mode Vocoder Service Option 56 for Wideband Spread Spectrum Communication Systems* (December 2001).

Table 1.1. *SMV operating modes and average bit rate (ABR)*

Mode	Full rate (FR) 8.5 kbps	Half rate (HR) 4.0 kbps	Quarter rate (QR) 2.0 kbps	Eighth rate (ER) 0.8 kbps	ABR
Mode 0	36%	4%	0%	60%	3.70[a]
Mode 1	20%	9%	10%	61%	2.75
Mode 2[b]	8%	20%	10%	62%	2.18
Mode 3	2%	26%	10%	62%	1.91
HRMax	0%	38%	0%	62%	2.02

Notes:
[a] 3.70 kbps is the average EVRC bandwidth requirement.
[b] Mode 2 is the EVRC equivalent voice quality.

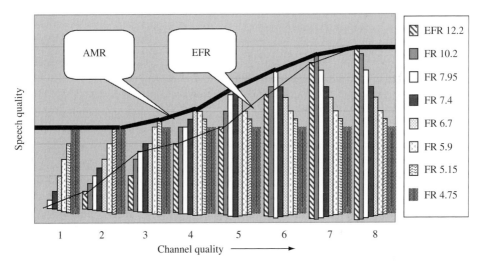

Figure 1.3 Codec performance at a given channel quality.

best universal codec. When compared with EVRC, it fashioned far better voice quality at the same bit rate, and equivalent voice quality at a lower bit rate (see Table 1.1).

The SMV is based on four codecs: full rate at 8.5 kbps, half rate at 4 kbps, quarter rate at 2 kbps, and eighth rate at 800 bps. The full rate and half rate are based on the extended CELP (eX-CELP) algorithm[22] that is based on a combined closed-loop–open-loop-analysis (COLA). In eX-CELP the signal frames are first classified as: silence/background noise, non-stationary unvoiced, stationary unvoiced, onset, non-stationary voiced, stationary voiced. The algorithm includes voice-activity detection (VAD) followed by an elaborate frame-classification scheme.

Silence/background noise and stationary unvoiced frames are represented by spectrum-modulated noise and coded at $\frac{1}{4}$ or $\frac{1}{8}$ rate. The stochastic (fixed) codebook structure is also

22 A. Yang Gao Benyassine, J. Huan-yu Su Thyssen, and E. Shlomot, Ex-CELP: a speech coding paradigm, *Acoustics, Speech, and Signal Processing Proceedings (ICASSP '01)*. IEEE International Conference, May 2001.

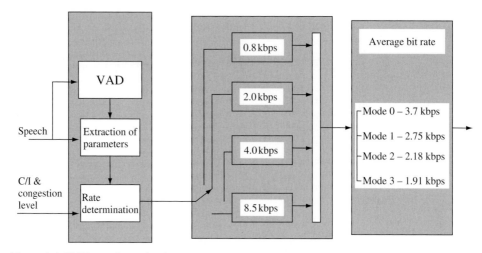

Figure 1.4 SMV rate-determination process.

elaborate and uses sub-codebooks each tuned for a particular type of speech. The sub-codebooks have different degrees of pulse sparseness (more sparse for noise-like excitation). The selectable-mode vocoder includes two types of noise-suppression (NS) algorithm, A and B, where B is a bit lighter than A. The noise suppression is applied as a preprocessor to the coder.

The coder works on a (20 ms) frame of 160 speech samples and requires a look-ahead of 80 samples (10 ms) if noise-suppression option B is used. An additional 24 samples of look-ahead is required if noise-suppression option A is used. So, the total algorithmic delay for the coder is 30 ms with NS option B enabled and 33 ms with NS option A enabled.

The selectable-mode vocoder rate-determination process is illustrated in Figure 1.4. After detecting and analyzing the signal, the codec determines the rate used for the next frame. The mode and the full/half-rate selection are pre-provisioned, since they govern the rate-determination process.

Both CDMA and GSM codecs employ a sophisticated voice-activity detection (VAD) procedure. In GSM the VAD procedure is used in conjunction with a feature labeled discontinuous transmission (DTX).[23]

Discontinuous transmission (DTX) is a method (used mostly in GSM) of powering down, or muting, a mobile or portable telephone set when there is no voice input to the set. This optimizes the overall efficiency of a wireless voice-communications system.

In a typical two-way conversation, each individual speaks slightly less than half of the time. If the transmitter signal is switched on only during periods of voice input it can be cut to less than 50 percent utilization. This conserves battery power, eases the workload of the components in the transmitter, and reduces traffic congestion and interference, while freeing the channel so that time-division multiple access (TDMA) can take advantage of the available bandwidth by sharing the channel with more speech and other signals.

---

[23] GSM 06.31, *Digital Cellular Telecommunications System (Phase 2+); Full Rate Speech; Discontinuous Transmission (DTX) for Full Rate Speech Traffic Channels.*
   See www.etsi.org/.

GSM

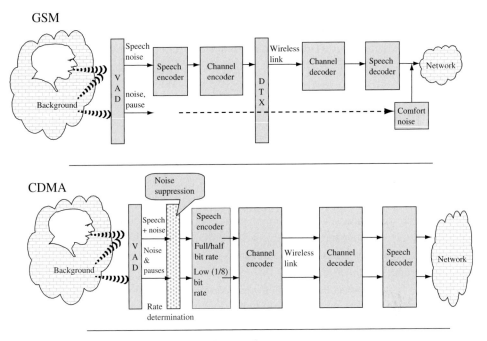

Figure 1.5 GSM and CDMA view of signal processing.

A discontinuous transmission circuit operates using voice-activity detection (VAD). Sophisticated engineering is necessary to ensure that circuits of this type operate properly. In wireless transmitters, VAD is sometimes called voice-operated transmission (VOX).

When the communications channel employs discontinuous transmission it replaces the effected silence with comfort noise matching. The transition from speech to noise matching is smoothed by having the noise matching and the background noise (during speech periods) at a noticeable level so as to mask potential bumpiness during the transition. Silence-descriptor (SID) frames are sent periodically to the receiver to update the comfort noise. In CDMA the VAD is used in conjunction with source control of the codec variable-bit rate. The encoding rate of the CDMA codec depends on the information content of the speech signal. Since voiced speech, unvoiced speech, noise and pauses are coded at different rates, the VAD output is used in determining the bit rate used for the next frame. Both EVRC and SMV employ a noise-suppression algorithm that preprocesses the digital speech signal prior to its encoding procedure. The noise-suppression algorithm in CDMA is applied during both speech and speech pauses. The GSM standard does not include a noise-suppression algorithm. However, since the NS algorithm is a preprocessor to the codec it may be included in the mobile handset as a particular cellphone feature rather than a feature of the codec standard. In general, since the GSM DTX feature is designed, among other things, to mask transitions between speech and reproduced matching comfort noise, the noise floor in GSM is bounded by the actual background-noise level measured by the DTX processor and experienced by the listener at the other end.

Figure 1.5 depicts the key differences between the latest GSM and CDMA voice-processing environments. It is worth pointing out that most of the functional blocks are equivalent except for the replacement of the GSM DTX by the CDMA NS. The GSM DTX ceases transmitting when the signal contains noise or silence, while the CDMA continues transmission at a lower bit rate. Global-systems mobile reproduces the background noise at the other end while CDMA transmits the actual noise-suppressed signal.

1.4 Summary

Digital speech-coding approaches have evolved through three major phases. Waveform coding was the initial phase where extensive sampling of the signal offered an accurate depiction en route for its reconstruction. The codecs associated with this approach were the PCM and ADPCM, which entailed bandwidth of 64 kbps and 32 kbps respectively. Waveform codecs were characterized as high quality, low complexity, low delay, and high bit rate. This bit-rate extent exceeded the feasible range that made the wireless spectrum available and economical. Accordingly, the emergence of wireless services spawned the second and third phases. In the second phase, coding approaches turned to source or parametric codecs that modeled white noise and voice tracts by simulating their response to pulse excitations. The transmitted parameters included gain, voiced/unvoiced decision, pitch (if voiced), and LPC parameters. The source or parametric codecs were referred to as vocoders. Although these vocoders could run under very low bit rates, with medium to high complexity, they were characterized as being too synthetic and inappropriate for quality-communication sessions. The third phase involved hybrid coding, combining waveform and source-coding techniques and utilizing codebooks – fixed, adaptive, algebraic, reduced, and extended – to generate a multipulse, CELP-type codec. The third phase and its evolving derivatives have produced high-quality codecs by means of low bit-rate coding, high complexity and moderate delay.

Figure 1.6 depicts the quality-versus-bit-rate trade-offs with the codec types.

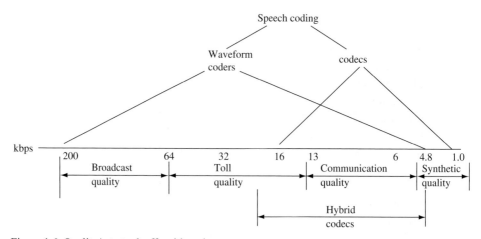

Figure 1.6 Quality/rate tradeoffs with codec type.

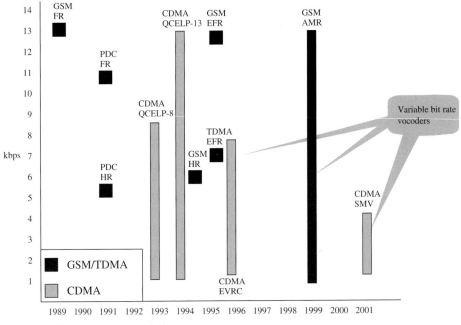

Figure 1.7 Wireless-codecs evolution.

As of this writing, wireless communication is stepping through the opening themes of its third generation. The first generation included transmission of analog speech. It was based on frequency-division multiple access (FDMA) a.k.a. AMPS, and used bandwidth equivalent to analog telephones. Speech quality was inconsistent, but complexity of implementation was low.

In 1989, Europe led the way to the second generation with the introduction of the RPE-LTP – the first digital codec for global-systems mobile. The second generation included the narrowband digital architectures embedded in GSM, TDMA (D-AMPS), and CDMA. The main advantages of the second generation over the first were significant bandwidth savings, more consistent and better speech quality, and as the second generation evolved, it became possible to make bandwidth/quality trade-offs.

The third generation emerged with the arrival of the AMR codec (in GSM) in 1999 and SMV (in CDMA) in 2001. These two codecs are compatible with second- and third-generation wireless architectures. The evolution of the wireless codecs is depicted in Figure 1.7, as it illustrates the changes in codec bit rate for GSM and CDMA over the 1990s and beyond.

Figure 1.8 illustrates the progression of CDMA voice quality with respect to average bit rate allocated to the transmitted speech. It demonstrates the pre-eminence of the latest SMV codec over its predecessors in terms of bit rate and speech quality.

Table 1.2 displays a summary of the major wireless speech-coding systems.

Table 1.2. *Summary of major speech coding systems*

System type	System name	Reference	Codec type	Codec rate (kbps)	Codec algorithmic delay (ms)[a]
TDMA	GSM/PCS full rate (FR)	GSM 06.10	RPE-LTP	13	20
TDMA	GSM/PCS half rate (HR)	GSM 06.20	VSELP	5.6	20
TDMA	GSM/US PCS 1900 enhanced full rate (EFR)	GSM 06.60	ACELP	12.2	20
TDMA	D-AMPS full rate (FR)	TIA/EIA IS-85	ACELP	8.0	28
TDMA	D-AMPS enhanced full rate (EFR)	TIA/EIA IS-641	ACELP	7.4	25
TDMA	PDC full rate	RCR-STD-27	VSELP	6.7	~30
TDMA	PDC half rate	RCR-STD-27	PSI-CELP	3.45	~30
CDMA	QCELP-8 half rate	TIA/EIA IS-96C	CELP	8/4/2/0.8	30
CDMA	QCELP-13 full rate	TIA/EIA IS-733	CELP	13.3/6.7/2.7/1.0	30
CDMA	EVRC	TIA/EIA IS-127	RCELP	8/4/0.8	30
TDMA/W-CDMA	AMR adaptive multi-rate	GSM 06.90	ACELP	12.2–4.75	25
CDMA2000	SMV multi-rate multi-mode	TIA/EIA IS-893	eX-CELP	3.7–1.91	30–33

Note:
[a] Algorithmic delay is the delay attributed to frame size only. These codecs require a complete frame of data to start the processing. Algorithmic delay is only one component in the total delay, which includes transmission and processing delays.

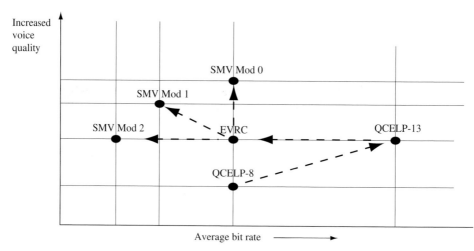

Figure 1.8 CDMA voice quality/average bit rate.

1.5   Transmission media

The transmission-media components that bear the voice signal from its wireless subscriber source to its final destination, whether it is a PSTN or another wireless subscriber, has a major effect on voice-quality performance.

The heaviest weight borne on voice-quality perception is the remarkable delay exacted by the wireless transport. The foremost part of the delay is incurred along the path segment stretching from the mobile phone to the transcoder rate adapter unit (TRAU). That round trip delay amounts to a range of 160 ms to 250 ms, depending on the particular codecs, and transport architectures employed by the service provider. A mobile to PSTN call includes a single mobile-to-TRAU round trip, but a mobile-to-mobile call may include two such delays unless the architecture enables a tandem-free operation (TFO)[24] (a.k.a. codec bypass) where the voice signal skips processing by the network codec functions and both TRAUs (see Figures 1.9 and 1.10), and moves through the network components in its coded form.

In addition to a limiting spectrum, and the PCM coding within, the speech channel may suffer from:

- attenuation distortion, where some frequency bands are attenuated more than others;
- envelop delay distortion, where some frequencies within the voice spectrum are phase shifted relative to others;
- high-level distortion, where overloaded amplifiers may set off cross-talk; or
- circuit noise, where thermal white noise[25] is the most common.

Distortions due to coding architectures can only be remedied by better designs. As technology becomes more advanced, more elaborate coding algorithms become practicable.

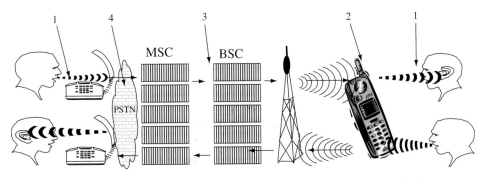

Figure 1.9 Functional view of speech transport process; mobile to PSTN view: (1) Analog or digitally coded signal to sound; (2) Codec function (wireless to sound); (3) TRAU function (wireless to PCM); (4) PCM codec function (PCM to analog).

---

[24]   Tandem-free operation is discussed in more detail in chapter 9.
[25]   White noise is a term used for noise power that is uniformly distributed across the entire frequency spectrum.

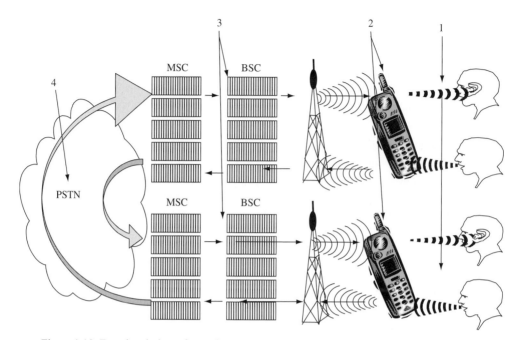

Figure 1.10 Functional view of speech transport process; mobile-to-mobile view: (1) sound; (2) codec function; (3) TRAU function; (4) when TFO is on, speech travels in its coded form; when TFO is off, speech travels as PCM.

Delay, power, and affordability budgets can be sustained, while algorithm complexity grows appreciably. At the same time, impairments caused by network architectures, handset limitations, and environmental interferences must be dealt with outside the codec infrastructure. Impairments such as electrical echo, acoustic echo, level management, and ambient noise are factors that affect voice quality. These aspects, when combined with particular coding architectures, are discussed in Part II and Part III of the book. They are referred to as voice quality assurance applications (VQA), and they deliver their functionality via voice-quality systems (VQS).

1.6   Audio illustrations

The associated website (www.cambridge.org/9780521855952) contains recordings of speech over a variety of codecs. The listener may find inconsequential differences in speech quality when comparing codecs with similar properties such as adjacent AMR modes. The differences in quality are much more prominent in the presence of noise and transmission errors. There are endless combinations of noise and error varieties, and it would be impossible to provide a comprehensive representation of these impairments. Nevertheless, the illustrations do provide a window into the speech performance of most codecs mentioned in this chapter. A careful listening would find perceptible differences when comparing codecs with greater dissimilarities such as GSM and CDMA types.

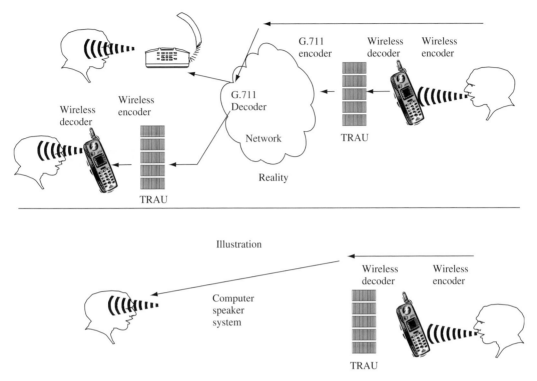

Figure 1.11 Contrasting reality with audio illustrations.

It should also be noted that in reality listeners are subject to codec combinations that involve several conversions before speech reaches its final destination ear. Conversions contribute to distortions in speech quality, and these deteriorations magnify otherwise minor differences in codec performance to make them much more striking.

Figure 1.11 illustrates the difference between the attached audio recordings and the harsh reality of transmitted speech.

Chapter 17 lists the recordings and gives more details concerning their specific identification and nature.

2 Quantitative assessment of voice quality

Chapter 2 kicks off the presentation with an overview of the standard metrics and methodologies followed by a description of specialized tools employed for obtaining subjective voice-quality scores through genuine opinion surveys and via computer modeling emulating human perceptive evaluation of speech quality. It then relates voice-quality scores obtained via surveys or computer evaluations to the perception of worth. It elaborates on the relationships between opinion scores and the potential return on investment in voice-quality technology. It examines results of voice-quality studies with reference to the three popular GSM codecs – full rate (FR), enhanced full rate (EFR) and half rate (HR). The presentation includes a discussion of the effect of noise and transmission errors on the relative performance of these codecs.

2.1 Introduction

It is widely agreed that the most vital element affecting the performance of voice quality in wireless networks is the type of codec used in the communication session.[1] It is also well known that spectrum is the most expensive (per channel) building block contributing to the viability of wireless infrastructure. Yet most European GSM wireless operators, throughout the nineties and the first half of the following decade, have opted to restrict their service offerings to full-rate (FR)[2] or enhanced full-rate (EFR),[3] rather than embracing the half-rate (HR)[4] option, which could cut their spectrum consumption in half and save them millions of dollars in operating expenses.

Prioritizing investment decisions that impact the quality of transmitted voice – particularly, decisions outside the scope of choosing a wireless codec – has always been challenging. The two key concerns affecting investment decisions have traditionally been:

---

[1] This is true under ideal radio-channel conditions. If the radio channel degrades, the C/I is low and the choice of speech compression is not relevant anymore.

[2] GSM 06.01, *Digital Cellular Telecommunications System (Phase 2+); Full Rate Speech; Processing Functions.* See www.etsi.org/.

[3] GSM 06.51, *Digital Cellular Telecommunications System (Phase 2+); Enhanced Full Rate (EFR) Speech Coding Functions; General Description.* See www.etsi.org/.

[4] GSM 06.02, *Digital Cellular Telecommunications System (Phase 2+); Half Rate Speech; Half Rate Speech Processing Functions.* See www.etsi.org/.

- how to interpret the outcome of voice-quality studies and relate mean opinion score (MOS) differences to a qualitative perception of the worth of improved voice quality, and
- how to use the MOS ratings and qualitative perception of worth to guide investment decisions about adding voice-quality systems to a telecommunications network.

To facilitate the investment decision process concerning voice-quality systems, this chapter explores the following topics:

- metrics and methodologies for measuring and quantifying voice quality in telecommunications networks,
- relating voice-quality metrics to perceived worth,
- results of studies demonstrating the impact of GSM codec selection on voice quality,
- assessing the potential impact of voice-quality systems on voice-quality scores,
- making economically sound investment decisions concerning voice-quality systems.

2.2 Measuring and quantifying voice quality

Ever since Alexander Graham Bell uttered the first words spoken via telephone in 1876, the measurement and quantification of voice quality in telecommunications networks has been an integral part of service qualification. However, traditional methodologies were largely subjective. For example, the interested party would solicit opinions from invited guests, who were asked to listen to transmitted speech on telephone handsets and then rate the various samples. The voters were subjected to various implementations of speech technologies and transmission media, and their mean opinion was then tabulated and used to drive investment decisions and strategy.

As telecommunication grew more pervasive and ever-larger networks were built, digital codecs (and later, lower bit-rate codecs) were introduced, exponentially increasing the possible number of combinations among various speech technologies. Correspondingly, the need for an international standard rating became vital. In response, the ITU published in August 1996 the P.800 standard recommendation[5] advocating a mean opinion-score (MOS) metric, employing five values representing the perceived voice quality of a phone call:

excellent $= 5$
good $= 4$
fair $= 3$
poor $= 2$
bad $= 1$

Since then, subjective mean opinion scoring (MOS-LQS)[6] has been obtained via extensive experimentation involving many human subjects (both male and female) who have rated

---

[5] ITU-T Recommendation P.800, *Methods for Subjective Determination of Transmission Quality* (1996).
[6] When a mean opinion score is derived subjectively, it is referred to as MOS-LQS, where LQS is an acronym for listening quality subjective. When an algorithm emulating human perception derives a MOS, it is referred to as MOS-LQO, where LQO stands for listening quality objective.

Table 2.1. *MOS ratings. These MOS ratings reflect clean transmission media, absent of errors and noise*

Quality of speech	Score
Toll-grade	4–4.6
Mobile-to-PSTN	3.5–4.2 depending on the specific vocoder employed
Mobile-to-mobile	3.1–4.1 depending on the specific codec pair and ordinary measurement variations

speech calls on the five-point scale. The data are processed statistically to obtain a mean value – the MOS. Since test conditions and human perceptions vary a bit, different studies produce diverse scores, but on the whole, conclusions have been fairly consistent, as reflected by the unofficial ratings given in Table 2.1.

When conducting subjective evaluations of voice quality, the standards call for the following requirements that need to be satisfied for the test to yield proper results:

- listening is in a quiet room with a controlled noise level,
- subjects listen through a telephone handset with a standard response,
- recordings are typically 8 seconds long and consist of a pair of unrelated sentences,
- tests are performed with speech from several different talkers (typically two male, two female) for each coding condition,
- subjects are non-expert.

Once the test is complete, the votes are summed up and a mean opinion score (MOS) is computed; the corresponding MOS gives the quality of each condition.

However, studying voice quality using subjective scoring was found to be both expensive and time consuming, revealing the need for an inexpensive laboratory tool that could emulate human perception of voice quality in order to assist in the design, testing, and fine-tuning of new codecs.

In response, in February 2001, the ITU issued a new standard for gauging the quality of transmitted speech: P.862 – perceptual evaluation of speech quality (PESQ).[7] The PESQ tool was designed to compute MOS-LQO values on voice samples consistent with subjective evaluations. To do this, PESQ derives a score from the difference between a reference signal and an output signal emerging from equipment in the signal path. In general, the greater the difference between the reference signal and the output signal, the lower the MOS value.

The PESQ tool proved to be a reliable, inexpensive, and quick appraiser of codec performance. In fact, it was verified by comparing its predictions to results obtained

---

[7] ITU-T Recommendation P.862, *Perceptual Evaluation of Speech Quality (PESQ): an Objective Method for End-to-end Speech Quality Assessment of Narrow-band Telephone Networks and Speech Codecs* (2001).

Table 2.2. *The DMOS five-point degradation category scale*

Quality of speech	Score
Degradation is inaudible	5
Degradation is audible but not annoying	4
Degradation is slightly annoying	3
Degradation is annoying	2
Degradation is very annoying	1

from traditional subjective studies, yielding statistical correlations (via least-squares analysis) greater than 0.9. However, PESQ is less accurate in judging performance of noise-reduction (NR) algorithms, primarily because its mathematical modeling and parameter tuning were based solely on codec studies exclusive of NR algorithms. As a result, it was branded inadequate for assessing echo cancelation and speech-level performance (see more detail in Chapter 9).

To remedy this weakness in NR measurement, ITU-T recommendation P.800 proposed a modified rating scale, referred to as the DMOS (degradation mean opinion score) metric. This measurement tool is equivalent to the MOS metric, but it is more specific as it provides for characterizations introduced by variations in noise level. Like the MOS scale, the DMOS metric is based on a five-point degradation category scale, as shown in Table 2.2.

When using DMOS to grade voice quality, a non-degraded call is chosen as a reference point. Noisy calls are then compared with the non-degraded one, and scores reflect that relative assessment. This is in contrast to MOS, which reflects an absolute rating. Other effective tools have been developed to assess voice quality in even greater detail. In recent years, ITU-T study group 12 and ITU-T study group 15, both of which specialize in voice-quality characterization, have proposed tools specifically designed for assessing the quality of noise-reduction techniques implemented as part of voice-quality systems. Some of these techniques are similar in approach to PESQ in that they employ intrusive procedures by contrasting clean speech signals with the same ones mixed with noise. However, they diverge from PESQ in that they measure performance twice, once before and once after the NR algorithm is applied. Unlike PESQ scoring, these techniques do not score based on relative deviations from the clean speech reference signal, but rather on relationships between speech and noise before and after the NR algorithm is applied.

The specific measurements include:

- total noise-level reduction (TNLR) during speech and speech pauses,
- noise power-level reduction (NPLR) during speech pauses,
- signal-to-noise ratio improvement (SNRI) during speech,
- signal power-level reduction (SPLR) during speech.

These measurements are then used to generate a quality index equivalent to the MOS-LQO, as it is a five-point scale reflecting equivalent categories to those of the ITU ACR[8] P.800 standard recommendation.[9]

2.3   Relating voice-quality metrics to perceived worth

The ability to make a telephone caller's voice sound somewhat better when it already sounds fine to the recipient is not very impressive. Making that voice sound somewhat better when it is almost unintelligible, however, can make a big difference. In other words, the value of a 0.5 increase in a MOS or DMOS score is heavily dependent on the point of degradation at which the score is measured. If the score improvement transforms the quality perception from "very annoying" to somewhere between "slightly annoying" and "not annoying," the increase in score would assume a steep value. Conversely, if the score improvement occurs when the call quality is already very high, the score's value is decreased significantly. To accommodate this, the DMOS scale is constructed so that any transition within the 2–3 range indicates a strong perception of worth, while transitions of equivalent magnitude within the 3–4 range are minute in comparison.

Figure 2.1 presents the non-linearity governing the relationships between the DMOS and the associated perception of worth. The range of 2–3 exhibits the steepest rise in worth, while the extreme ranges 1–2 and 4–5 are fundamentally levelled.

2.4   Mean opinion scores of wireless codecs

The choice of codec can have a substantial impact on the voice quality of wireless networks. This fact was recognized by the 3rd Generation Partnership Project (3GPP), a body formed in December 1998 to expedite the development of specifications for 3G networks.[10] Included in two 3GPP specifications are GSM 06.85 and 3GPP TR–46.085,[11] tests that provide voice quality results obtained from subjective evaluations of HR, FR, and EFR codecs.

Figure 2.2 depicts MOS for HR, FR, and EFR encoding during mobile-to-PSTN calls (single encoding), where C/I is carrier-to-interference ratio and GBER is average gross bit-error rate. EP0 represents no errors, while EP1 and EP2 represent different types of error

---

[8]   ACR is the acronym for absolute category rating. It is used by the MOS metric; DCR is the acronym for degradation category rating and it is used by the DMOS metric.

[9]   ITU SG-15 question 6 delayed contribution, SQuad *on Objective Method For Evaluation of Noise-Reduction Systems*, SwissQual, SQuad, Geneva, Switzerland (April 2002).
        ITU-T SG-12 question 9, contribution 32, *Cause Analysis of Objective Speech Quality Degradations Used in Squad*, SwissQual (August 2001).

[10]   See www.3gpp.org/About/about.htm for further details.

[11]   ETSI TR 101 294 V8.0.0 (2000–06), *Digital Cellular Telecommunications System (Phase 2+); Subjective Tests on the Interoperability of the Half Rate/Full Rate/Enhanced Full Rate (HR/FR/EFR) Speech Codecs, Single, Tandem and Tandem Free Operation*, GSM 06.85 version 8.0.0 (release 1999). See www.etsi.org/.

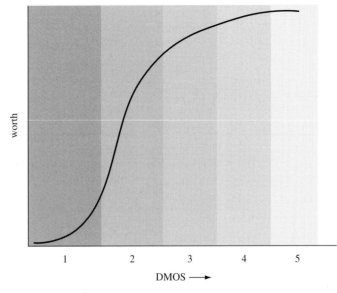

Figure 2.1 DMOS versus worth.

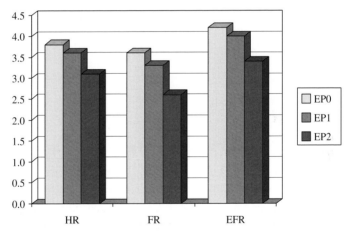

Figure 2.2 Mobile-to-PSTN MOS performance.

conditions. In particular EP0 represents no channel errors, EP1 represents C/I = 10 dB; 5% GBER (well inside a cell) and EP2 represents C/I = 7 dB; 8% GBER (at a cell boundary).

The interesting and somewhat ironic result is that HR yielded higher MOS than FR under single encoding, with or without errors. The MOS scoring of EFR was consistent with common understanding.

While the results indicate a slight advantage (within the margin of error) for HR over FR for single encoding with or without errors, the addition of noise on the mobile end (street noise or vehicular noise) reveals a noteworthy transformation in the relative MOS, as illustrated in Figure 2.3.

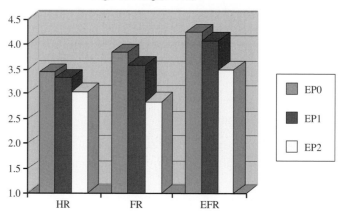

Figure 2.3 Mobile-to-PSTN DMOS performance with added noise (SNR = 10 dB). Single encoding (DMOS).

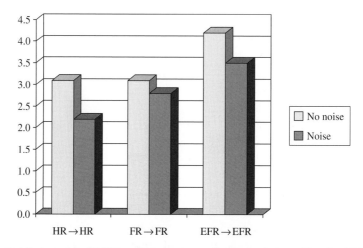

Figure 2.4 Mobile-to-mobile DMOS performance; error-free transmission with noise (SNR = 10 dB).

Figure 2.3 depicts DMOS scores for HR, FR, and EFR during mobile-to-PSTN calls (single encoding) with vehicular noise added to speech (SNR = 10 dB) and under various error conditions. Again, EP0 represents no errors, while EP1 and EP2 represent different types of error conditions.

Although Figure 2.3 presents an acceptable performance (even for HR), the picture changes considerably for the mobile-to-mobile case as illustrated in Figure 2.4. Because of the relatively large number of codec combinations, a summary explains the outcome better than multiple detailed graphs.

Most codec combinations involving HR (HR–HR, HR–FR, HR–EFR) – with vehicular noise added to speech yielding SNR = 10 dB – reduce the DMOS score to the 2+ region, with the lowest score involving HR–HR and the highest involving HR–EFR. Codec combinations excluding HR result in average scores ranging from slightly to distinctly

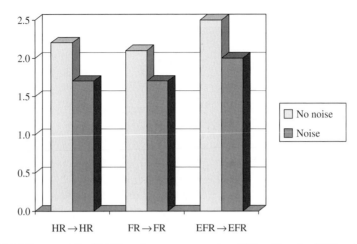

Figure 2.5 Mobile-to-mobile DMOS performance; error type EP2 with noise (SNR = 10 dB).

above three. Errors added to the noisy conditions result in consistently worse scores for all possible codec combinations.

Considering the fact that DMOS scores below three reflect an ever-increasing level of annoyance, it is understandable why, despite the resulting expense of double spectrum, many European carriers have opted for an FR standard as opposed to HR. The HR codec deteriorates a great deal faster than FR or EFR in the presence of noise, and this deterioration is accelerated when a mobile-to-mobile session is conducted between two HR-equipped handsets.

It should be noted that the DMOS score has been shown to deteriorate rapidly (more so when HR is involved) as the SNR ratio is reduced to levels below 10 dB.

Figures 2.4 and 2.5 exhibit GSM 06.85 study results for the three GSM codecs. Figure 2.4 illustrates how background noise instigates diverse stimuli on the different GSM codecs' performances in a mobile-to-mobile (double encoding) situation. Figure 2.5 demonstrates the additional impact caused by transmission errors. The results reflect the earlier conclusion that the HR is more susceptible to noise than the other codecs. Half rate is no worse than FR in the absence of noise. Consequently, if a HR codec is selected, elimination of the noise element carries a significant savings potential since expensive spectrum is better utilized.

Evidently, in the presence of transmission errors and noise, the overall score drops to a DMOS region where any minor improvement due to noise reduction is highly valued. Under these conditions, even the small increase in DMOS due to noise reduction is of significant worth for every codec, even EFR.

The emergence of the wireless UMTS architecture[12] has been accompanied by the adaptive multi-rate (AMR) codec.[13] As was already discussed in Chapter 1, the key benefit

[12] Peter Rysavy, Voice capacity enhancements for GSM evolution to UMTS, White Paper, July 18 (2002).
H. Holma and A. Toskala, *WCDMA for UMTS*, John Wiley and Sons (2004).
[13] 3GPP TS 26.190, *AMR Wideband Speech Codec: Transcoding Functions*, 3GPP technical specification, see www.3gpp.org/.
Alan D. Sharpley, *3G AMR NB, Characterization Experiment 1A – Dynastat Results*, 3GPP TSG-S4#14 meeting, Bath, UK (November 27–December 1, 2000).

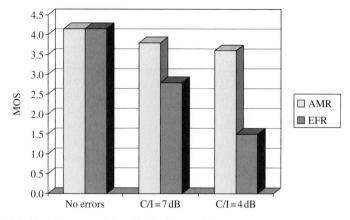

Figure 2.6 MOS of AMR versus EFR with single encoding.

provided by AMR is its stability in the face of high channel interference. Under optimal channel conditions, the performance of AMR is equivalent to EFR. But as the channel to interference ratio (C/I) (in dB) grows, AMR exhibits continuously stable performance while quality of speech under EFR deteriorates rapidly in comparison. Several studies[14] concerning MOS ratings of AMR in comparison with EFR concluded the comparable outcomes. Their consensus is exhibited in Figure 2.6.

Code-division multiple-access codecs delivered better voice quality at lower bit rates as they progressed through time from QCELP13, through EVRC, toward SMV. When comparing CDMA codecs with the popular GSM EFR and AMR, and taking readily available technology into account, one conclusion stands out clearly. While the CDMA and GSM codecs have delivered equivalent voice-quality performance under channel interference and background-noise environments, the CDMA codecs starting with the EVRC required considerably less bandwidth in comparison.

The summary results presented in Figure 2.7 are derived from numerous studies[15] that reflect the above conclusion. Although not reflected directly in the figure, it should be noted that noise tends to impact the mean opinion score of every codec. Street noise is generally worse than car noise because it possesses more dynamic changes in color and level. Street noise yielding 15 dB SNR may drop the MOS of SMV and EVRC by about 0.3 to 0.5 points on the average, while it would push down the MOS of AMR and EFR by about 0.5 and up to a full point respectively.[16]

When increasing the relative noise level to 10 dB SNR, the MOS penalties associated with SMV, EVRC, and AMR are comparable. They all tend to reduce their MOS by about

[14] 3GPP TS 46.008 V4.0.0 (2001–03), see www.3gpp.org/.

[15] Huan-yu Su, Eyal Shlomot, and Herman K. Nakamura, *Selectable Mode Vocoder Emerges for CDMA2000 Designs*, Mindspeed Technologies, Inc., CommsDesign.com (Jan 07 2003).
   Skyworks Solutions, *Selectable Mode Vocoder* (SMV), *Enhanced Voice Quality and Capacity for CDMA Networks*, Skyworks Solutions (May 2004).
   Sassan Ahmadi, *Tutorial on the Variable-Rate Multimode Wideband Speech Codec*, Nokia, CommsDesign.com (Sep 02, 2003).
   S. Craig Greer, *Speech Coding for Wireless Communications*, Nokia (November 27, 2001).

[16] Skyworks, *Selectable Mode Vocoder, Enhanced Voice Quality and Capacity for CDMA Networks*.

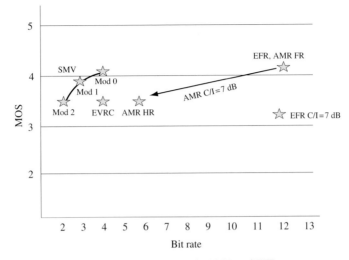

Figure 2.7 MOS versus codec bit rate of SMV, EVRC, AMR and EFR.

one point. The main reason for the SMV and EVRC better stability, under relative street noise of 15 dB SNR, is their built-in noise reduction function. At higher noise levels, however, the noise reduction feature loses some of its potency, as will be seen in Chapter 5. At the same time, both SMV and AMR exhibit high stability in the face of channel errors. At C/I = 7 dB either SMV or AMR MOS drops by about 0.3 points. At C/I = 4 the point drop for either is about 0.5 points from their 4+ highs, while EFR drops to about 1.5 on the mean opinion MOS scale.[17]

2.5 Standard computerized tools for speech-quality assessment

Subjective evaluations resulting in MOS are expensive and time consuming. They require proper setup, supervision, people, and disruption of normal work around the MSC. Fortunately, there are computerized tools that make simulations of voice-quality appraisal possible. Some of these computerized tools generate a MOS-LQO that is designed to correlate well with a MOS-LQS obtained via subjective study involving people's opinions. Some other tools generate a pass/fail score based on particular pre-determined criteria. Some tools are intrusive. The equipment and systems must be taken out of normal service and run in a lab where various signals are fed through their path. Output signals are then compared with input signals or judged against an objective criterion to determine a quality score. Some tools are not intrusive. They may be applied in the field under normal traffic conditions. These tools pass judgment on the signal quality by devising or projecting an internal model of an ideal output signal and contrasting it with the actual output signal.

---

[17]   J. Makinen, P. Ojala, and H. Toukamaa, *Performance Comparison of Source Controlled AMR and SMV Vocoders*, Nokia Research Center, www.nokia.com/library/files/docs/Makinen2.pdf.
ETSI/SMG11, *Speech Aspects*, Presentation of SMG11 activities to Tiphon, http://portal.etsi.org/stq/presentations/smg11.pdf#search='speech%20aspects'.

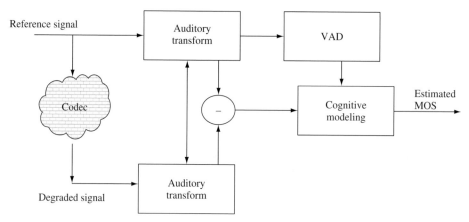

Figure 2.8 PSQM architecture.

This segment introduces the most popular tools for appraising voice quality with some of the history behind the latest versions. The discussion takes account of their scope, methodology, advantages, and limitations. Additional details of the specific procedures as they apply to particular test methodologies are available in Chapters 9 and 12, which deal with testing and evaluation of voice-quality products.

Perceptual speech-quality measure (PSQM)

Perceptual speech-quality measure[18] was developed by Beerends and Stemerdink.[19] Its main purpose was the assessment of speech codecs,[20] and it was the first to evaluate and score voice on a quality scale by using the architecture depicted in Figure 2.8. When emulating human perceptions PSQM utilizes key psychophysical processes in the human auditory system by computing an internal representation of loudness in time and frequency.

Cognitive-modeling schemes were the breakthrough that PSQM brought to the VQ assessment methodology. There were early approaches to estimating a single quality score based on the average distance between the transforms of the reference and degraded signal,[21] but PSQM was able to interpret the difference between the transforms while improving correlation between objective score and subjective mean opinion score by treating asymmetry and different weighting for active speech and silent periods.

[18] ITU-T Recommendation P.861, *Objective Quality Measurement of Telephone-band (300–3400 Hz) Speech Codecs* (1998).

[19] J. G. Beerends and J. A. Stemerdink, A perceptual audio quality measure based on a psychoacoustic sound representation. *Journal of the Audio Engineering Society*, **40**: 12 (1992), 963–974.
    J. G. Beerends and J. A. Stemerdink, Measuring the quality of audio devices. 90th AES Convention, pre-print no. 3070 (1991).

[20] J. G. Beerends and J. A. Stemerdink, A perceptual speech-quality measure based on a psychoacoustic sound representation. *Journal of the Audio Engineering Society*, **42**: 3 (1994), 115–123.

[21] J. Karjalainen, A new auditory model for the evaluation of sound quality of audio systems, IEEE ICASSP, 608–611 (1985).
    Beerends and Stemerdink, Measuring the quality of audio devices.
    S. Wang, A. Sekey, and A. Gersho, An objective measure for predicting subjective quality of speech coders. *IEEE Journal on Selected Areas in Communications*, **10**: 5 (1992), 819–829.

Perceptual speech quality measure considers the asymmetry effect caused by the codec distortion[22] of the input signal. It weights disturbances that occur during speech active periods more than those occurring during silent intervals, since disturbances during active speech tend to affect perception of voice quality more than during inactive periods.

In 1996, the ITU-T P.861 standard recommendation adopted PSQM as the standard methodology for objective quality measurement of narrowband speech codecs. In its description of the procedure, P.861 limits its scope to exclude live network testing or conditions in which variable delay (time warping) can occur. It places requirements on time alignments of the reference and degraded signal and calls for speech-level equalization in line with the standard level used in subjective listening tests. By 1999, PSQM was revised to include end-to-end testing of real networks.

Perceptual analysis measurement system (PAMS)

Hollier proposed a perceptual model[23] for interpreting what was termed "the error surface." The perceptual analysis measurement system (PAMS) is an enhanced version of this model. The key enhancement focuses on end-to-end measurements utilizing time and level alignment and equalization respectively.[24]

The first step initiated by the model divides the signals into utterances, aligning and equalizing them. It identifies delay changes due to transmission and processing architecture. Equalization is used to set the listening level to 79 dB SPL[25] before performing the auditory transforms. It follows with auditory transforms of both reference and degraded signals. The perceptual layer is performed by quantifying the errors and using the measured distortions as data for regression analysis (see Figure 2.9). The results are mapped to two quality scores – a standard MOS based on the ACR listening-quality opinion scale and the other on the ACR listening-effort opinion scale.[26]

The perceptual analysis measurement system can be used in end-to-end measurement applications[27] as well as testing monophonic wideband telephony at 16 kHz sample rate.[28]

---

[22] J. G. Beerends, Modelling cognitive effects that play a role in the perception of speech quality, in *Proceedings of the International Workshop on Speech Quality Assessment*, Bochum (November 1994), 1–9.

[23] M. P. Hollier, M. O. Hawksford, and D. R. Guard, Characterisation of communications systems using a speech-like test stimulus, *Journal of the Audio Engineering Society*, **41**: 12 (1993), 1008–1021.

    M. P. Hollier, M. O. Hawksford, and D. R. Guard, Error activity and error entropy as a measure of psychoacoustic significance in the perceptual domain. *IEE Proceedings – Vision, Image and Signal Processing*, **141**: 3 (1994), 203–208.

[24] A. W. Rix and M. P. Hollier, The perceptual analysis measurement system for robust end-to-end speech quality assessment, IEEE ICASSP (June 2000).

[25] ITU-T Recommendation P.830, *Subjective Performance Assessment of Telephone-Band and Wideband Digital Codecs*, August 1996.

[26] ITU-T P.800, *Methods for Subjective Determination of Transmission Quality*.

[27] A. W. Rix, R. Reynolds, and M. P. Hollier, Perceptual measurement of end-to-end speech quality over audio and packet-based networks. 106th AES Convention, pre-print no. 4873 (May 1999).

    A. W. Rix and M. P. Hollier, *Perceptual Analysis Measurement System*.

    Psytechnics limited, *PAMS, Measuring Speech Quality Over Networks*, www. psytechnics.com, white paper (May 2001).

[28] A. W. Rix and M. P. Hollier, Perceptual speech quality assessment from narrowband telephony to wideband audio, 107th AES convention, pre-print no. 5018 (September 1999).

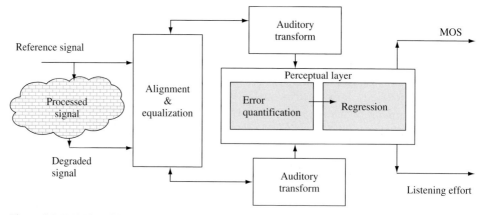

Figure 2.9 PAMS architecture.

PESQ: perceptual evaluation of speech quality – ITU P.862 recommendation

Perceptual evaluation of speech quality (PESQ) emerged in the year 2000 following the realization that correlation between P.861 PSQM and subjective opinion scores were poor. In 1998, another procedure, measuring normalizing blocks (MNB), was proposed as an optional appendix to P.861 to remedy some of PSQM shortcomings. However, neither P.861 PSQM nor MNB were found to be suitable for end-to-end measurement of voice quality in networks. Neither PSQM nor MNB were able to handle variable delay or background noise conditions.

In 2001, ITU-T approved PESQ as recommendation P.862.[29] PESQ combined the best parts of PAMS and PSQM99 to yield a significantly higher correlation with subjective studies than any of its predecessors.

Figure 2.10 presents the PESQ architecture. The procedure starts the ball rolling by aligning both reference and degraded signals to a pre-determined loudness level. Following the level adjustment, the signals are then processed by a fast Fourier transform (FFT) with an input filter that emulates distortions brought about by processing in a telephone handset. The two signals, the reference and the degraded signals, are then aligned in time and are processed through an auditory transform, a psycho-acoustic model which maps the signals into a representation of perceived loudness in time and frequency, similar to that of PSQM, including equalization of the signals for the frequency response of the system and for gain variation. The disturbance, which is the absolute difference or the gap between the reference and the degraded transforms, is processed in frequency and time. When the first time alignment yields large errors with incorrect delay, the signals are realigned and the disturbance is recalculated. The process is repeated until it yields a lower disturbance value, and only then is it moved forward to the next step of aggregation where short intervals and the whole signal are considered.

---

[29]   Beerends and Stemerdink, *Measuring the Quality of Audio Devices.*

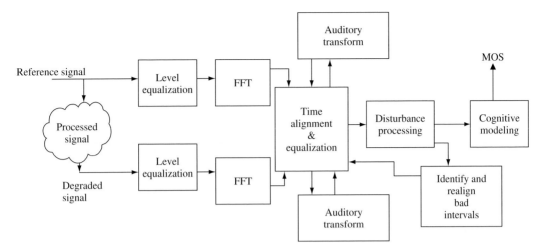

Figure 2.10 PESQ architecture.

Mapping to an estimate of subjective mean opinion score on the ACR scale is performed by using a combination of two parameters – symmetric and asymmetric disturbance.

The mapping used in PESQ is given by

$$MOS = 4.5 - 0.1 \times \text{symmetric disturbance} - 0.0309 \times \text{asymmetric disturbance}$$

The coefficients 0.1 and 0.0309 were obtained from a least-squares regression analysis performed on a database consisting of 30 separate studies involving subjective tests.

Various studies of wireless codecs, fixed networks and VoIP types comparing subjective wireless tests, fixed network tests, and VoIP networks tests with results obtained via PESQ, PSQM, and MNB[30] under a variety of background noise and language conditions, have demonstrated the superiority of PESQ over the rest. The tests covered several different languages and various background noise conditions. Perceptual evaluation of speech quality scores have been found to correlate well with scores obtained via subjective evaluations. They yielded an average correlation with subjective tests exceeding 0.95, 0.94, and 0.92 for wireless, fixed networks, and VoIP respectively. About 95% of the studies yielded correlations better than 0.9. At the same time, comparisons with PSQM produced correlations exceeding 0.92, 0.85, and 0.70 for wireless, fixed, and VoIP networks respectively with several worst-case correlations falling behind 0.7. Measuring normalizing blocks produced significantly lower correlations than PSQM.

Correlations less than 100% are subject to residual errors. The ITU P.862 reports that for 22 known ITU benchmark experiments the average PESQ residual-error distribution showed that the absolute error was less than 0.25 MOS for 69.2% of the conditions and less than 0.5 MOS for 91.3% of the conditions. The derived conclusion from this study is that when comparing voice-processing algorithms via PESQ, caution must be placed before declaring a winner. A mean opinion score obtained via PESQ may be within the

30   ITU-T COM 12- C46-E, *Mapping the PESQ Algorithm's Score on the Subjective Scale*, L. M. Ericsson, November 2002.

margin of error if the score reflects a small (less than 0.25 or even less than 0.4) improvement or degradation in speech quality due to different processing algorithms.

The residual error brought on by PESQ – with respect to a potential subjective evaluation – is understated because subjective evaluations are susceptible to variations just as well. Consequently, a way to remedy the drawback and stem decisions is to treat PESQ as "just another opinion" while relying on some form of subjective judgment concurrently.

Perceptual evaluation of speech quality is an intrusive tool designed specifically for evaluating codec performance. It evaluates the signal quality by contrasting it against an unprocessed original. Its output is a mean opinion score based on the absolute category-rating method with the listening-quality opinion scale.

Perceptual evaluation of speech quality is less suitable for assessing remedies designed for reducing noise and should never be used in assessing echo cancelation. It should only be applied to gauging performance of one-way communications (rather than interactive or conversational speech). The P.862 standard recommendation clearly states that:

> . . . the PESQ algorithm does not provide a comprehensive evaluation of transmission quality. It only measures the effects of one-way speech distortion and noise on speech quality. The effects of loudness loss, delay, sidetone, echo, and other impairments related to two-way interaction (e.g., center clipper) are not reflected in the PESQ scores. Therefore, it is possible to have high PESQ scores, yet poor quality of the connection overall.

It should also be noted that the P.862 standard recommendation states clearly that the PESQ provides inaccurate predictions when used in conjunction with varying listening levels and loudness loss. Consequently, it becomes a less effective tool for evaluating noise reduction algorithms. Unlike PESQ, noise reduction tools utilize selective and uneven attenuation of frequency bands within a given signal. As a result, they may attenuate frequency bands containing speech and affect speech distortion. Since PESQ is not designed to evaluate loudness loss, it is less suited for evaluating performance of noise-reduction algorithms.

ITU P.563 draft recommendation: a non-intrusive procedure for evaluating vocoder performance

The ITU P.563 recommendation[31] presents a non-intrusive procedure that can be run in live network monitoring without affecting service. It is similar to PESQ in concept, but very different in execution since it is lacking a reference signal that can only be used under intrusive measures. Live networks may include background noise, filtering and variable delay, as well as distortions from channel errors and speech codecs. The procedure was designed for single-ended non-intrusive measurement applications, taking into account the full range of distortions occurring in public switched telephone networks. It was intended for assessing the speech quality on a perception-based scale MOS-LQO according to ITU-T Recommendation P.800. P.563 is not restricted to end-to-end measurements; it can be used at any arbitrary location in the transmission chain. The calculated score is highly correlated with subjective tests with people listening with a conventionally shaped handset.

---

[31]   ITU-T Recommendation P.563, *Single Ended Method for Objective Speech Quality Assessment in Narrow-Band Telephony Applications* (2004).

The P.563 recommendation approach should be treated as an additional (rather than a sole) expert opinion. The quality score estimated by P.563 is its computed perceived quality of the voice signal when taking into account a conventional handset by applying a pre-processing which models the changes to the signal inflicted by the receiving handset. Pre-processing the signal involves applying a set of basic speech descriptors such as active speech-level adjustment, and application of voice-activity detection (VAD). The process moves on through three independent functional blocks that correspond to the main classes of distortion:

- Vocal tract analysis and unnaturalness of speech including male and female voices, and identification of robotization,[32] where speech is modeled for the purpose of extracting signal parts that could be interpreted as voice, separating them from the non-speech parts. Ratings are computed for the unnaturalness of speech for a sex-independent strong robotization, as well as repeated speech frames, which occur in packet-based transmission systems employing error-concealment methods using packet (frame) repetitions that replace lost packets with previously successfully trans-mitted packets. And to top it off, a more general description of the received speech quality compares the input signal with a pseudo reference signal generated by a speech enhancer.

- Analysis of strong additional noise including low static SNR (background noise floor) and low segmental SNR (noise that is related to the signal envelope). The noise analysis computes two key parameters determining the type of noise – static residing over most of the signal (at least during speech activity) where the noise power is not correlated with the speech signal, or one dependent on the signal-power envelope. When noise is static, several detectors try to quantify the amount of noise "locally" and "globally." "Local" noise is defined as the signal parts between phonemes, whereas "global" noise is characterized by the signal between utterances or sentences. Distinguishing between noise types is key to mobile communications shaping of comfort noise.

- Interruptions, mutes, and time clipping turn out a separate distortion class. Signal interruption can occur either as temporal speech clipping or as speech interruption. Either one leads to a loss of signal information. Temporal clipping may occur when voice-activity detection is used in conjunction with DTX or DCME, when the signal is interrupted and replaced by comfort noise. Clipping of this type cuts off a tad of speech during the instant it takes for the transmitter to detect its presence. The P.563 recommendation distinguishes between normal word ends and abnormal signal inter-ruptions as well as unnatural silence intervals in a speech utterance.

In summary – the P.563 recommendation is able to construct a hypothetical reference accompanied by several criteria when gaging the quality of the scrutinized processed signal. This recommendation has the same objectives and limitations as PESQ when it comes to case studies. It is best used for evaluating codec performance. It is less effective in judging quality of noise-reduction algorithms and should not be used for evaluating

---

[32] Robotization is caused by a voice signal that contains too much periodicity.

interactive conversations. For these reasons the methodology should not be employed in evaluating performance of echo cancelation or noise reduction algorithms. The P.563 recommendation clearly states that "although correlations between objective and subjective scores in the benchmark were around 0.89 for both known and unknown data, the algorithm cannot be used to replace subjective testing but it can be applied for measurements where auditory tests would be too expensive or not applicable at all."

Like PESQ, the P.563 algorithm is not intended to provide a comprehensive evaluation of transmission quality. It is confined to the effects of one-way speech distortion and noise effects on speech quality. The P.563 algorithm removes the effect of a common receiving terminal handset while equalizing the speech level to a SPL of 79 dB. What is more, the effects of loudness loss, delay, side-tone, talker-echo, music or network tones as input signal, delay in conversational tests, and other impairments related to the talking quality or two-way interaction are not reflected in the P.563 scores. It should be noted that P.563 has not been fully validated, at the time of the standardization for:

- amplitude clipping of speech,
- singing voice and child's voice as input to a codec,
- bit-rate mismatching between an encoder and a decoder if a codec has more than one bit-rate mode,
- artificial speech signals as input to a codec,
- artifacts from isolated noise-reduction algorithms,
- synthetic speech used as input to a speech codec.

Therefore, when including any of the above inputs or characteristics, it is possible to have high P.563 scores, yet non-optimal quality of the connection overall.

To evaluate the reliability of the P.563 algorithm, its results were compared with subjective scores obtained via 24 known ITU benchmark experiments. In that specific study, the average correlation was 0.88. For an agreed set of six experiments used in the independent validation – experiments that were unknown during the development of P.563 – the average correlation was 0.90.

When employing the P.563 test procedure it is recommended that a variety of suitable conditions and applications be manipulated and used as test factors. Examples of these conditions and applications are listed below.

- Characteristics of the acoustical environment (reflections, different reverberation times);
- mobile and conventional shaped handsets as well as hands-free terminals, as in P.340 test-set-up in office environments;
- environmental noise at the sending side;
- speech input levels to a codec;
- transmission channel errors;
- a variety of CELP codecs and a combination of packet loss and packet-loss concealment;
- a variety of transcoding combinations;
- a variety of transmission delays.

The P.563 recommendation was validated with unknown speech material by third-party laboratories under strictly defined requirements.

ITU G.107 recommendation: E model

The ITU-T G.107 recommendation[33] provides a rating factor, R, used to estimate customer opinion. The E model is a planning tool designed to estimate projected (rather than existing) conversational (rather than listening, as in P.862 and P.563) overall quality. In that sense, it is a proper procedure for evaluating the quality of improvements dispensed due to echo cancelation, noise reduction, level optimization, and packet-loss concealment.

The E model was first submitted to standards bodies in 1993 although its origins date back to the models first developed in the 1960s by BT, Bellcore and others.

The basic principle behind the model treats degrees of various impairments as additive factors. The overall equality estimate is derived by adding objective ratings of noise, echo, delay, codec performance, jitter, etc., to obtain an overall objective rating of quality. The overall quality, (R factor), is calculated by estimating the signal-to-noise ratio of a connection (Ro) and subtracting the network impairments (Is, Id, Ie) that, in turn, are offset by any expectations of quality of the caller (A).

The basic formula for the E model is

$$R \text{ factor} = Ro - Is - Id - Ie + A,$$

where the R factor is the overall network quality rating (ranges between 0 and 100) and:

Ro: signal-to-noise ratio
Is: impairments that occur more or less simultaneously with the voice signal
Id: impairments caused by delay
Ie: quality impairments due to codecs effects and packet losses of random distribution
A: advantage factor (attempts to account for caller expectations)

The term Ro and the Is and Id values are subdivided into further specific impairment values.

The R factor ranges from 0 to 100.[34] It can be mapped onto an estimated mean opinion score where:

- for $R = 0$ (estimated conversational quality), MOS $= 1$
- for $0 < R < 100$ (estimated conversational quality),

$$\text{MOS} = 1 + 0.035R + R(R - 60)(100 - R)7^*106$$

- for $R = 100$ (estimated conversational quality), MOS $= 4.5$

---

[33] ITU-T Recommendation G.107, *The E-model, a Computational Model for Use in Transmission Planning* (1998).
[34] Although R is a value between 0 and 100, there is a slim theoretical possibility that a particular implementation of the E model may yield R values lower than 0 or higher than 100. Under these circumstances the value of R should be set at 0 or 100 respectively.

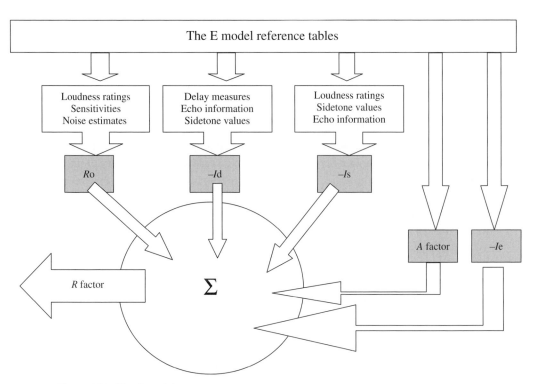

Figure 2.11 The E model process.

The interpretation of the *R* factor is as follows:

- $R = 90$, MOS lower limit $= 4.34$ (users very satisfied)
- $R = 80$, MOS lower limit $= 4.03$ (users satisfied)
- $R = 70$, MOS lower limit $= 3.60$ (some users dissatisfied)
- $R = 60$, MOS lower limit $= 3.10$ (many users dissatisfied)
- $R = 50$, MOS lower limit $= 2.58$ (nearly all users dissatisfied)

The E model process is illustrated in Figure 2.11.

ITU G.168 recommendation: appraisal of echo-cancelation performance

The ITU-T G.168 recommendation[35] suggests a series of tests, run in a lab environment, measuring the dynamics of speech and echo-power levels after a pre-recorded standard signal passes through the echo-canceler processing algorithms. The recommendation provides acceptance thresholds yielding a pass/fail grade. Tests are conducted by sending pre-recorded source and echo signals through the opposite input (receive and send, respectively) ports of the echo canceler and measuring the power level of the resulting

---

[35]   ITU-T Recommendation G.168, *Digital Network Echo Cancellers* (1997, 2000, 2002).

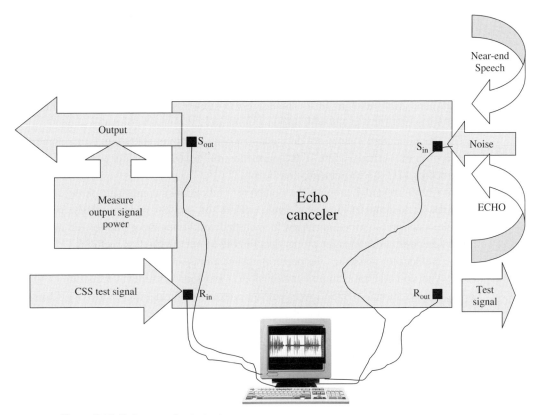

Figure 2.12 Echo-canceler test set-up.

echo signal at the send output port (see Figure 2.12). When measuring particular character-istics of the echo canceler, it becomes necessary to disable certain capabilities and isolate others. Tests require user control of some internal parameters so that manipulation of the various components during the test is feasible.

Measurements at the send output port assess the amount of signal (echo) power loss resulting from the echo canceler processing. Signal loss is a combination of three components:

- loss due to network conditions independent of the echo-canceler performance. This loss is referred to as the echo-return loss (ERL),
- loss due to linear processing by the echo canceler,
- the combined loss due to non-linear processing (NLP) and the linear processor.

(See Chapter 9 for further details).

ITU G.160 recommendation – appendix II

This recommendation suggests a series of intrusive tests, run in a lab environment, measuring the quality and trade-off nature of noise-reduction algorithms. Although the scope and purpose of the standard document is broader, as it encompasses voice-quality systems as a whole while defining certain performance constraints for them, its chief contribution to voice-quality assessment methodology is sheltered in its second appendix. The ITU-T G.160[36] recommendation covers a description of the general characteristics of voice quality systems indicating what characteristics are important to provide acceptable telecommunications performance. The document discusses issues relating to the interaction of voice-quality systems with other network and subscriber equipment as well.

The recommendation does not define a standard algorithm, applied to either electrical echo control (governed by ITU-T Recommendation G.168[37]) or automatic level control (governed by ITU-T Recommendation G.169).[38]

The G.160 appendix presents an objective methodology comprising three objective measures for characterizing the basic effect of noise-reduction (NR) methods. Recommendation target value ranges in specified conditions are given for the measures to serve as basic guidelines for proper functioning of noise reduction methods.

The methodology assesses the performance of a noise-reduction application by measuring signal-to-noise ratio improvement (SNRI) and total noise level reduction (TNLR) and a delta measurement (DSN). While SNRI is measured during speech activity, focusing on the effect of noise reduction on speech signal, TNLR estimates the overall level of noise reduction, experienced both during speech and speech pauses. The delta measurement is computed by taking measurements of noise power-level reduction (NPLR) during speech pauses and subtracting it from SNRI. The improvement during speech and during speech pauses should be almost identical, and the value of DSN should be equal to 0 under ideal conditions. The actual negative or positive value of DSN is then determined to reveal speech attenuation or undesired speech amplification respectively caused by a noise reduction solution.

The proposed methodology is a further development of one included in the 3GPP GSM Technical Specification 06.77,[39] *Minimum Performance Requirements for Noise Suppresser*[40] *Application to the AMR Speech Encoder* and the TIA TR45 specification[41] TIA/EIA-IS-853.[42] A detailed description of the earlier methodology is incorporated in the named specifications. A more detailed discussion of the performance assessment proposed by the G.160 is presented in Chapter 9, which deals with testing and appraising of voice quality systems.

---

[36]   ITU-T Recommendation G.160, *Voice Enhancement Devices*, draft 14 (2004).
[37]   ITU-T Recommendation G.107, *The E-model*.
[38]   ITU-T Recommendation G.169, *Automatic Level Control Devices*, 1999.
[39]   3GPP TS 06.77 V8.1.1, *Minimum Performance Requirements for Noise Suppresser Application to the AMR Speech Encoder* (1999), see www.3gpp.org/.
[40]   "Noise suppression" in the named standard specifications refers to methods that reduce the background noise impact in the desired speech signal, with no limitation as to their capability of reducing the noise level either only outside of or also during speech activity. The terms "noise suppression" and "noise reduction" are used interchangeably in this book.
[41]   TR-45, *Mobile and Personal Communications Standards*.
[42]   TIA/EIA IS-853, *TDMA Third Generation Wireless; Noise Suppression Minimum Performance for AMR*.

Part II

Applications

3 Electrical echo and echo cancelation

Chapter 3 provides an overview of echo in telecommunications networks, its root causes, and its parameters. It follows the presentation with the methods used for controlling electrical echo, including network loss, echo suppression, linear convolution, non-linear processing, and comfort noise injection. The chapter covers the application of echo cancelation in wireless communications. And, in view of the fact that today's wireless networks include long-distance circuit switched, VoIP, and VoATM infrastructures (specifically as part of third-generation architectures), the chapter covers echo cancelation in long-distance and voice-over-packet applications.

3.1 Electrical echo

Many people have never experienced echo on a telephone call. They either never made calls outside their vicinity, or never encountered a malfunctioning echo canceler on their long-distance or wireless calls. In reality, echo does exist in the network. It accompanies every call involving an analog PSTN phone, but in most cases it either gets canceled before reaching its listening ear or it arrives too quickly with little delay, so it can sneak in undetected.

To hear an echo, we must first generate a sound. Then the sound must travel over a substantial distance or a slow terrain (a.k.a. complex processing) to accumulate delay. Next, the sound must be reflected and then travel back towards us. Finally, when the reflected sound reaches our ear, it must be loud enough to be heard. In addition, the amount of delay that the original sound signal incurs influences our perception of "reverberated echo" (i.e., increasing delay produces a greater echo effect). These basic concepts are the foundation for explaining the echo effect that occurs in both our natural surroundings (e.g., a stadium, cavern, or mountainous region) and telecommunications networks.

For historical and economic reasons fixed (PSTN) telephone sets are usually connected to a local switch with only two wires. The two wires form the local loop, and carry signals in two directions simultaneously. This technique works fine for short distances when signal loss is minimal. However, in situations when greater distances are covered the signal loss becomes unbecomingly large. In this case, the send and receive signals are separated, and amplifiers or repeaters are inserted in the path for each direction of transmission. These "long-distance circuits" are called four-wire sections, because one pair of wires carries the receive signal and a separate pair of wires carries the transmit or send signal.

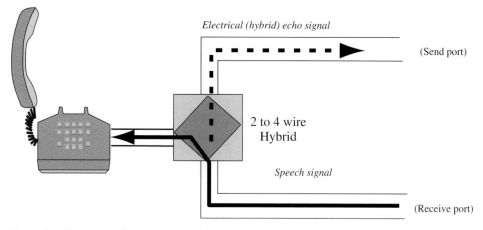

Figure 3.1 The origin of electrical (hybrid) echo.

Two-wire to four-wire hybrid

The device that connects the two-wire local loop to the four-wire long-distance circuit is called a hybrid. Ideally, a four-port network element splits the receive signal power equally across the two-wire ports. However, theoretically, this only happens if the lumped balancing impedance (Z-bal) exactly matches the distributed line impedance of the local loop. In all other cases, a portion of the receive signal will "leak through" the hybrid, and will appear at the send port. From here, it will be transported back to the speaking subscriber, who may perceive this signal (which is an attenuated and delayed version of his or her own voice) as an echo. Since a hybrid can be switched between many local loops that have different impedances, the Z-bal value is typically set to an average.

Furthermore, because of the large number of hybrids deployed in a network, the Z-bal circuitry must be inexpensive. In most cases the Z-bal function is realized with a minimum number of standard components, and the resulting hybrids are far from perfect. Therefore, signal leakage is usually present in hybrids, even if an exact impedance match is achieved. Figure 3.1 illustrates a typical hybrid configuration, and the corresponding electrical (hybrid)-echo condition that is created.

Echo as a function of delay and return loss

All telephone calls involving at least one PSTN party give off echo as a by-product. Echo is the reflection of the talker's speech to his/her own ear; for this reason it is referred to as talker echo. In a scenario where a wireless caller speaks to a PSTN caller, it is the wireless caller (not the PSTN caller), who perceives his or her own voice reflected back away from the PSTN hybrid. The signal originates as speech on the wireless side of the call. When the signal reaches the hybrid on the PSTN side, it is reflected back to the speech-originating wireless end. The wireless caller would experience that echo at a delay equal to the time it takes the signal to travel from the mobile phone to the hybrid and back to the mobile phone.

This round-trip signal delay affects the perception of echo. The longer the delay the more annoying the echo is.

Echo signals reflected at round-trip delays shorter than 25 ms are not perceived as echo but rather as side tones on the telephone earpiece. As the delay grows longer, exceeding 50 ms, the reflected signal may be perceived as a singing side tone. At delays greater than 100 ms, the reflected signal would be perceived as a clear echo.

Delay is a function of intensive signal processing, propagation, or a combination of both. Nearly all of the delay incurred in digital wireless communications is due to signal processing. Very-long-distance wireline communications incur sufficient propagation delay to have echo perceived by the talker. Echo reflected during wireline communications involving short distances is not perceived by the talker, and may be ignored by telecommunications engineers. The round-trip delay threshold accepted by telecommunications planners as the point where echo becomes bothersome is 25 ms.

Since signal-processing delays in wireless communications far exceed the 25 ms threshold, it is evident that every wireless session involving a PSTN end is accompanied by perceivable echo regardless of the distance between the endpoints of the call. On the other hand, perception of echo in wireline communications depends on distance. The 25 ms threshold kicks in once the length of the round-trip path of a given connection exceeds 3000 miles (or 1500 miles one way). This threshold distance can be calculated by taking into account the speed of signal propagation.

Since the time physics became a veritable science, the speed of light has been recognized as its most celebrated constraint. Light and electrons travel at an absolute speed of 299 792 458 meters (approximately 187 000 miles) per second (in a vacuum). This is a high speed (i.e., "light speed"), but when the metric is changed, the number and one's perception of what construes the swiftness of "light speed" shrinks considerably.[1] For example, light and electrons travel at approximately 300 kilometers per millisecond. Therefore, even a photon needs approximately 133 milliseconds to encircle the earth when following a path around the equator (assuming it remains in a vacuum).

Furthermore, "light speed" is reduced considerably when photons travel inside a medium such as glass fiber. The fiber medium slows light transmission to just about 200 kilometers per millisecond (125 miles per millisecond or 1000 miles per 8 milliseconds). Hence, a photon traveling in a fiber would need almost 200 milliseconds to circle the earth's equator.

Although the deployment of optical fiber technology has had an enormous impact on the overall telecommunications infrastructure, network planners and engineers must still consider the effect of signal propagation delay in the transmission path. This is especially significant given the rapid expansion of telecommunications networks that service a widely dispersed global community of subscribers.

Propagation delay is a less significant component of an overall signal delay in wireless communications. Nearly all of the signal delay is due to heavy processing within the wireless-access portion of the transmission path. Long distances between the end parties make the situation worse, and unless the signal is carried via satellite communications, the

---

[1] To be exact, the figure "shrivels" into 1/1000th.

relative weight of a delay associated with signal propagation is trivial in comparison with the processing delay.

Delay is not the only factor affecting echo perception. Two other important aspects are crucial in shaping awareness of this bothersome effect. They are the echo-return loss (ERL) associated with the hybrid, and the level of the source signal.

Hybrid echo carries a very clear signature of the original source signal that started it off. In fact, hybrid echo is a close reproduction, an attenuated offspring, of the source signal. The main difference between the two is the energy level, as the source signal spawning the echo is generally louder, and the power-level difference (in dB) between the source and its echo, the echo-return loss (ERL), governs its signal strength relative to the source. Standard transmission plans require that communication networks must adopt a minimum ERL of 6 dB marking a clear level distinction between source and echo signals. Evidently, a hybrid reflecting a weaker echo is a better hybrid with a larger ERL. In general, a large ERL makes echo less apparent.

The most important ingredient governing the echo power is the one driving the source signal. A louder source produces a louder echo and vice versa. By and large, the entire relationships between the two are linear, and can be expressed mathematically as a linear function. In fact, the entire echo signal, its amplitude, spectral content, and all other audio characteristics can be calculated and constructed ahead of its emergence once the source signal and the hybrid characteristics (ERL) are known. This important observation serves as a key to treating and canceling hybrid echo as discussed later in this chapter.

Figure 3.2 illustrates how perception of echo is augmented as ERL becomes smaller and delay grows larger for a given source signal. It points out the fact that perception of echo is sharper when ERL is low, and fainter when ERL is considerable. The same is demonstrated on the website (see Chapter 17).

As shown in Figure 3.2, when a voice signal originating at a wireless phone encounters an electrical hybrid in the vicinity of its PSTN destination phone (at the far end of a network), the impedance mismatch returns an electrical (hybrid) echo signal back to the wireless caller. Since the round-trip delay (i.e., the signal delay between the two ends) is approximately 200 m, the resulting echo signal would be perceived clearly by the wireless caller.

A typical wireless to (local) PSTN connection sustains approximately 200 ms of round-trip delay, almost all of which is incurred between the mobile handset and the base-station controller (BSC). The delay associated with a wireless application is primarily a result of codec equipment that is used to process voice signals for transmission over the air interface. The total delay may exceed 200 ms if the end-to-end path includes additional propagation delay due to its long range, or if transmission is carried out over IP or ATM.

Delays in long-distance telecommunications with no wireless component are mostly due to signal propagation. A propagation delay of 25 ms would account for a round trip of 5000 kilometers (2500 km in each direction). This is approximately equivalent to 1600 miles, which is the distance (air miles) between New York City and Denver. However, economic routing of telephony traffic often results in calls being bulk concentrated and redirected through major arteries before reaching specific geographic distribution points. For example, the traffic route taken by a voice signal traveling from New York

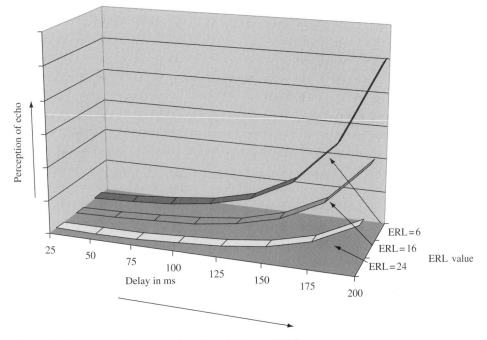

Figure 3.2 Perception of echo as a function of delay and ERL.

to Chicago can easily double the airline mile distance, thereby exceeding the "critical-echo round-trip" distance of 1600 miles.[2]

If propagation delays were the only factor affecting echo characteristics, then echo would not be perceived over routing distances shorter than 1600 miles. However, most modern-day networks (especially wireless and voice-over-packet [VoP] applications) introduce excessive delays during signal-processing activities (see Figure 3.3), hence the physical distance in these cases becomes almost a secondary consideration.

A typical voice-over-internet protocol (VoIP) connection incurs about 200 ms of round-trip delay. The delay associated with a VoIP application is distributed among codec operations, packet assembly and disassembly (including variations associated with packet sizing), buffering (and buffer sizing), and routing (including network-congestion issues) functions. In comparison, a typical voice-over-asynchronous-transfer-mode (VoATM) connection experiences about 50 ms of round-trip delay. The delay associated with a VoATM application is distributed among vocoder operations, buffering and cell assembly, and disassembly functions.

Even under highly-favorable conditions (e.g., a well planned 3G network implementation), when blending a 2G wireless mobile-to-mobile call (typical processing delay 400 ms) with 3G VoATM services packaged within a VoIP infrastructure, the round-trip voice-signal delay could approach 650 ms or more when distance and signal propagation are considered as part of the overall network topology.

---

[2] The air-mile distance between New York and Chicago is about 800 miles.

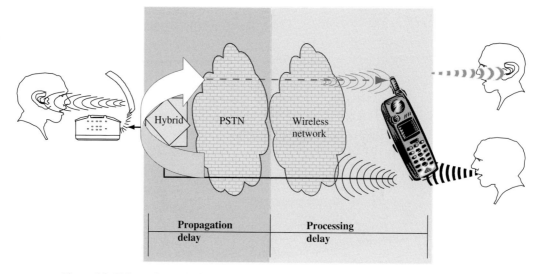

Figure 3.3 Echo and round-trip delay.

Impulse response and echo dispersion

Just like distinct musical instruments, which maintain particular timbre and frequency responses to stimulants, different network hybrids generate impulse responses, or echo, assuming unlike characteristics. The particular distinctions are put across in terms of dispersion and frequency response. Dispersion is a direct result of the fact that hybrids do not reflect the entire impulse at a single instant. Some frequencies are delayed, and are reflected later than others, generating an impulse that includes a drawn-out fading effect that is reflected over several milliseconds, as shown in Figure 3.4, while, at the same time, imparting a minor distortion onto the reflected source due to a non-uniform frequency response, as illustrated in Figure 3.5.

The ITU-T G.168[3] standard provides details for several hybrid models recommended for the purpose of testing and experimentation. The fading spell of these test hybrids ranges from about 3 ms to roughly 10 ms. The frequency responses take many different shapes.

3.2   Echo in wireless and long-distance networks

The average echo-return loss (ERL) is about 12 dB in most local-loop applications. This means that a considerable portion of the received signal is fed back to the subscriber who is speaking. Whether this condition is perceived as an objectionable echo depends upon the returned signal's amplitude (loudness) and the amount of delay. If the signal delay is less than 20 ms, the talker cannot distinguish the delayed speech signal from the original voice. However, the talker may describe the delayed signal as background noise. If the

---

[3]   ITU-T Recommendation G.168, *Digital Network Echo cancellers* (1997, 2000, 2002).

Time-domain representation

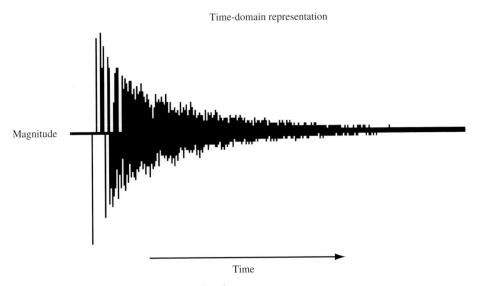

Magnitude

Time

Figure 3.4 Impulse response of an echo signal.

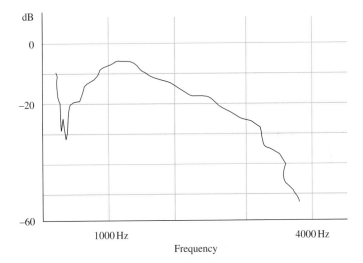

Figure 3.5 Frequency response of an echo signal.

signal delay is between 20 and 30 ms, the talker still cannot distinguish the delayed speech as echo, but may perceive sound distortion in the voice path. If the signal delay exceeds 40 ms, a distinct reverberation effect is perceived by the talker, which is commonly described as echo. If the signal delay exceeds 100 ms, separate syllables (and at 200 ms, even entire words) may be distinguished by the talker. If the returned signal level is lower by more than 50 dB (relative to the received speech signal) the talker would not perceive any echo. However, if the returned signal is loud it can confuse the talkers to the extent that they stop speaking entirely. Obviously, uncontrolled echo in telecommunications networks is undesirable because it increases subscriber complaints, and it reduces the average length of time subscribers remain on a call. Both of these situations impose economic penalties on the service provider. That is, complaints about voice quality may cause subscribers to change service providers or place fewer calls that are of shorter duration.

3.3   Echo control

In the early part of the twentieth century, echo control was needed for any circuit exposed to excessive signal delays. Circuit delays existed in long-distance wireline communications only, since digital wireless was not yet invented.

The initial technique devised to reduce (control) echo used a method called "via net loss." This approach was based on the deliberate insertion of loss, designated R, in both the send and receive voice paths. The overall objective was to improve the "signal to echo ratio" (i.e., to ensure that speech signals were louder than echo signals). In applying via net loss, the echo signals are attenuated by $2R$ dB (see Figure 3.6), but valid speech signals are only attenuated by R dB. Therefore, the improvement in the signal to echo ratio is R dB. An obvious drawback of this approach is the attenuation of the valid speech signals.

Via net loss is not the most effective way for controlling echo. Provided that the theoretical difference between the speech and its echo is R, then a greater R yields a greater echo return loss (ERL). This is good. At the same time, since R affects the speech signal as well, a greater R yields a greater loss to the actual speech. This may not be good, since too much loss may reduce signal quality.

Echo suppression

Advancements in telecommunications technology led to "echo suppression" techniques. The first echo suppressor (ES) was introduced in 1925 as an improved method for eliminating (controlling) the electrical (hybrid) echo in long distance applications.

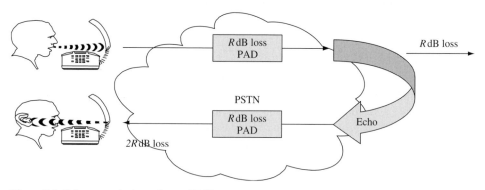

Figure 3.6 Echo control via net loss of R dB.

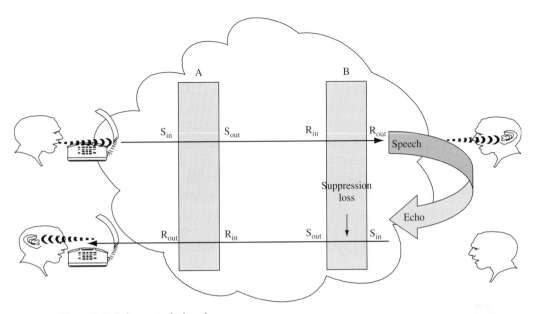

Figure 3.7 Echo control via echo suppressors.

Echo-suppressor technology is based on the concept of a speech-activated switch. Figure 3.7 illustrates a typical echo-suppressor configuration for a wireline long-distance application. The ES contains four interface ports – send in, abbreviated as S_{in}, send out, abbreviated as S_{out}, receive in, abbreviated as R_{in}, and receive out, abbreviated as R_{out}. As depicted in Figure 3.7 the send path is the path that contains the signal transmitted from the near-end of the communication session, while the receive path contains the signal originated at the far end of the communication session. The signal always flows from the "in" side to the "out" side of the echo suppressor in either the receive or send directions.

Three types of echo suppressors were defined in the ITU-T G.164 standard:[4]

a full-echo suppressor, in which the speech signals on either path control the suppression loss in the other path,

a half-echo suppressor, in which the speech signals of one path control the suppression loss in the other path, but in which this action is not reciprocal, and

a differential echo suppressor, in which the difference in level between the signals on the two speech paths control the operation of the echo suppressor.

In Figure 3.7 the speech detector located in echo suppressor B continuously monitors the send and receive paths to determine the signal levels at the echo suppressor on the B side. If ES B identifies a valid speech signal on the receive side of B, and, at the same time, it determines that the signal on the send side is not speech (the signal is determined to be weak either in relative[5] or absolute terms) then it inserts loss or clips the signal on the send side of ES B. The loss insertion or the signal cut off prevents subscriber A's echo

---

4   ITU-T Recommendation G.164, *Echo Suppressors* (1988).
5   That is, relative to the receive-side signal strength.

(which leaked through the hybrid at B) from reaching back to subscriber A. Although not in the figure, it should be noted that subscriber B presents a mirror image (top and bottom are swapped) of subscriber A, and the equivalent mirror logic works in turn when subscriber B speaks.

Naturally, echo-suppressor technology effectively eliminates echo, provided only one subscriber speaks at a time and they do not interrupt each other. However, if the subscribers A and B speak simultaneously, echo-suppression technology becomes much less effective. When employing the one-way speech logic both echo suppressors insert suppression loss in their send path, and communication in each direction is temporarily disconnected. This condition can be avoided if the subscribers use disciplined procedures, such as saying the word "over" whenever they have finished speaking to indicate it is the other subscriber's turn to speak (similar to the approach used for military radio transmissions). However, natural conversation is rarely conducted in this manner, and studies have shown that there is a 20% probability of interruptions in a typical telephone call, a condition commonly called "double-talk."

To address double-talk issues, echo suppressors were enhanced with near-end input (NI) speech level monitors. If the NI speech level exceeds a preset threshold for a specific time interval a state of double-talk is declared (typically the NI threshold setting in each echo suppressor is based on the level of the far-end input (FI) subscriber's speech signal). When double-talk occurs, both echo suppressors cease their loss insertion in the send path (see Figure 3.7), thereby fully connecting the voice path between subscribers. Accordingly, echo is not suppressed during double-talk. As a compromise, some echo suppressors insert loss in the receive path to reduce echo during double-talk. The rationale for inserting loss is that the higher amplitude of the incoming speech signals masks the lower-amplitude echo signal returned to each subscriber during double-talking. Of course, the best results are obtained if these actions are taken at both ends of the voice path simultaneously.

The characteristics of the speech detectors are crucial to the performance of echo suppressors. If the speech detectors react instantaneously to incidental threshold crossings (e.g., signal spikes and valleys), the echo-suppressor loss insertion would start and stop in rapid succession. This may cause multiple abrupt level changes that severely chop speech signals (commonly called speech clipping). Therefore, speech detectors are usually designed with "inertia" that allows settling time before determining whether the input signals have actually exceeded the preset NI thresholds. This characteristic is called "hangover time," which can occasionally cause the echo suppressor to continue suppressing when it should have stopped, and vice versa. As a result, the first syllables of double-talk speech may be "chopped off" or the last syllables of a single subscriber's speech signal (i.e., during single-talk) may leak through the hybrid as echo. These effects are intensified when there is a large amount of delay in the voice path (e.g., international long distance calls).

Long delays in the communication path bring about diminished synchronization between the two communicating ends. Reduced synchronization escalates incidences of interrupting double-talk, stopping and starting. Echo suppressors do not control echo very well during double-talk episodes, and they tend to clip first syllables after a speech pause.

These attributes produce more double-talk interruption: "What did you say?" and more stop and go situations surrendering to a cycle of increasingly poorer and poorer voice quality.

Echo-suppressor technology is no longer used in telecommunications networks because of its obvious drawbacks. Specifically, increased network delay (e.g., long-haul international calls, satellite transmission, wireless communications, voice-over-packet) rendered echo-suppressor technology ineffective, and encouraged the development of echo cancelation equipment.

Echo cancelation

In the early 1960s, the first commercial communications satellites were launched. These satellites were (and still are) located in equatorial geo-stationary (GEO) orbits at 36 000 km above the earth's surface. In this arrangement, radio signals traveling at the speed of light (300 000 km/s) accumulate a one-way delay of approximately 240 ms between satellite to earth stations. With the introduction of satellites as a transport mechanism for intercontinental calls, round-trip signal delays of 300 ms (or more) became prevalent. New methods of echo control had to be developed to accommodate this condition because echo suppressors could not support natural conversations in the presence of excessively long signal delays (i.e., subscribers were required to resort to the highly disciplined speech procedures described earlier).

A new concept for echo control was invented at Bell Laboratories in 1964, commonly called echo cancelation. An echo canceler is a device that employs two major processes – linear and non-linear – whereas echo suppressors rely solely on the non-linear part. The linear process analyzes the received speech signal, its corresponding echo, and the relationships between the two. It then generates a predictive model of the next estimated echo signal. It follows by subtracting the estimated echo from the signal(s) that is returned from the far-end of the connection.

The echo estimate is obtained via the following linear mathematical function

$$\text{new echo} = \text{linear function (previous and present signal and echo pairs)} + \text{residual echo}$$

$$\text{new echo estimate} = \text{linear function (previous and present signal and echo pairs)}$$

$$\text{new echo estimate} = \text{new echo} - \text{residual echo}$$

The linear function is chosen by a method that minimizes a form of the residual echo. There are several different algorithms for selecting the most proper form to be minimized.[6]

---

[6] ITU-T Recommendation G.165, *Echo Cancellers* (1993).

M. M. Sondhi, An adaptive echo canceler, *Bell System Technical Journal*, **46** (March 1967), 497–510.

Mun Keat Chan, Adaptive signal processing algorithms for non-Gaussian signals, Ph.D. dissertation, Engineering Faculty, Queen's University Belfast (2002).

Perry P. He, Roman A. Dyba, and F. C. Pessoa, *Network Echo Cancellers; Requirements, Applications and Solutions*, Motorola Inc. (2002).

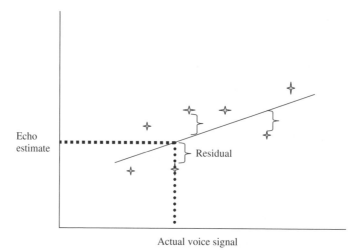

Figure 3.8 Echo estimate via a linear function of actual voice signal.

The most popular ones are forms that minimize a weighted sum of the mean-squared residuals.[7] A residual error may be either positive or negative. When adding up residual errors, the sum must take into account either the absolute or the squared error.[8] The squared error is the preferred method, since the second-degree exponent tends to weigh larger errors more than smaller ones.

In Figure 3.8, the linear function is drawn as a straight line. When a new signal point enters the system, the new echo estimate is computed as a corresponding point on the line, and the actual echo may be positioned a small residual away.

The linear process is capable of canceling much of the echo, but the estimate and signal summation do not eliminate it in its entirety. It leaves a trace of a weak echo residual that, by and large, can still be heard. Accordingly, a non-linear process (NLP), similar to the echo suppressor course, follows the linear one with the task of removing the residual portion of the echo.

Figure 3.9 provides a high-level illustration of an echo-canceler functional process. During single-talk originating at the far end, a speech signal enters the EC on the receive (R_{in}) port. Once in, the EC computes an estimate of expected consequent echo and stores it in the EC memory as the signal progresses through the R_{out} port on its way to the final destination. No echo processing is carried out by the EC on the speech signal while on route to its destination within the receive side. Signal processing begins when the ensuing echo arrives after some delay at the send (S_{in}) port. Once the echo is in, the EC subtracts the echo estimate from the actual echo signal. The resulting residual is then used internally as an additional data point for updating the next echo estimate; while

---

[7] D. L. Duttweiler, Proportionate normalized least-mean-squares adaptation in echo cancelers, *IEEE Transactions on Speech and Audio Processing*, **8**:5 (2000), 508–518.

[8] A square is always a positive number.

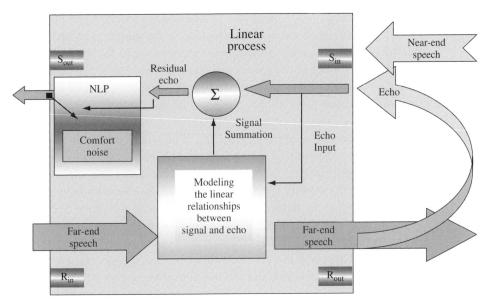

Figure 3.9 Echo canceler functional architecture.

at the same time the signal progresses toward the non-linear processor (NLP). Once in the NLP, the residual echo is attenuated or broken off as long as there is no near-end speech.

The theory used to analyze echo cancelation is based on an assumption that the echo signal entering the S_{in} port can be represented as a linear expression of the far-end subscriber's speech signal (R_{in}). In reality, the far-end subscriber's speech signal may encounter coder/decoder (CODEC) equipment, a hybrid, loaded and non-loaded network sections, and several local switches before it reaches the S_{in} port of the echo canceler. As a result, the S_{in} signal is delayed, distorted, dispersed, de-coded, re-coded, and may have had significant noise added to it. Fortunately, the non-linear and stochastic components in the S_{in} signal are relatively small, and a well-designed linear echo canceler can reduce the echo signal level by about 30 dB relative to its ancestor speech signal. However, when an echo signal is delayed more than several hundred milliseconds, it can be perceived at levels of −50 or −60 dBm0,[9] depending on the far-end subscriber's ambient (surrounding) noise level. Therefore, a non-linear processor (NLP) is incorporated into most echo cancelers to handle the residual echo condition. Although non-linear processors effectively eliminate residual echo, the on/off NLP switching operation may cause clipping of speech signals. Consequently, many NLP implementations have replaced the abrupt on/off operation with a smoothing procedure, applying gradual introduction of loss that attenuates the residual echo signal while reducing the likelihood of speech clipping.

In addition to cutting the residual echo off the send path, the NLP process includes a comfort-noise injection feature that is turned on as soon as the NLP cuts off the actual

[9] An explanation of dB, dBm, dBm0, dBr, etc., is given in Section 6.1.

signal from the send path. Comfort-noise injection is a crucial feature in echo cancelers. It is designed to match the noise accompanying the speech in an attempt to maintain smooth continuity once the NLP cuts off the signal and the background goes silent. Without proper comfort-noise injection, the NLP operation becomes evident to the listener, and its on/off action is quite disturbing. It creates a feeling of heavy breathing, noise pumping, and noise clipping. (Check the audio illustration on the website. See Chapter 17.)

Complete elimination of circuit noise is a drawback of the switching function used in both NLP echo cancelers and echo suppressors. Typically, the partially analog tail-end circuit has a higher background-noise level compared with a fully digital long distance or wireless circuit. As a result, when the echo-canceler NLP or echo-suppressor switch does not block the signal the far-end subscriber hears the tail end of the analog-circuit noise. When the NLP blocks the signal, the noise level suddenly drops down to the lower noise level of the digital circuit. If the difference between noise levels is extreme, the far-end subscriber may perceive the call to be disconnected (i.e., the far-end subscriber may assume that the near-end subscriber has hung up). The comfort-noise injection overcomes this problem by smoothing out the background noise perceived during speech and speech pauses.

To implement the comfort-noise feature, the tail-end circuit noise is characterized during silent periods and stored in a memory. When the NLP switch blocks the connection to remove residual echo, a matching noise signal (i.e., comfort noise) is inserted at the S_{out} port of the echo canceler. The background noise heard by the far-end subscriber remains constant throughout the call, thus significantly contributing to the perception of high-grade speech quality.

Noise characterization involves two key parameters, level and color. When the background-noise level is relatively low, as is true in most PSTN circuits, level matching is sufficient. Since echo cancelers are deployed at the tail of the PSTN where noise levels are relatively low, comfort-noise matching may employ white noise that matches the level of the actual noise. A more advanced procedure referred to as "colored noise matching" enhances the actual noise approximation by adding matched spectral characteristics to the simulated comfort noise signal. The topic is given a comprehensive analysis as part of the discussion on acoustic echo control in Chapter 4 where background noise is louder, more colorful, and more perceptible by the listener, and where comfort noise becomes even more vital to the quality of the communication.

Wireless speech codecs that utilize predictive speech interpolation techniques require continuity of voice path signals to insure optimum performance. It is obvious that both "on/off NLP switching" and "on/off comfort-noise injection" introduce discontinuity in the voice path. As an alternative, smooth operation of NLP with a coordinated inverse ramp up/down operation of comfort-noise insertion can maintain voice path continuity and improve the performance of the wireless-codec equipment.

Although the echo-canceler NLP operation is equivalent to an echo suppressor, its effectiveness is a great deal superior. The key to its utility is its far improved echo-activity detection (EAD). The residual echo left over after the linear process completes its task is much weaker than a likely speech signal; it can better be characterized as to what it is, so that it can be more easily separated from speech. For that reason, the NLP becomes less prone to erroneous decisions, which increase speech clipping and bursts of echo due to signal misclassification.

One of the most challenging tasks associated with an echo canceler is the elimination of echo during double talk. Under these conditions, the signal entering the S_{in} port is a blend of echo and near-end speech. The linear relationship between the far-end signal and the near-end one is no longer valid because the far-end and the near-end speech signals are uncorrelated due to double talking, i.e., the signal entering the R_{in} has nothing in common with the one entering the S_{in} port, even though the S_{in} port contains an echo component. Any update to the echo estimate employing the S_{in} echo plus near-end speech signal could deform the existing estimate. Consequently, the proper way to operate the linear portion of the EC algorithm is to freeze its adaptation operation, while prolonging the application of the existing estimate on all new S_{in} signals as long as double talk conditions persist. At the same time, the NLP must cease its operation and let the signal leaving the linear filter pass through its gates intact.

It should be understood that operation of the NLP function during double-talk conditions does not intrude on the voice path. This is because the bulk of the echo signal is removed by the linear convolution function, and the NLP removes only the leftover residual echo for the duration of single-talk. During double-talk, signal degradation may occur, since the NLP lets the signal through, allowing the residual echo signal to return to the far-end subscriber. However, since the residual echo has been reduced by 30 dB, it may be effectively masked by the near-end subscriber's much louder speech signal during double-talking (i.e., the residual echo may not be detected by the far-end subscriber). In comparison, an arrangement using an echo suppressor would allow the far-end subscriber's entire echo signal to be returned, which would cause a much higher level of signal degradation and annoyance for the far-end subscriber throughout the double-talk condition.

As seen in the figure, the major and most crucial element absent from the echo suppressor is the linear component of the algorithm.

Figure 3.10 illustrates the EAD decision process while contrasting an echo suppressor with an echo canceler process. In the figure, the near-end signal is either speech or echo. If it is echo, the linear process attenuates it by just about 30 dB, and in the process, it downgrades the signal level seen by the NLP. The amount of cancelation achieved by the linear process is referred to as the echo return loss enhancement (ERLE), and the echo signal strength is reduced by that amount. If the near-end signal is pure speech then the linear process contributes nothing, yielding ERLE $= 0$, and the signal strength reflects the speech level. If it is speech $+$ echo, then the signal strength is still reduced by ERLE to become speech $+$ echo $-$ ERLE, to reflect the speech level and any residual echo.[10] When the signal is classified as speech or speech $+$ echo, the NLP neither inserts loss, nor does it clip the signal. When the signal is classified as pure echo, the NLP either clips it or inserts sufficient loss to reduce its level all the way to irrelevance.

---

[10]   Residual echo $=$ echo $-$ ERLE.

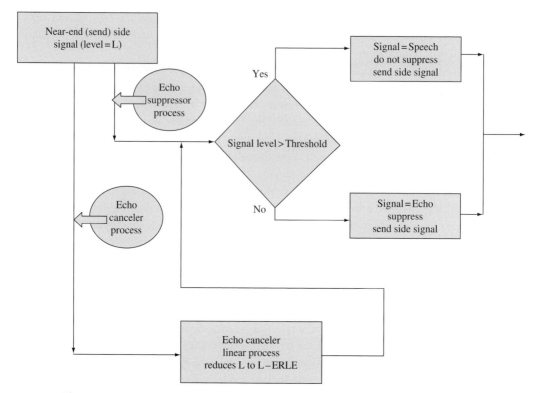

Figure 3.10 Echo control algorithms logical flow chart.

3.4   Placement in the network

Figure 3.11 illustrates a typical echo-canceler configuration for a long-distance wireline application and for the wireless application, respectively.

Network echo cancelers are usually deployed next to a long-distance toll switch, an international gateway, a wireless switch, or a voice-over-packet gateway. The first general rule calls for deployment of an echo canceler next to a gateway confronted with a long transmission delay. The second general rule positions the echo canceler in such a way that its tail (S_{in} and R_{out} ports) faces the PSTN hybrid. Although these rules appear to be one and the same among the different network types, caution must be applied to insure that the application is put into service the way it is supposed to.

When it comes to echo cancelers, wireless networks appended to a PSTN network exhibit a non-symmetric connection.[11] The non-symmetry with respect to echo is due to the digital-wireless access, which does not include a hybrid, and, therefore, does not produce

---

[11]   A non-symmetric condition comes about when only one end of the connection contains a hybrid emitting echo, whereas the other side comprises a digital (wireless, VoIP or ISDN) phone with no need for a hybrid.

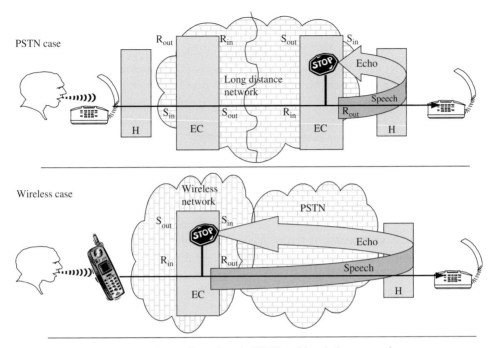

Figure 3.11 Echo-canceler configurations in PSTN and in wireless networks.

electrical echo in the PSTN direction. The only electrical echo is the one experienced by the wireless caller. On the other hand, toll networks terminating on the PSTN at either end are symmetric with respect to hybrid echo.[12] Voice-over-packet networks may or may not be symmetric. Their symmetry depends on whether any end point terminates on the PSTN or on a digital PBX.

An important architectural feature of toll networks is that echo cancelers installed next to a toll or an international gateway switch are designed to protect callers from the far end. If a PSTN subscriber from the USA calls a country in a different continent, and the American hears her own voice reflected back to her ear at a disturbing delay, she might call her long-distance service provider and complain. Unfortunately, the US-based service provider is clean and innocent. The disturbing echo is the foreign service provider's fault. The echo heard on the US side of the connection has been reflected back from the foreign land. It should have been canceled in the foreign country before crossing the ocean back to the USA.

The main reason for the architectural feature of long-distance networks has been, and still is, the need for minimizing the tail delay, thus improving cost and performance of the

---

[12] Symmetry is brought about by the fact that both sides of the connection face an identical echo problem. Accordingly, each side deploys an echo canceler that cancels echo originating on its far end.

echo cancelers. Tail delay is defined as the delay incurred to the speech-turned-echo signal on its journey from the R_{out} port through the hybrid and back to the S_{in} port of the echo canceler as seen in Figure 3.11. This measure reflects the minimum time interval that an echo canceler must incorporate into its processor when correlating a speech signal to its corresponding echo. Long delays require the processing of long intervals. Long intervals require more memory or more computing resources, or both. At the same time, the performance of echo cancelers is inversely related to the maximum amount of tail delay that they are designed or configured for controlling.

Since the long-distance connection between two PSTN ends is symmetric, both sides must cancel hybrid echo, and both sides must incur the expense of buying, installing, and managing echo cancelers to protect subscribers from the other end who may be served by another administration. This recognition drove the international community to set standards (ITU-T G.164,[13] G.165,[14] and G.168[15] recommendations) benefiting both sides of the long-distance connection, by recommending that each network segment protects the other, rather than its own, from hybrid echo, while reducing costs and boosting performance for both.

While echo cancelation responsibility for toll networks is managed by the network segment servicing the other end point, it is not so when it comes to wireless networks. Echo cancelation responsibility protecting the wireless subscriber is the wireless-network service provider's duty. The three main reasons for this requirement are history, economics, and lack of symmetry. Digital-wireless networks were built years after the PSTN was already in place. They were managed by service providers, different, in most cases, than their contiguous PSTN administration. Local PSTN networks have not incorporated echo cancelation equipment within their infrastructure because round-trip delay is short and echo has not been an issue. The appendage of the wireless network, and the added delay caused by the wireless codec made the echo large enough to be perceived by the wireless subscriber on PSTN connections. The PSTN subscriber does not experience electrical echo on calls going to wireless end points, simply because there is no hybrid in that particular receive path. Obviously, echo is the wireless-network problem. And the delay that brings it to the surface is instigated by the wireless codec that has been imposed on top of an existing PSTN infrastructure. The simple attitude: "You've caused it. You fix it," prevails here as well.

Lack of symmetry implies no change to a local PSTN infrastructure as wireless networks are adjoined to its neighborhood. Naturally, the add-ons must take care of the interface and the associated performance issues that may come about. And consequently, no disturbance to the PSTN is called for as the new neighbor moves in. What's more, even combined economical reasoning justifies placement of echo cancelers inside the wireless network within its gateway to the PSTN, where every call requires echo to be filtered out. Placing echo canceling on local PSTN switches may become vastly impractical owing to

---

[13]   ITU-T G.164, *Echo Suppressors.*
[14]   ITU-T G.165, *Echo Cancellers.*
[15]   ITU-T G.168, *Digital Network Echo Cancellers.*

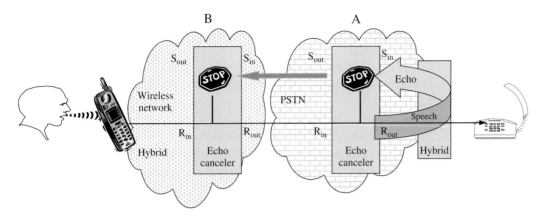

Figure 3.12 Wireless to PSTN (long-distance) call configuration.

the very poor utilization affected by the large volume of calls that originate and terminate on the local PSTN with no need for echo cancelation.

It should be noted that tail delay, a.k.a. echo-path delay, is independent of round-trip delay. The echo tail is typically very short relative to the round trip. As illustrated in Figure 3.11 the bulk of the delay and the distance in the long-distance network is outside the tail, and the bulk of the delay in the wireless network is caused by processing inside the wireless network prior to the signal encounter with the echo canceler.

Figure 3.12 illustrates a call configuration between a wireless and a long-distance PSTN subscriber. In this case, the wireless caller is protected by two echo cancelers working in tandem. Tandem configuration is not always a preferred configuration. Although the echo should be canceled out by the long distance EC, the situation may become complex because, at the very start of the call, the relationship between the speech signal and the echo inside the wireless echo canceler is non-linear and time variant. It may become non-linear throughout the call if the long-distance echo canceler performs less than a perfect job. A more detailed discussion of tandem configuration is presented in the following section.

3.5 Convergence and adaptation processes

Earlier in the chapter, the methodology for computing the next echo estimate was established. The estimate was shown to be a linear function of past speech and corresponding echo pairs. However, at a start of a call, there is either no history or there is irrelevant history,[16] and the echo estimate is either null or erroneous

---

[16] Irrelevant or even misleading history may exist inside the echo canceler at the start of a new call, since any historical data may belong to a call that has just terminated. This situation is very common since many echo-canceler implementations do not reset their internal memory at the start of a new call. Many echo cancelers may not even be aware of call boundaries.

respectively. A proper echo estimate is constructed over time, as more and more relevant data keep moving in. The relevant data required for establishing a proper echo estimate are far-end speech[17] and its corresponding near-end echo. The echo estimate improves as it builds upon the streaming input up to a point where it becomes stable, realizing a relative steady state. Once that point is attained, the echo canceler has converged.

The process of convergence may take several seconds of far-end speech input. In practice, the far-end speaker may experience a short burst of echo at the beginning of a call before the echo fades away. The process of convergence applies only to the linear portion of the echo-canceler algorithm. Although it may take ten seconds (or more) of far-end speech for completing the convergence process, the entire echo may be eliminated much sooner. The ongoing algorithmic progress towards a steady state boosts up the ERLE such that the non-linear processor threshold sensor crosses the echo-activity detection point; triggering a blocking action even before full convergence is attained. Although echo may be blocked off early on by the NLP action, the echo canceler continues its convergence process in the background, as additional far-end speech input keeps entering the system.

The convergence process inherent in an echo canceler at the beginning of a call makes its impact on a second echo canceler that is placed in tandem along an echo path. Tandem configurations of echo cancelers lead to time variant and non-linear echo tail input at the S_{in} port of the second echo canceler in the sequence. The process of convergence within the first echo canceler in the sequence contributes to consistent reduction in echo seen by the second echo canceler. In the absence of a stable relationship between signal and echo, it becomes impossible for the second echo canceler (EC B in Figure 3.12) to construct a viable echo estimate. Instead, it may generate an erroneous estimate that once added to the actual signal at the S_{in} port, might cause some unpleasant artifacts at the very beginning of the call. The distortion effect may last seconds before the first echo canceler (A) in the sequence attains full convergence. At that point, as long as the first EC (EC (A) in Figure 3.12) works properly, the second EC (EC (B) in Figure 3.12) may not see echo at all for as long as the call is up. If the first echo canceler is less than perfect, than the signal arriving at the S_{in} port of the second EC may present inconsistent relationships between the R_{in} speech and the S_{in} echo signal. Inconsistency may lead to the construction of erroneous echo estimates and erroneous EC behavior due to misinterpretation of signal input at the S_{in} port. Erroneous behavior by an echo canceler may set off strange noises, echo, or other interfering sound effects, which severely impair the quality of the voice during the phone call.

A well-designed echo canceler maintains several sensors that provide status information on a number of fronts. Generally, condition details are required when a change

---

[17]   The process of convergence may take ten or more seconds of far-end speech input. If far-end speech is paused or interrupted, the amount of real time required for full convergance may take longer than that. Far-end speech is speech entering the EC at the R_{in} port. Near-end speech is speech entering the EC at the S_{in} port.

in status may trigger a change in behavior of the algorithm. The most notable examples are as follows.

- Rapid convergence at the start of a call or at some point of instability requires measures different from an ordinary adaptation once in steady state. When the ERLE is deemed insufficient or when the difference between the actual echo and the computed estimate – denoted as the "error" or S_e – is relatively large, it is an indicator that rapid convergence must take place to close the gap. The EC algorithm may trigger a faster convergence process by assigning much more weight to the most recent input taps relative to past input. It may even increase the leak rate of past data. That way, the EC boosts the gain until it is satisfied that the error has become smaller or the ERLE has strengthened and attained a minimum threshold value. The opposite takes place as the EC senses a steady state. When the error or S_e is small, and the ERLE appears to be high and stable, past speech and echo input pairs would be valuable, and recent input ought not force any major change in the echo estimate. Under these circumstances, the gain is lowered so that a sudden (short) burst of atypical noise does not cart the estimate off its steady-state course by diverging and yielding a larger S_e and a poorer ERLE.
- Following single talk, double-talk conditions trigger a significant shift in the algorithm behavior. The algorithm makes the NLP cease its blocking action; it ends the adaptation carried out by the linear convolution process, while continuing to apply the frozen estimate. Consequently, double-talk conditions must be sensed quickly before the echo estimate gets corrupted and before the near-end speech gets clipped.

Sensing double-talk conditions may be performed on several fronts. The first indicator is steered by the near-end speech detector. In most echo cancelers, the chief sensor detecting near-end speech is based on absolute or relative signal-level thresholds. When the near-end signal level exceeds a specified threshold, the echo canceler assumes it to take the form of speech. Nevertheless, relying upon signal level without any supporting evidence may lead to a higher probability of classification errors, and some EC implementations have resorted to a forward–backward type procedure,[18] employing a background shadow canceler that keeps on updating the echo estimate without committing the result to the near-end signal. The shadow canceler is employed for the sole purpose of comparing the resulting error of the frozen estimate with the updated one. If the updated estimate scores better[19] than the frozen estimate, then the sensing of double talk must be erroneous, otherwise the classification is accurate.

Allocating resources to a shadow canceler for the sole purpose of verifying double talk conditions may prove to be an expensive proposition. Methods that go half way do exist. One such procedure (see Figure 3.13) checks the S_e magnitude after adaptation. If the S_e moves in the wrong direction, while signal level at the S_{in} port draws near to the double-talk threshold, the adaptation is called off, and the frozen estimate is applied instead.

---

[18]   J. Hennbert, C. Ris, H. Bourlard, S. Renals, and N. Morgan, Estimation of global posteriors and forward–backward training of hybrid HMM/ANN Systems, *Eurospeech*, (1997), 1951–1954.
[19]   If the error is smaller, the score is higher.

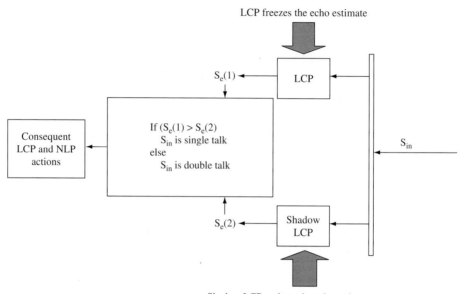

Figure 3.13 Shadow linear convolution.

3.6   Tail delay and its impact on performance

Echo cancelers are designed with a particular maximum tail delay capacity. As an example: when tail delay capacity is set to 64 ms, it is said to have a 64 ms h-register window. The h-register is the term used for the echo-canceler engine that maintains and processes the echo estimate. If the actual tail delay of the echo is less than 64 ms, then its impulse response shows up somewhere within the 64 ms window, and the echo canceler should be able to hunt it down at the start of its convergence process. If the actual delay exceeds 64 ms, the impulse response is outside the echo canceler range and the echo cannot be canceled.

With developments in voice-over-packet technologies that inserted end points, network links, and components inside the echo tails, delays have increased substantially, and the echo-tail capacity for echo cancelers has risen in turn to cover the extended delays.

As tail delays became longer, early implementations of echo-canceler algorithms, like the plain normalized least-mean-square (NLMS),[20] faced fresh performance challenges as the unvarying allocation of the algorithm effort over the extended window yielded slower convergence and slower reaction to changes in single to double-talk states.[21] It became evident that smarter algorithms, which concentrate the processing effort on the narrow region occupied by the impulse response (rather than the extended interval), were required.

---

[20]   Sondhi, Adaptive echo canceler, 497.

[21]   The NLMS algorithm allocates its resources equally throughout the h-register window, even though most of the h-register window is irrelevant to the echo-cancelation process because the impulse response occupies a very small part of the window space.

The next generation of echo canceler algorithms combined more pre-processing and cleverer math. The pre-processing effort focuses on locating the region (or regions) occupied by the impulse response ahead of the recursive process. Once the region is located, the recursive algorithm that computes the echo estimate can operate over the smaller, though relevant, interval. Mathematically, some implementations of the NLMS algorithm were upgraded to a variation of the proportionate NLMS (PNLMS),[22] which assigns heavy weights to data points located in the region of the impulse response, thus speeding up convergence and reaction time to state changes by allocating resources more efficiently. Because the PNLMS algorithm has shown instability under certain conditions, versions of PNLMS, which alternate between NLMS and PNLMS, are known to exist in several implementations of the least-squares algorithms for canceling echo. Although the NLMS part of the set slows down the convergence process, it introduces stability that pays off handsomely.

Another way to combat very large tail delays without over-extending the size of the h-register window has been applied to situations where tail delays are known to inhabit specific ranges. When remote regions connect to the main network via satellite communications, the delay range between a particular endpoint and the echo canceler, i.e., the echo path (or tail) delay may vary by as much as 40 ms among end points. Accordingly, the total delay is always equal to a satellite delay plus X ms, where X is a number between 0 and 40. Under these circumstances an h register window size of 64 ms may be time shifted by Y ms since the impulse response would never appear earlier than Y ms on a window occupying the full range. The time shift is implemented by having the speech signal stored for Y ms before applying it into the linear convolution process.[23]

3.7   Far-end echo cancelation

The typical transmission plan for echo control between two international administrations is for each administration to deploy an echo-control device at its international switching center (ISC). As illustrated in Figure 3.14A, each ISC controls the echo generated on its domestic network. This way, the echo heard by administration A's customers is controlled by administration B's echo control device and vice versa.

With the diverse levels of deployed echo control devices between international administrations, several administrations deploy a long-haul echo canceler as shown in Figure 3.14B.

As illustrated in Figure 3.14B, administration A deploys two echo cancelers. One controls the echo generated within A's network and the second one controls the echo generated in B's network. This solution is practical when the long haul is within the maximum tail delay of the echo canceler. If, however, the long-haul facility consists of

---

[22]   Duttweiler, Proportionate normalized least-mean-squares adaptation in echo cancelers, p. 508.
[23]   See a more detailed discussion in the following section.

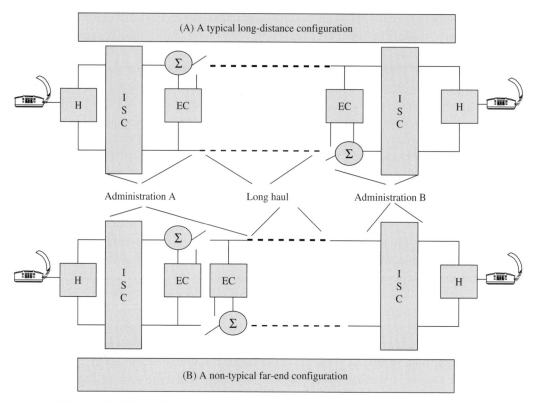

Figure 3.14 EC long-distance configurations.

7000 km of fiber cable and administration B's maximum practical facility mileage is 1000 km, the round-trip delay for the echo canceler is calculated using the formula below:

$$\text{round trip long haul delay (LHD)} = \text{total one-way distance (km)}$$
$$\times\ (0.01\,\text{ms}/1\,\text{km})$$

Using the formula above the LHD = 7000 km × (0.01 ms/1 km) = 70 ms, meaning that the long-haul echo canceler must provide a minimum of 70 ms tail capacity. In the case where the long haul route is accommodated by a geostationary satellite approximately 30 000 km up in space, the facility distance is 60 000 km (comprising up-link and down-link). The LHD (60 000 km × 0.01 ms/km) amounts to 600 ms. Unfortunately, the tail capacities of most commercial echo cancelers do not generally accommodate delays longer than 200 ms, and there is a need for a more creative approach that utilizes existing technology and its limitations to solve the far-end echo-cancelation situation.

The solution to the far-end echo cancelation utilizes a shift of the tail-capacity window. Normally, an echo canceler's tail-capacity window is set to cancel echo that turns up at its S_{in} port at delays between 0 ms and its maximum tail capacity (64, 128, or even 192 ms). When delays are known to be permanently longer, the same window may be shifted to cover the same range, but at a different time slice.

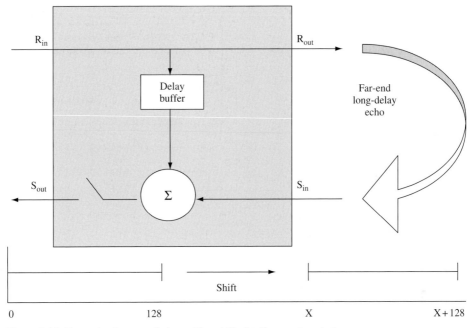

Figure 3.15 Far-end echo cancelation with a shifted tail-capacity window.

Figure 3.15 illustrates how signals arriving at the R_{in} port are buffered for the proper amount of time before being introduced to the convolution processor. In this case, the echo canceler operates as if the signal arrives later on. Once released from the delay buffer, the signal information is introduced in the convolution processor, which employs its standard tail capacity (128 ms in the example of the figure) for canceling the echo.

In some cases, actual far-end delay may be variable. Trunks may be routed over a satellite when a terrestrial route fails or when terrestrial capacity is exhausted due to high traffic volume. Under these conditions, the echo canceler must resort to a variable delay buffer. The procedure involves a nearly instantaneously change in the canceler's long-haul delay from terrestrial short to satellite long (600 ms). However, the echo canceler must be able to detect the change in facility delay automatically and without special equipment in administration B's ISC. The change in delay is detected by resorting to detection of signaling tones present on ITU-T no. 7 or no. 5 signaling systems.

ITU-T no. 7 solution

ITU Signaling system 7 operates by generating a 2000 Hz continuity check (CC), a.k.a. voice path assurance (VPA), tone for every call between ISCs. Figure 3.16A illustrates a block diagram of an echo canceler with the CC1 detector monitoring the S_{out} port and the CC2 detector monitoring R_{in}.

With detection of the outgoing CC tone at CC1, a timer inside the delay processor (DP) starts counting. When the returning CC tone is detected by CC2, the timer is stopped.

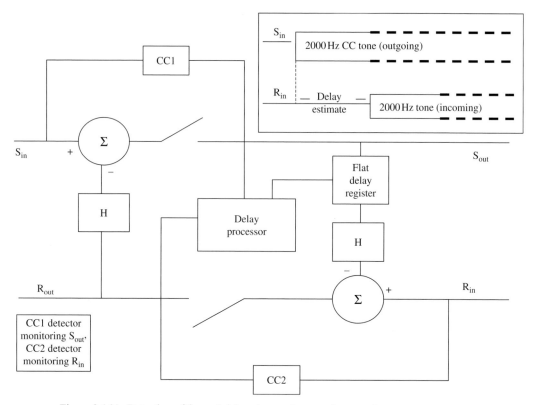

Figure 3.16A Detection of far-end delay changes by an echo canceler.

The value calculated by the DP is the round-trip delay between ISC's A and B administrations. This value is compared with the current one. If the delay is found to be different from the one in the flat-delay (FD) register, the new value replaces the current one or an entire trunk group. A range or window of time (a $\pm 5\%$ window) around the present FD register value must be established. The flat-delay register may not be updated when the CC tone is first detected by CC2. Any delay-processor value outside this window may update the flat-delay register to the new DP value.

ITU-T no. 5 solution

A similar delay-monitoring concept is utilized to update the value in the FD register for changes in the long-haul facility delay for facilities where ITU-T no. 5 signaling is deployed. Table 3.1 provides an overview of the supervisory-signaling tone sequences.

The algorithm is based on detecting the "seizure–proceed to send" of signaling no. 5, which occurs at the start of every call and is completed before the call begins to ring. The only risk to this choice is the fact that the F1 and F2 detectors shown in Figure 3.16B must operate within 40 ms. The algorithm is effective only on calls from ISC A to ISC B.

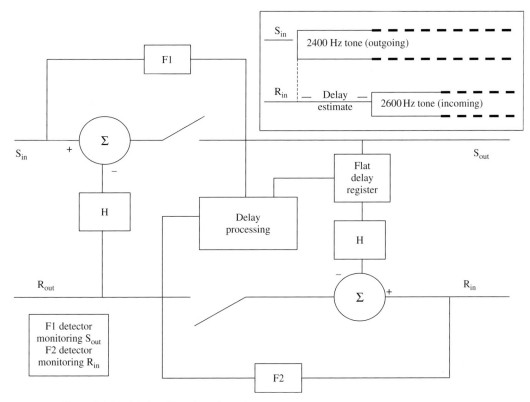

Figure 3.16B C5 signaling-detection scheme.

Table 3.1. *An overview of the supervisory-signaling tone sequences*

	Signal	Frequency	Sending duration	Recognition time
1	Seizure	F1	Continuous	40 ± 10 ms
	Proceed to send	F2	Continuous	40 ± 10 ms
2	Busy flash	F2	Continuous	125 ± 25 ms
	Acknowledgment	F1	Continuous	125 ± 25 ms
3	Answer	F1	Continuous	125 ± 25 ms
	Acknowledgment	F1	Continuous	125 ± 25 ms
4	Clear back	F2	Continuous	125 ± 25 ms
	Acknowledgment	F1	Continuous	125 ± 25 ms
5	Forward transfer	F2	850 ± 200 ms	125 ± 25 ms
6	Clear forward	F1 & F2	Continuous	125 ± 25 ms
	Release guard	F1 & F2	Continuous	125 ± 25 ms
	Forward signals		$F1 = 2400$ Hz	
	Backward signals		$F2 = 2600$ Hz	

The shaded area is preset at the start of every outbound call.

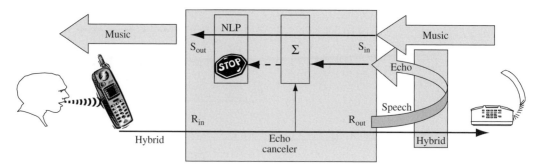

Figure 3.17 Music on hold – uninterrupted.

3.8   Background music – music on hold

When calling a busy PSTN operator who asks the caller to "please hold" while letting her "enjoy" the background music, it is quite common for the music to get clipped in reaction to noise or speech originating at the caller's end. Choppy background music may become an annoying matter, and special measures have been taken by echo-canceler algorithms to smooth the broken-up audio and let it flow uninterrupted. (see Figure 3.17)

The main reason for having chopped music in reaction to far-end noise or speech is inherent in older NLP algorithms. Older implementations based their NLP action trigger on absolute or relative signal levels. The threshold setting was provisioned in advance, and was derived from the rule that contrasts the R_{in} to the S_e level.[24] It blocks the near-end signal from moving forward when the R_{in} level is noisier by x dB[25] than the NLP, otherwise the NLP lets the signal through. In the case of music on hold, when the far-end caller speaks, the R_{in} signal reflects a high level. If, at any instant, the level difference between the speech (R_{in}) and the music level, i.e., ($R_{in} - S_e$) happens to move below the threshold level setting, the NLP blocks off the music. It chops a few notes away, until the speech gets weaker relative to the music.

Improvements in the NLP design allow music to sneak through the NLP uninterrupted even when the far-end speaker is active. The enabling technology takes advantage of the intelligence gathered by the linear convolution process. It takes account of the noise floor at the near end, the residual-echo amplitude, and of the exact echo-path delay.

Real-time awareness of the constantly changing noise floor lets the NLP pass through samples comprising lower amplitude. When the lower-level samples are musical audio, the far-end speaker can hear it uninterrupted. When they are echo, they are masked by the noise. This NLP process is referred to as a center-clipper transfer function.

Since the echo return can occur with a delay from zero to the maximum tail length (typically 64–192 ms), older implementations of the NLP had it operating over that entire

---

[24] The S_e level is the signal level measured after being processed by the linear convolution processor and before entering the NLP.

[25] Where x is a user's provisioned value.

window when significant energy was detected at R_{in}. This process guaranteed the complete elimination of the residual echo while leading to excessive clipping during times for cases where there was no actual residual echo but rather an undetected near-end speech, as well as for cases where the near-end speech level was lower than the R_{in} speech level.

Awareness of the echo delay and amplitude provides precise timing and better signal-characterization information. It minimizes the probability of applying the NLP to non-echo signals. As a result, the non-linear treatment is only applied at instants when residual echo is present. When applied, the NLP uses the information to fine-tune the amount of time it stays in a blocking state, and consequently, near-end speech signals pass through with little distortion.

3.9   Voice-response systems – barging

Unlike human beings, who perceive echo only if it is delayed by a minimum of 50 ms relative to their own speech, computer or voice-response systems (VRS) perceive echo regardless of the delay. These systems are designed to respond to voice barging, and echo may seem like barging but for an echo canceler that classifies it correctly. Accordingly, every voice-response system must be protected from echo. Because of the particular nature of the VRS the tail-delay capacity of the echo canceler that protects it may be short since local calls do not normally pass through an echo canceler while long distance calls do.

3.10   Hot signals

Hot signals at the S_{in} port must be treated as special. The μ law and the A law[26] that govern the signal representation within the transmission channel limit the peaks of any signal once they exceed +3 dBm0 in power. Peak limiting affects the perceived or computed level difference between the R_{in} signal and its corresponding S_{in} echo. If the R_{in} signal is severely peak limited, then the difference between the R_{in} and the S_{in} gets trimmed, and the stock echo-activity detector, which relies on signal-level difference for its signal characterization, may misclassify a hot echo at the S_{in} port and call it near-end speech. Consequently, during periods of hot signals and hot echo, an echo canceler that does not bear special treatment would freeze its linear adaptation, let the residual echo through the NLP, and take up a very long time before convergence is achieved. It should be noted that hot signals produce hot echo, and hot echo is louder and quite more disturbing than soft echo, hence echo cancelers must pay special attention to the treatment of hot signals.

A standard procedure for treating hot signals is based on anticipation of the trimmed difference between the R_{in} and the S_{in} levels. When sensing a hot R_{in} level the EC must

---

[26]   ITU-T Recommendation G.711, *Pulse Code Modulation (PCM) of Voice Frequencies* (1988).

adapt (lower) its ERL threshold level[27] setting to compensate for the limited peaks. Setting the ERL threshold level to a lower value carries the trade-off risk of misclassifying double talk as echo. Accordingly, after convergence is achieved, the EC must return the ERL threshold setting to its provisioned value, thus reducing the likelihood of quick divergence in the face of double talk.[28]

3.11   Voice-band data and fax

Switching systems must disable (i.e., turn off) echo cancelation (or suppression) during certain in-band signaling sequences, when carrying medium-speed and high-speed modem calls, and for configurations where several network sections are equipped with echo cancelers (i.e., tandem echo-canceler arrangements). Typically, an in-band 2100 Hz tone (as described in International Telecommunications Union – Telecommunications [ITU-T] Standard G.164) is generated by a switching system or voice-band modem to disable echo cancelers (or suppressors). This technique is commonly called "straight tone disable."

All echo-canceler (and suppressor) equipment must be capable of detecting a 2100 Hz tone and subsequently disabling the echo cancelation (or suppression) function for the respective channels carrying the tone. Similarly, when an echo cancelation (or suppression) is disabled by a 2100 Hz tone, its disabling state lasts as long as the signal energy level does not drop below a preset threshold for a specific time interval. The echo canceler (or suppressor) function is automatically re-enabled (i.e., the echo canceler or suppressor is "turned back on") when the 2100 Hz tone is absent.

It is often incorrectly assumed that echo-cancelation techniques are only applied to voice transmission facilities. In reality, a variety of other traffic types are also routed through the voice network, and these signals encounter echo-canceler equipment. A major portion of global non-voice connections is fax traffic. As previously indicated, field experiences[29] have shown that improperly controlled echo can corrupt fax protocol. This causes customer complaints associated with lower fax-transmission speeds and longer call-hold times.

The same studies have shown that well-designed echo cancelers can improve network performance, even for voice-band data calls (i.e., for some voice-band modems and particularly facsimile [FAX] transport applications). As a result, the ITU-T standard recommendation G.165[30] released in 1993, explicitly states that echo cancelers must remain enabled during fax calls, and must ignore 2100 Hz tones that do not have phase reversals in their signal patterns (i.e., ignore straight tone disable). For that reason, modern

---

[27]   The ERL threshold level is the minimum difference setting designed to detect and separate speech from echo. The setting is the mechanism used by an echo canceler as part of its process for detecting an echo. A common default setting in many echo cancelers is 6 dB. Under these conditions, when the perceived level difference between speech and echo is less than 6 dB, the EC may interpret the echo as double talk. For that reason, the ERL threshold setting must be lower in the presence of hot signals, thus reducing the likelihood of misclassification.

[28]   When the ERL threshold setting is too low, a double-talk signal may assume a higher likelihood of being classified as echo. Under these conditions, the EC may continue converging (rather than freezing its operation) on the wrong signal. It would naturally diverge, and set off disturbing noises and artifacts.

[29]   ITU-T G.165, *Echo Cancellers*. ITU-T G.168, *Digital Network Echo Cancellers*.

[30]   ITU-T G.165, *Echo Cancellers*.

switches and high-speed modems generate an in-band 2100 Hz tone with periodic phase reversals (as per ITU-T G.165) to disable echo cancelers, while echo cancelation remains enabled for fax calls that employ 2100 Hz tones without phase reversal.

Echo cancelers should also be disabled in four-wire digital-data applications (i.e., services that require 64 kbps clear-channel operation). To accommodate 64 kbps clear channel operation, some echo cancelers are equipped with per-call control (PCC) or signal processor control (SPC) capabilities. For global applications, switches that support PCC or SPC protocol (specified in the ITU-T Recommendation Q.50[31]) can control (i.e., enable or disable) echo cancelation on an individual-call basis by using specific signaling bits contained in channel 16 of the E1 signal format. In North America, several methods are used to implement the PCC/SPC capability. These methods are "vendor-specific features" (i.e. non-standard techniques) that rely on external local-area network (LAN) connections, common-channel signaling (CCS), rob-bit signaling, extended-super-frame (ESF) signaling, or a variety of proprietary asynchronous physical serial-channel protocols as the control mechanisms. In this case, the echo-canceler equipment must be specifically designed to interface properly with the proprietary PCC/SPC protocol used by the switching system. A primary advantage of PCC/SPC is that it allows service providers to mix voice-band and digital data services on the same facilities, thereby increasing efficiency and improving the utilization of resources.

3.12   Signaling systems nos. 5, 6, and 7

Signaling systems nos. 5, 6, and 7 make use of tones while communicating at the start of a call. Echo cancelers installed on either or both sides of the communications path must be disabled for the duration of the tone communications. If not disabled, the echo cancelers may treat signals communicated from one end as the echo of the initiating end, since the tones are alike. Signaling system 7 makes use of a 2000 Hz tone referred to as voice path assurance (VPA). Signaling system 5 (C5) makes use of 2400 Hz, 2600 Hz, and a combination of both in its protocol. Echo cancelers must be equipped with tone detection and a disabling mechanism that detects either the VPA or the C5 signaling.

Although VPA detection and disabling is essential in wireless communications, C5 detection is necessary only in conjunction with certain international gateways. It is not a requirement for the digital-wireless environment.

3.13   Conclusions

Echo control has gone through tremendous technological developments since its original inception. The first digital echo canceler was the size of a double-door refrigerator and

---

[31]   ITU-T Recommendation Q.50, *Q.50 Signalling Between Circuit Multiplication Equipments (CME) and International Switching Centres (ISC)* (1993).

dissipated well over 2000 W. Today the physical size and power consumption of echo-canceler equipment has decreased by factors of 1000 and 100 000 respectively.

Continued trends in technology will support ever-increasing levels of integration and density. Echo cancelers are presently designed to process more than 1000 channels (64 kbps DS0 or equivalent signals) on a single moderate-size printed-circuit board. These also include specialized functions that further enhance voice quality and provide new network services. Consequently, more and more implementations of echo-canceler and voice-enhancement features move away from a stand-alone design and into an integrated architectures. Echo cancelers change from being a product to becoming a feature within a larger system such as a mobile switch, VoIP gateway, digital cross-connections system, etc.

Although higher levels of concentration are emerging (e.g., SDH, ATM, IP), the majority of subscribers, nevertheless, continue to access telecommunications networks via two-wire local-loop connections. Therefore, echo will be generated by the hybrids that terminate these local loops, and consequently echo-canceler products will continue to be deployed in the foreseeable future.

4 Acoustic echo and its control

4.1 Introduction

This chapter examines the sources and the reasons for the existence of acoustic echo in wireless networks. It explains how acoustic echo is different from hybrid or electrical echo, and how it can be diagnosed away from its hybrid relative. The chapter follows the description of the impairment by examining the present methods for properly controlling acoustic echo in wireless communications. It also gives details of how background noise makes it more difficult to control acoustic echo properly. It describes those particular impairments that may be set off by some acoustic-echo control algorithms, specifically those built into mobile handsets, and it describes how they can be remedied by proper treatment brought about by means of voice-quality systems (VQS) in the network.

Acoustic echo and its derivatives are common problems facing customers of wireless communications services. Unlike electrical echo, these problems are rarely anticipated during initial deployments of wireless infrastructures because they are supposed to be addressed by handset manufacturers and controlled inside the mobile phone.

Unfortunately, many handset solutions do not control acoustic echo properly. In fact, a considerable number of mobile handsets introduce solutions that exacerbate the impairment by spawning severe side effects in the form of noise clipping and ambiance discontinuity. These side effects are, in most cases, more damaging than the impairment affected by the acoustic echo on its own.

Unlike hybrid echo, which is triggered by impedance mismatch associated with the two-wire to four-wire conversion inside the hybrid, acoustic echo is provoked inside the network by means of sound waves bouncing away from either signal-reflecting objects, acoustic leaks inside the telephone, or direct feeds from speaker phones. It is triggered outside the network. The reflected signal returns to the far-end talker via the mouthpiece of the receiving end of the connection.

Acoustic echo is, more often than not, set off by the mobile phone, the mobile environment, or both. Nonetheless, instigation of acoustic echo is not confined to a wireless end point. Wireline fixed phones may give rise to acoustic echo when conditions bear a resemblance to those of wireless phones. In particular, wireless phones and the environment in which they are used are more prone to triggering acoustic echo. Wireless phones are small, so the distance between their speaker and their receiver is shorter, giving rise to a potential signal leak. Wireless phones are used inside cars where the listener's receiver is in close proximity to the car frame. Wireless subscribers are more likely to use speaker

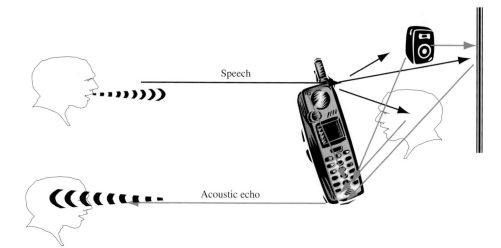

Figure 4.1 Sources of acoustic echo.

phones, especially inside their cars, setting off multiple reflections, direct and indirect, that enter the mouthpiece of the mobile phone. And, finally, the listener's head may reflect the speech signal back to the far-end talker when the mouthpiece is rested at a particular angle facing the reflected signal.

It may sound amusing, but some users of mobile phones wrap their prized instruments with protective jackets, which may trigger acoustic echo by having the proximate envelope serve as a sound conductor from the receiver to the mouthpiece.

Figure 4.1 illustrates the potential sources of acoustic echo.

In Chapter 3 I discussed the relationships between hybrid echo and its originating speech signal. I have shown that the echo signal was predictable once we have had information concerning the impulse response, ERL, and delay. Predictability of hybrid echo is achievable and straightforward because these relationships do not change throughout a particular telephone call. They are stationary over time, and maintain high linear correlation between the source signal and its hybrid echo.

On the other hand –

Acoustic echo in digital-wireless applications is non-stationary

Since acoustic echo is the result of sound waves bouncing off solid objects, changes in the positions of these solid objects relative to the microphone (mouthpiece) of a wireless phone produce changes in acoustic-echo characteristics. For example, head or hand movements, walking while talking, and the movement of nearby people or objects in the course of a phone conversation, all create different, non-stationary, acoustic echo effects. Being non-stationary implies that the impulse response (i.e., the echo) associated with a speech signal is, in essence, a moving target. Acoustic echo may also behave intermittently. As echo-emitting objects keep moving their relative positions, acoustic echo may change in intensity, and may fade away entirely at times, so that delay measurements become inconsistent. Therefore, the relationships between the source signal and its acoustic echo may not be modeled as a deterministic linear function like the one used for hybrid echo. The relationships keep changing over time, and a learning process that tries to keep track of impulse response, ERL, and delay, would be perpetually chasing the changing echo characteristics throughout a particular call.

Acoustic echo in digital-wireless applications is non-linear

The acoustic echo injected into the mouthpiece of a wireless phone is subsequently applied to a wireless codec, which processes the acoustic-echo signal in the same manner as a valid speech signal. That is, the acoustic-echo signal is subjected to speech compression techniques that modify the signal characteristics, while creating a non-linear association between the source signal and its acoustic-echo offspring.

For these reasons, techniques for the elimination of acoustic echo resort to different treatments than the ones applied for canceling hybrid echo.

4.2 Round-trip delay and echo dispersion

Acoustic echo is created by sound waves originating in the receiver (earpiece). These waves enter the microphone (mouthpiece) of a handset or a speaker phone via reflections bouncing off solid objects in the sound path, and return to the speaker on the far end as acoustic echo (see Figure 4.2).

Acoustic echo reaches its destination at a round-trip delay consisting of one additional component not present in the case of hybrid echo. While both hybrid and acoustic echo incur delays associated with codec processing and signal propagation through the wireline network, the acoustic echo travels an extra distance – from the receiver to the

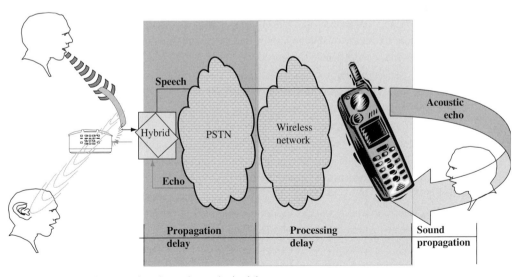

Figure 4.2 Acoustic echo and round-trip delay.

mouthpiece – at the speed of sound rather than the speed of light. Consequently, the additional delay associated with acoustic echo reflected off an object 0.3 meters, 1 meter, 5 meters, or 10 meters away takes more than 2 ms, 6 ms, 30 ms, or 60 ms, respectively. It should be noted, however, that the power of the acoustic echo is trimmed down exponentially as the distance from the source receiver grows longer. Therefore, in the absence of a speaker phone, most acoustic echoes reflected off objects farther than 30 cm away from the phone receiver are negligible, or practically non-existent, and may be ignored.

While most acoustic echoes that reflect a telephone-receiver output consist of a relatively short acoustic feedback, the use of a speaker phone changes the type and characteristics of acoustic echo. The changes are not limited to a significant increase in the round-trip delay, but also affect the makeup and extent of the impulse responses. Speaker-phone output may spawn multiple reflections, which bring about a swelling of the impulse response as the surrounding walls spread farther away from the signal source.

Acoustic-echo impulse responses formed from within a car may persist over 250 samples or just over 30 ms. Impulse responses formed from inside a 4 m^2 room may persist over 560 samples or just over 70 ms. Impulse responses formed from inside a small conference room of about 10 m^2 size may persist for as long as 1600 samples or 200 ms.

When digital-wireless standards were first laid out, the only echo problem requiring infrastructure attention was the electrical echo. Acoustic echo was not considered an infrastructure issue. Manufacturers of mobile handsets were charged with the responsibility of controlling this potential impairment and making it disappear.[1] It did not. Nowadays, with the continuing expansion of the worldwide wireless-subscribers base, and with the ongoing growth in the relative frequency of mobile-to-mobile communications, the phenomenon of acoustic echo and its noise-clipping derivative have drawn closer to the forefront of voice-quality impairments that have yet to be fully conquered.

In contrast to electrical echo generated by PSTN hybrids, acoustic echo instigated by mobile terminals (handsets or speaker phones) is more difficult to characterize mathematically because of its inconsistency, non-linearity, and dynamic-delay attributes. These peculiarities make it difficult to reproduce an exact duplicate or a close estimate that models a particular acoustic-echo signal. The first part of this chapter discusses the nature and sources of acoustic echo, its main attributes, its detection algorithm, and the associated control technology. The rest discusses proper and improper treatments designed to eliminate the annoying impairment.

The comprehensive acoustic echo and treatments described in this chapter are not universal. There are various implementations employing similar, partial, or different methodologies. At the same time, methodologies reported herein have proven themselves to be highly effective when deployed and experienced under various field conditions including ones containing noise, ring tones, ring back, background music, and, of course, pure acoustic echo with little or no added flavors.

---

[1]   ITU-T Recommendation G.167, *Acoustic Echo Controllers* (1993).

4.3   Reasons for perceiving acoustic echo in wireless communications

Acoustic echo can be present in both wireline and wireless applications. Most wireline configurations that are exposed to echo conditions are equipped with electrical echo cancelers, therefore acoustic echo in wireline applications is controlled via standard echo-cancelation algorithms and associated NLP techniques. During the early nineties, digital-wireless applications deployed echo cancelers to nullify echo (i.e., electrical echo) generated within the public switched-telephone network (PSTN) end only. In a digital-wireless scenario, the PSTN subscriber is not expected to experience either acoustic or electrical echo. This assumption is based on the following rationale: wireless-phone standards require equipment manufacturers to engineer sufficient loss in the acoustic echo path.

The two- to four-wire hybrids present in the PSTN network are absent in the digital-wireless environment, hence there should be no electrical (hybrid) echo generated by digital-wireless equipment.

Unfortunately, some wireless-phones do not comply with the official standards that specify adequate isolation between the receiver (earpiece) and the microphone (mouthpiece) of digital mobile phones. As a result, acoustic echo is a concern for wireless-service providers. Furthermore, the acoustic-echo problem is more pronounced in the case of digital-wireless applications because of long processing-delay times ($\sim$200 ms round-trip delay) introduced by speech-compression techniques and the non-linearities of speech-compression algorithms.

4.4   Controlling acoustic echo in digital-wireless networks – main considerations

The preferred or the most effective technique for controlling non-linear, non-stationary, acoustic echo is via non-linear methods. Non-linear processing works by employing suppression techniques upon detection of an acoustic-echo signal. Obviously, the key to high performance is proper detection. Poor detection capability may lead to errors setting off suppression and clipping of speech segments rather than echo. The echo-activity detection (EAD) algorithm relies on the echo-return loss (ERL) settings. The acoustic-echo ERL is the difference in level (dB) between the source speech signal and the resulting echo signal. The greater the ERL, the greater is the difference between speech and echo, and, therefore, the smaller the likelihood of signal misclassification.

In a traditional hybrid-echo canceler, the linear convolution algorithm of the standard linear processor within the echo canceler facilitates the subsequent detection and decision-making process of the non-linear processor (NLP). The linear processor cancels a hefty chunk of the echo before letting the NLP take care of the residual part. Consequently, the ERL seen by the linear EC part is enhanced as measured by the ERLE (where E stands for enhancement), and the distinction (in levels) between the source speech signal and the echo seen by the NLP is widened as reflected by the measure of ERL + ERLE. Accordingly, when the ERLE is significant, then ERL + ERLE provides sufficient level disparity between speech and echo, yielding a lower likelihood of mistaking one for another.

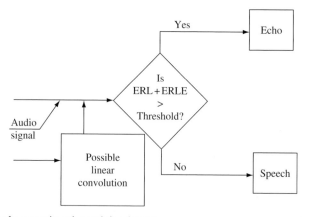

Figure 4.3 A simple acoustic-echo activity detector.

In the absence of a linear convolution processor, the ERLE is equal to 0, and the ERL is the only indicator setting speech and echo apart.

Although the acoustic echo in wireless communications assumes non-linear relationships to its speech origin, a linear convolution process, that is, a close copy to the one used for canceling linear echo, may be employed as a pre-processor to the foremost non-linear algorithm. The realized ERLE born out of the linear process is trivial and it all but borders on irrelevance. The resulting sum ERL + ERLE tends to yield acoustic-echo detection quality quite comparable to the one employing ERL alone. Accordingly, many algorithm designers may conclude that embracing computing resources for placing a linear convolution process in front of a non-linear one may not be worth the cost of inclusion, and skipping it may not significantly impair the overall acoustic-echo control performance carried out by a lonely non-linear process.

Figure 4.3 illustrates a common and popular acoustic-echo activity-detection algorithm that relies on level discrimination as the chief ingredient in determining whether the signal is speech or echo. A decision is critical as to whether or not there should be a call for action that suppresses the signal. A linear convolution process that acts only as a pre-processor without actually applying a loss may be used to compute a potential ERLE.[2] In the absence of a computed ERLE the decision rests on the ERL alone.

A study prepared by Cristian Hera,[3] concerning the impact of GSM wireless codecs on acoustic-echo cancelation performance, estimated the ERLE of a linear convolution pre-process of an acoustic-echo signal by running a test as described in Figure 4.4. A speech signal was applied at R_{in} and was than passed through an encoding decoding process, before being applied to the hybrid where an echo was reflected back towards the echo canceler. This echo was coded and decoded again and then passed to the echo canceler. The signal at S_{in}, compared with the signal at R_{out}, was distorted twice by the speech codecs in the path.

The echo path was simulated using the codecs defined in the ITU-T G.168 recommendation. All these echo paths correspond to electrical echo.

---

[2] Details of the specifics are discussed in the next section.
[3] Cristian Hera (private communication).

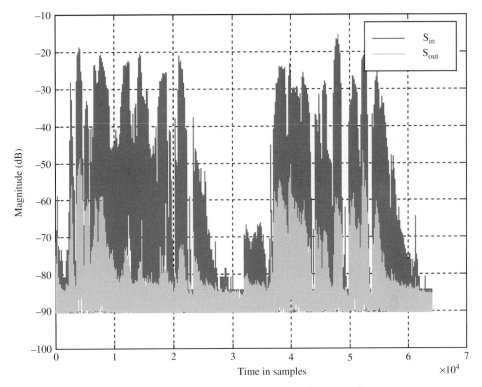

Figure 4.4 Acoustic-echo signals before and after echo cancelation; no codecs.

The study considered an echo canceler with a window length of 32 ms (number of taps, M = 256). The input signal was speech with the active speech level of 26 dB from the overload point of 16 bits.

The study compared the amount of echo cancelation pulled off by the LCP after subjecting the echo signal to the GSM codecs EFR, FR, and HR, as well as the PCM G.711. For the sake of comparison, echo-cancelation processing was applied to the unprocessed linear representation of the speech signal.

The main findings affirmed that for most echo paths the ERLE realized by a commercial echo canceler was highest for the unprocessed signal, where the average ERLE amounted to about 45 dB for signal peaks. The second best was the G.711 codec pre-processed signal, where the average ERLE amounted to about 27 dB. The third best was the EFR codec pre-processed signal, where the average ERLE amounted to about 15 dB for signal peaks. The FR and HR codecs pre-processed signals exhibited equivalent performance. For both, the average ERLE for signal peaks amounted to about 8 dB. In spite of the non-linear components introduced by the GSM codecs, the average peak performance of the LCP appeared to reduce the GSM codecs pre-processed echo by a small amount. These results, although minor in significance, may still be helpful in enhancing the performance of the echo-activity detector and reduce the probability of misclassification.

Figures 4.5 to 4.8 present a sample of the findings for unprocessed, G.711 pre-processed, EFR, FR, and HR pre-processed signals that were subjected to a computer simulation of an echo-canceler LCP.

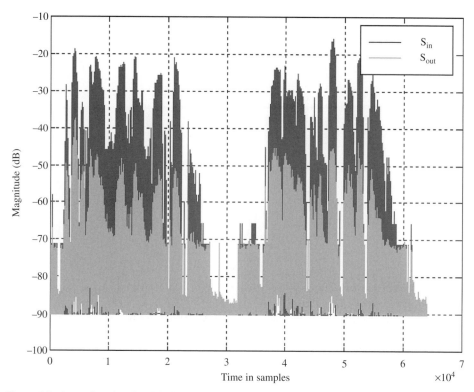

Figure 4.5 Acoustic-echo signal before and after echo cancelation; codec = G.711.

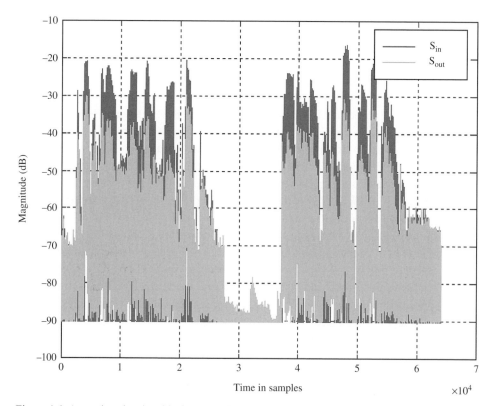

Figure 4.6 Acoustic-echo signal before and after echo cancelation; codec = GSM-EFR.

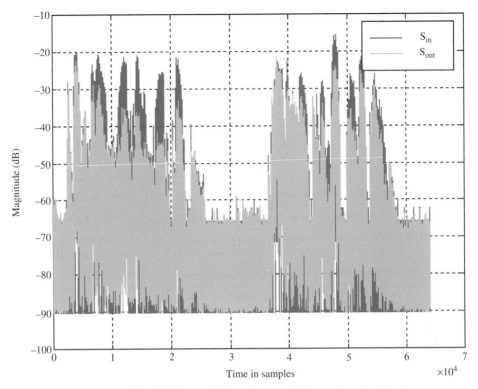

Figure 4.7 Acoustic-echo signal before and after echo cancelation; codec = GSM FR.

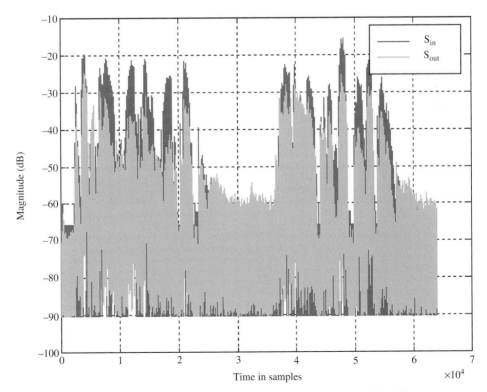

Figure 4.8 Acoustic-echo signal before and after echo cancelation; codec = GSM HR.

4.5   Acoustic-echo activity detection

There are two major issues associated with the logical approach for distinguishing acoustic echo from a valid speech signal. These issues are phrased in the form of the following questions:

- What is the appropriate threshold power level that should be used to minimize the probability of incorrect classification?
- How can the speech or acoustic-echo signal power level be measured without introducing measurement deformities?[4]

The answer to question 1: create a dynamically adaptive ERL. The procedure involves making continuous measurements and updating speech and noise levels during spurts of single talk. These measurements are used to update the ERL threshold. The algorithm computes an adaptive ERL by taking into account the difference in power (as computed under single-talk conditions) between the two ends, and it incorporates that difference into the ERL threshold figure.

The answer to question 2: employ noise reduction (NR) ahead of AEC. Noise-reduction algorithms not only bring listening comfort to users but, at the same time, may also be very beneficial to the acoustic-echo control (AEC) algorithm performance if they precede the AEC function. An untreated variable noise level distorts the power measurements made by the VQS. In the absence of a noise-reduction function, this distortion may breed erroneous decisions by the AEC algorithm, triggering more frequent clipping of speech, and more frequent bursts of acoustic echo.

Classification errors can be minimized when employing the two procedures above. Some of the trade-offs can be summarized by referring to an error type.

Error type I (with and without noise reduction)

This type of error mistakes acoustic echo for speech and lets acoustic echo return to the speaker. The error may occur as a consequence of setting the ERL at a relatively high level, requiring a sizeable safety margin separating speech and echo levels. When the echo is louder than expected it impersonates speech, and by so doing it gets through the gate. In the example shown in Figure 4.9, the echo-activity detector (EAD) ERL threshold is set to 15 dB. Actual ERL without line noise is 18 dB, which would have classified the signal as echo given an ERL threshold of 15 dB. However, due to line noise of 6 dB, the measured ERL is 12 dB, crossing the threshold with a 3 dB margin. It becomes apparent that proper noise reduction could have prevented error type I from occurring in this specific case.

---

[4] Measurement deformities are brought about by background noise, which tends to exaggerate the perceived loudness (power) of the signal level.

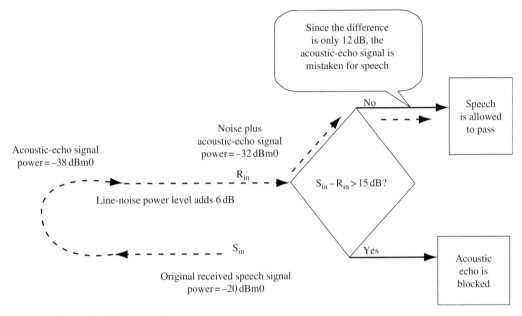

Figure 4.9 Error type I.

Error type II

This type of error mistakes speech for acoustic echo. It clips low-level bits of speech, particularly if the speaker exercises an adequate amount of dynamics. The error may occur as a consequence of setting the ERL threshold of the EAD at a relatively low level, requiring a thin margin separating speech from echo. When speech at the R_{in} is transmitted at a lower level than expected, it fails the ERL threshold. It mimics echo, and by so doing it is blocked from getting across the gate. In the example shown in Figure 4.10, the speech activity is at R_{in} input -30 dBm0, the speech activity at the S_{in} input is -15 dBm0. This mismatch in speech levels contributes to a greater frequency in tripping over error type II, particularly when the echo activity detector (EAD) ERL threshold is set to 14 dB. The derived difference between the speech signals at the S_{in} and the R_{in} is $-15 - (-30) = 15$ dB, which would have classified the signal as echo given an ERL threshold of 14 dB.

In conclusion, the dynamic setting of the ERL threshold is crucial in minimizing the probability of going down the error path for either error type, while the noise-reduction algorithm may add additional benefit by avoiding misclassification due to error type I.

4.6   Residual acoustic-echo control in the frequency domain

In addition to a relatively high likelihood for signal misclassification, one of the most notable weaknesses of relying solely or mostly on an NLP approach for suppressing

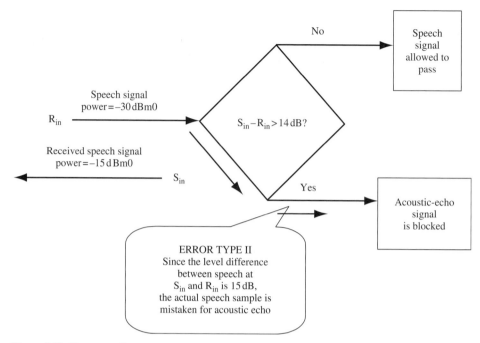

Figure 4.10 Error type II.

acoustic echo is the inability to employ it successfully during double-talk periods. When applying a blocking action during a double-talk event, the near-end speech is blocked from moving forward to the far end. If the NLP remains in a non-blocking state, then the echo is not canceled and it accompanies the near-end speech. Since nearly all the acoustic echo in wireless communications is non-linear, a linear convolution is ineffective on its own (as we have seen earlier) for controlling it. It is possible to enhance the NLP action by resorting to a sub-band NLP, where attenuation is applied only to frequency bins that contain echo energy without affecting or blocking the rest of the spectrum.[5]

Time-domain versus frequency-domain representation

Most people can understand the concept of representing an object moving across time. A typical illustration of a signal is depicted by having its power plotted relative to time as illustrated in Figure 4.11. There, a signal and its corresponding echo are represented in a stereo depiction where one channel carries the speech and the other (in the opposite direction) bears the corresponding acoustic echo.

Audio signals can also be represented in the frequency domain, where a short time slice of signal information is collected into a buffer. There it is submitted to a mathematical

---

[5] Gobu Rasalingham and Saeed Vaseghi, *Subband Acoustic Echo Cancelation for Mobile and Hands-Free Phones in Car*, http://dea.brunel.ac.uk/cmsp/Home_Gobu_Rasalingam/Gobu_new_way.htm.

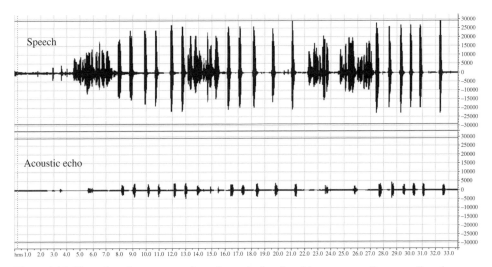

Figure 4.11 Time-domain representation of speech signal and its corresponding acoustic echo.

procedure – fast Fourier transformation (FFT) – where its spectrum is broken up into its components, and collected into separate frequency bins. The concept is analogous to a rainbow of sound equivalent to light displayed in the sky after a rainy episode.

Once the separate frequency bins are known to contain signal-power information, algorithms can manipulate each bin by either amplifying or attenuating their contents. The power management of separate frequency bins is equivalent to having a graphics equalizer operate on an incoming sound.

When applying a frequency-domain analysis to acoustic echo in wireless communications, it is evident that most of the non-linear acoustic-echo energy is concentrated in a relatively narrow range of the 4 kHz spectrum.

As an example, the signal and its acoustic echo in Figure 4.12 are converted to a frequency-domain representation, as depicted in Figure 4.13. It is interesting to note that although the speech-signal power is distributed almost uniformly across the entire spectrum, the acoustic-echo power concentrates most of its energy in the 1700 Hz–2700 Hz range.

An acoustic-echo algorithm that applies its sub-band NLP to the 1700–2700 Hz range would attenuate the signal in these frequency bins without affecting any other spectral contents. It would follow its discriminatory attenuation process by converting the signal back to the time domain via an inverse FFT (IFFT), and sending the outcome on to its destination. The sub-band NLP process is illustrated in Figure 4.13.

Sub-band NLP requires much more processing than a full spectrum NLP. Although it commands superior performance in comparison, the price does not always justify the enhancement, lest the sub-band NLP frequency-domain processing shares resources with other applications such as noise reduction (See Chapter 5). Since both noise reduction and AEC operate on the same side of the communications channel (see next section) a sub-band NLP implementation may be less expensive when considered as a by-product to the noise-reduction algorithm.

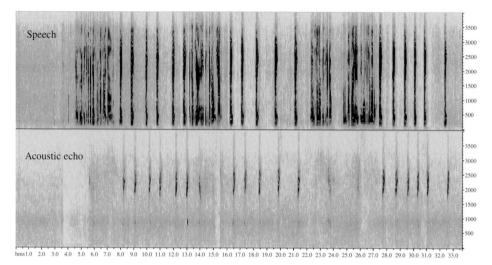

Figure 4.12 Spectral representation of speech signal and its corresponding acoustic echo.

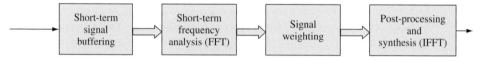

Figure 4.13 Sub-band NLP process for controlling acoustic echo.

4.7   The impact of comfort-noise matching on the performance of acoustic-echo control

One of the most important features of a voice-quality and echo-cancelation system is its ability to generate noise and transmit it to the speaker over the listening track of the voice-communications path.

Comfort-noise matching is generated by an echo-canceler system upon the blocking of a transmission path (listening track) by a non-linear processor (NLP). This operation is triggered by the algorithm as one of the steps designed to block echo from sneaking through to the talker end point of the transmission path. A signal-blocking NLP must be accompanied by the injection of matching background noise, without which the blocking and unblocking of the path would be perceived as heavy breathing since the background noise would shift between actual noise and complete silence – a notable contrast.

Note: the comfort-noise matching becomes more essential for the case of acoustic-echo control (AEC) as compared with hybrid-echo cancelation since the background noise surrounding a mobile listener is a great deal louder than the average noise surrounding a PSTN endpoint. In the absence of a matching comfort noise, the breathing effect perceived by the opposite end (the ear in Figure 4.14) would be intolerable due to the disparity (in background-noise level) between silence and genuine background noise. Glaring contrasts between louder actual noise and comfort noise (or lack of it) are quite apparent, and would

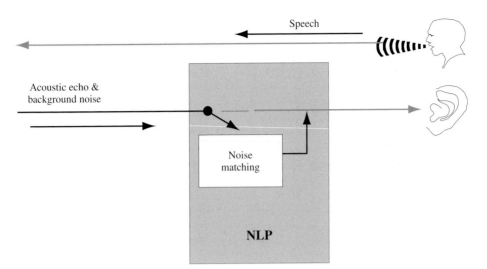

Figure 4.14 Acoustic-echo NLP and its corresponding comfort noise matching.

be perceived as serious signal clipping. That fact is underlined in view of the fact that the AEC employs an NLP as its critical algorithmic component. Figure 4.14 illustrates the NLP architecture as part of the acoustic-echo control process.

The ideal comfort noise is a duplicate of the bona-fide background noise. It provides a perfect match in level and spectrum. A less than ideal but reasonable match would match the levels and a proximate (most of the times) spectrum. A slighter conformity would match the levels but would fall short spectrally. Evidently, as the noise match becomes less accurate, the listening effort becomes more strenuous, and voice quality degenerates considerably.

A weak comfort-noise-matching algorithm may not cause severe degradation of voice quality when background noise is low. It would simply be less audible, and hence, less perceptible. This conclusion implies that a noise reduction (NR) algorithm preceding an NLP operation would be accommodating not only as a consequence of a lesser amount of transmitted noise, but also as a mitigating factor minimizing the perceived contrast between the NLP on/off mismatched noise states.

Speech-detection algorithms distinguish between noise and speech by relying on differences in signal levels. Specifically, speech, in most cases, is supposed to assume higher energy levels than background noise. A proper algorithm for deriving the matching (comfort-noise) level is to train the acoustic-echo control algorithm on low-level signals passing through it, while the AEC/NLP is in the off position.[6] The training distinguishes between low-level speech and background noise by using signal levels. It imposes an upper bound on levels classified as noise so as not to train on low-level speech.

---

[6] The AEC/NLP is in the off position when there is silence on the send side. When there is silence on the send side there is no acoustic echo on the receive side. A signal traversing through may not contain acoustic echo. It may either contain background noise or background noise + speech. If the algorithm imposes a low upper bound on the noise level, then the distinction between background noise without speech and background noise with speech would have a notable margin and the likelihood of signal-classification error would be minor.

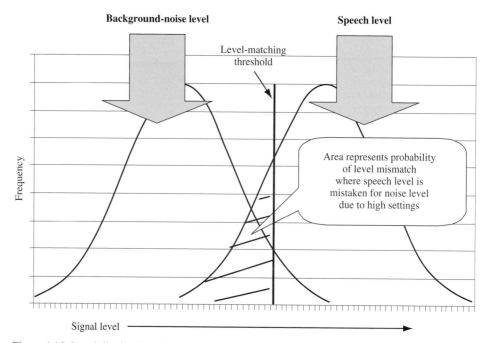

Figure 4.15 Level distribution of speech and background noise.

If no bound is imposed on classification of noise, and comfort noise is trained on low-level speech, mistaking it for the noise level, there is a reasonable probability that the matching noise level would be too high in comparison with the true background noise. Under these circumstances, the algorithm would insert a high comfort-noise level, matching the low speech level, while mismatching the background-noise level (see Figure 4.15). For this reason, adding the noise-reduction (NR) feature in front of the acoustic-echo control (AEC) would help greatly in minimizing any erroneous classification of incoming signals.

When the background noise is loud and the spectrum of the injected comfort noise is less than a perfect match with the actual noise, the spectrum mismatch is highly noticeable, and it interferes with voice quality. When a noise-reduction algorithm precedes the acoustic-echo control, it reduces the background noise entering the acoustic-echo control algorithm and the end user. Accordingly, a less than perfect spectrum matching would not be greatly noticeable since the noise level would not be as high as otherwise.

As illustrated in Figure 4.15, moving the level-matching threshold to the left (i.e., limiting the comfort-noise matching to low levels) would result in a lower probability of misclassifying speech as noise, but higher probability of misclassifying noise as speech. A signal classified as noise would be blocked by the NLP, while a signal classified as speech would be let through. When relying on level alone as a differentiator between background noise and speech, errors are inevitable. Trade-offs must be made between error types I and II. Consequently, thresholds are set to favor a higher likelihood of classifying noise as speech, while yielding a lower probability for classifying speech as noise.

The different types of comfort noise are white, colored with a fixed filter, and colored with an adaptive spectrally matching filter.

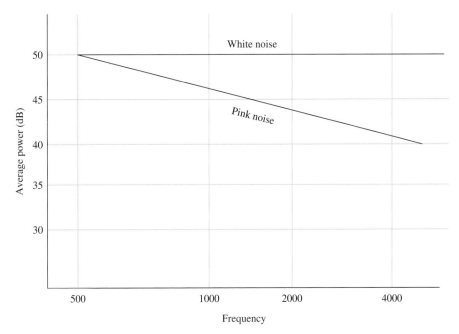

Figure 4.16 Spectral distribution of white and pink noises.

White noise is equivalent to the color white. The main characteristic of white is the uniform distribution of power along the entire (limited 4 kHz narrowband) audio spectrum as illustrated in Figure 4.16.

Colored noise, just like painted color, comprises non-uniform distribution of power along the entire relevant (4 kHz) spectrum. Most ambient noises that surround mobile environments are colored. They may resemble pink noise, where power density decreases by 3 dB per octave with increasing frequency over a finite frequency range, which does not include DC, and where each octave contains the same amount of power. In other words, pink-noise power is higher at low frequencies, and lower at high frequencies.

There have been attempts to define noises other than white and pink. For example: red noise has been associated with oceanic ambient noise (i.e., noise distant from the sources). It has been described as red due to the selective absorption at higher frequencies.

Orange noise has been defined by some as a quasi-stationary noise with a finite power spectrum with a finite number of small bands of zero energy dispersed throughout a continuous spectrum. These bands of zero energy are centered about the frequencies of musical notes in whatever system of music is of interest. Since all in-tune musical notes are eliminated, the remaining spectrum could be said to consist of sour, citrus, or "orange" notes. A room full of primary school students equipped with plastic soprano recorders most easily generates orange noise.

Black noise is defined by some as silence, and by others as the inverse (the cancelation signal) of white noise.[7]

---

[7] Joseph S. Wisniewski, *The Colors of Noise*, www.ptpart.co.uk (Oct 1996).

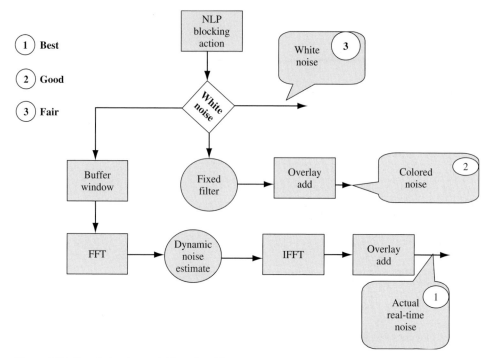

Figure 4.17 Comfort-noise matching algorithms.

The highest quality of comfort matching noise is colored noise with an adaptive filter. This type of comfort noise replaces the actual noise with a computer-generated version that matches the spectrum and the power of the actual noise. The spectral match is achieved by resorting to the process depicted in Figure 4.17.

When noise levels are high, the quality of comfort-noise matching becomes critical to the overall transmission quality. For low background noise,[8] white comfort noise that matches the noise power level is sufficient.

When noise levels are medium level,[9] the contrast between the comfort white and the actual ambient noise is quite noticeable, and the matching noise needs color to reduce the contrast.

Loud background noise[10] requires spectrally adaptive comfort noise matching. Otherwise, the noise contrast between actual and computer-generated noise is noticeable.

Typical power spectra of the above are depicted in Figures 4.18–4.24.

When combining the different ambient noises into a single blend that captures the essence of loud surroundings, it is possible to compose a fixed comfort-noise filter that comprises low contrast in comparison with an actual background. The fixed filter would maintain an average signal-power peak at low frequencies; it would reach a relative plateau to a slower decline at mid frequencies, but would further decline at high frequencies.

---

[8]   Noise level below –40 dBm0.
[9]   Noise power levels between –30 and –40 dBm0.
[10]   Loud background noise is defined as noise with power level above –30 dBm0.

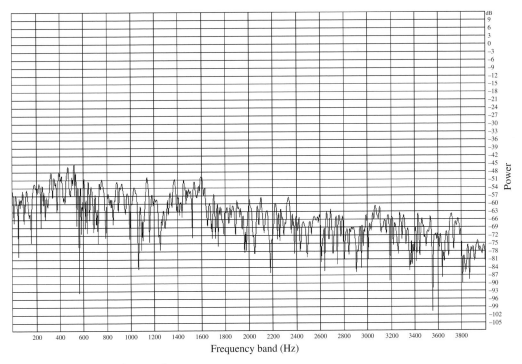

Figure 4.18 Airport noise.

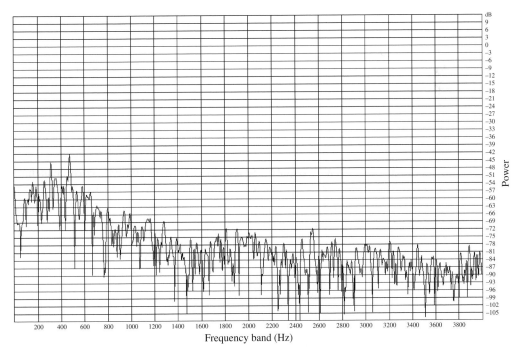

Figure 4.19 Subway noise.

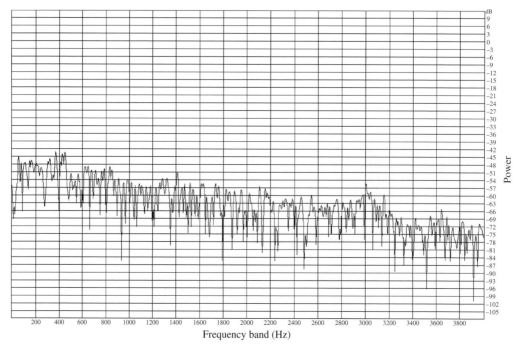

Figure 4.20 City street noise.

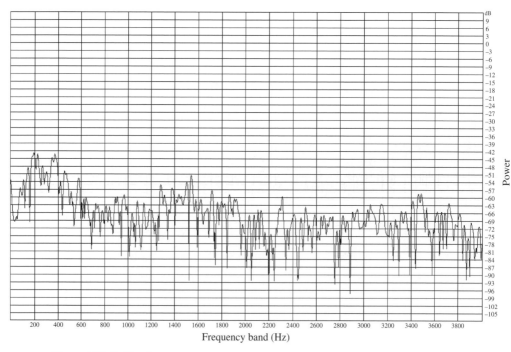

Figure 4.21 Café noise.

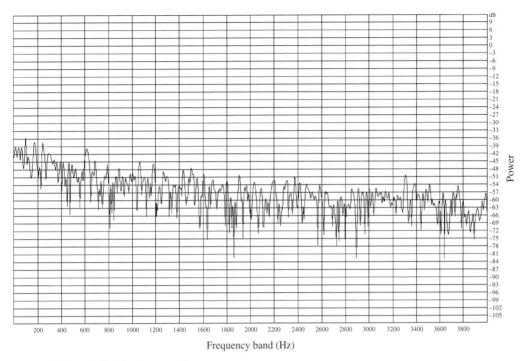

Figure 4.22 Moving car noise.

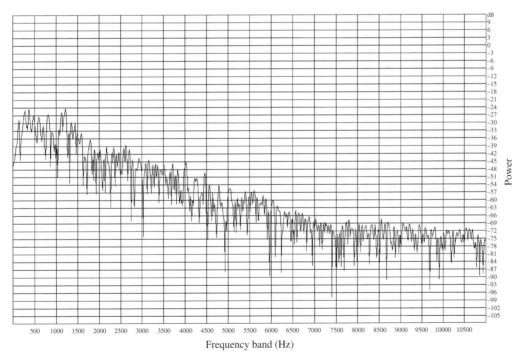

Figure 4.23A High wind noise.

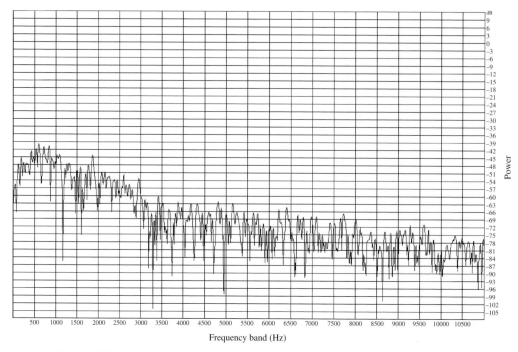

Figure 4.23B Large crowd noise.

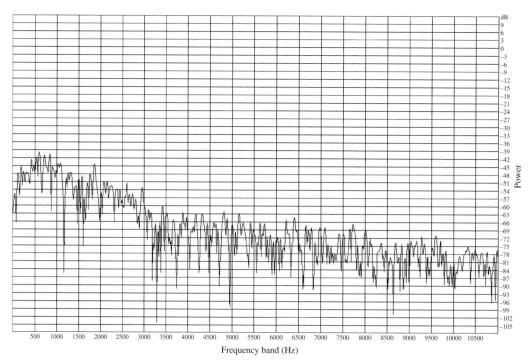

Figure 4.24 Beach noise.

The difference between peak and bottom power measures would entail about 30 to 40 dB across the spectrum.[11]

The different noise sounds (train noise, subway, bus, car (tire and wind), airport, café, high wind, beach, and large crowd noises) are illustrated on the website (www.cambridge. org/9780521855952). In addition, the website contains illustrations that depict five scenarios of acoustic-echo control with various forms of comfort-noise injection replacing the suppressed echo (see Chapter 17).

The scenarios are:

- input speech signal,
- a case of acoustic echo without acoustic-echo control,
- a case of acoustic-echo control with no comfort-noise match,
- a case of acoustic-echo control with white-noise match,
- a case of acoustic-echo control with a fixed filter and colored-noise match, and
- a case of acoustic-echo control with an adaptive-filter noise match.

1. All the scenarios above are activated by an input speech signal

This specific signal is demonstrated on the website.

2. A case of acoustic echo without acoustic-echo control

In recording 17.2.2 all signals including echo and actual noise are passed through to the talker side (on his listening track). Although background noise is not interrupted and stays continuous, the uninterrupted transmission of echo is a primary voice-quality drawback.

3. A case of acoustic-echo control with no comfort-noise match

In recording 17.2.3 echo is canceled but there is no comfort-noise matching. The listening track sounds choppy and it interferes with voice quality.

4. A case of acoustic-echo control with white comfort-noise matching

In recording 17.2.4 echo is canceled and noise level is matched. Because of the loudness of the background noise and the strong disparity in spectrum the match is better than silence but the transitions are very audible.

5. A case of acoustic-echo control with colored comfort-noise matching

In recording 17.2.5 echo is canceled, the noise level is matched, and the color (spectrum) match is closer, but transitions are still heard, although they are not as disturbing as before.

---

[11] In comparison, the difference between peak and bottom across a 4 kHz spectrum for a standard pink noise is about 10 dB.

6.   A case of acoustic-echo control with dynamically adaptive comfort-noise matching

In recording 17.2.6 echo is canceled and noise is matched both in level and in spectrum. Transitions are almost completely removed.

The dynamically adaptive noise matching is realized by employing the real-time noise estimate obtained through processing in the frequency domain via fast Fourier transforms (FFT) and their inverse as depicted by the Figure 4.17.

4.8   Correcting impairments caused by some acoustic-echo control algorithms built into mobile handsets

The most significant part of the acoustic-echo treatment involves signal suppression. The suppression operation leaves a hole in the signal flow that must be filled with comfort noise, designed for maintaining ambiance continuity. Making comfort noise comfortable is not a trivial task.

Many mobile handsets incorporate an acoustic-echo suppression function implemented as an NLP that is triggered by speech originated at the other end (the other mobile or PSTN phone entering the S_{in} port). When the mobile subscriber is surrounded by high noise and mobile handsets treat acoustic echo by suppressing it without filling the gap with proper comfort noise,[12] the signal arriving at the network voice quality system (VQS) terminal may already contain silent holes in place of the acoustic echo. This phenomenon occurs because the internal handset acoustic-echo suppressor tends to clip the noisy signal, replacing it with short intervals containing relative stillness. These silent intervals exacerbate the echo impairment by spawning severe side effects in the form of noise clipping and ambience discontinuity; the VQS algorithm must be able to remedy this derivative problem by resorting to complementary technology coupled with the acoustic-echo control algorithm. The website provides a demonstration of the phenomenon.

Proper treatment of the above impairment requires a coordinated operation of three functions.

- An intelligent noise estimate
- Noise reduction incorporated in the network VQS
- Acoustic-echo control incorporating spectral-noise matching in the VQS equipment

Intelligent noise estimate

When it comes to estimating noise, the two most important parameters are level and spectrum. Estimating level is relatively easier. However, in the case where noise is being clipped by a pre-processor (e.g., echo suppression in the handset, see Figure 4.25), the level

---

[12]   Unfortunately, this type of acoustic-echo control in handsets is quite common.

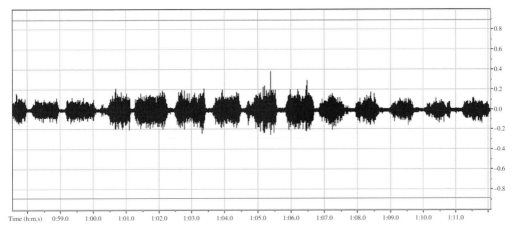

Figure 4.25 Noise clipping resulting from handset echo suppression.

estimate may be subject to distorted input as it would apply the exact signal-power level[13] it is exposed to, resulting in a much lower power-level estimate than the actual power level of the background noise. Furthermore, the spectrum would also be infected by the spectral content of the intervals containing the clipped noise. Consequently, a noise-reduction algorithm (if turned on) and the matching noise generated by the VQS AEC would use a distorted (a considerably undervalued) noise-level estimate, and erroneous spectrum, and would not perform properly.

A proper noise estimate should take the expected distortion into account when estimating the noise spectrum. By so doing the noise estimate would only embrace the actual noise as input for the calculation.

It should be noted that the NLP in the handset and the one in the VQS would only open whenever the noise level on the send side[14] is higher than the provisioned threshold for the noise floor. Under these circumstances, the voice-activity detector (VAD), built into the AEC in both gear (mobile phone and VQS), would consider the signal to be speech, would open the NLP, and replace the actual background noise with matched comfort noise. Figure 4.26 illustrates the adjustment made to the noise estimate when mobile phones are likely to have an acoustic-echo suppressor built in.

The AEC function within the VQS operates in parallel with the echo suppression function in the handset. When applying its suppression by reacting to the far-end (send-side) speech, the VQS AEC function replaces the post NR noise signal[15] with its own comfort matching noise. Thus, the combination of NR and the spectrally correct noise matching associated with the AEC should produce a low-level noise fill, while not interrupting the noise signal, and maintaining complete continuity.

---

[13] An average of the actual power level and the silent holes would produce a considerable bias downward and cause a significant underestimation of the actual power level.

[14] PSTN or another far-end wireless.

[15] This statement assumes that a noise reduction process is applied to the signal before it leaves the R_{out} port.

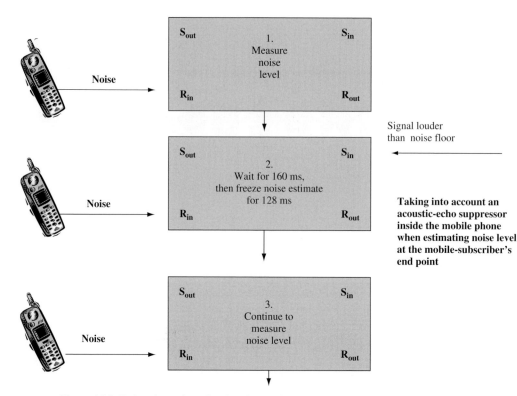

Figure 4.26 Estimating noise when handset includes an acoustic-echo suppressor.

4.9   Network topology of acoustic-echo control

In today's wireless networks, equipment designed to cancel or suppress acoustic echo must operate on an uncompressed signal, one that is coded in a 64 kbps channel, and is ITU-T G.711[16] compatible. Canceling or suppressing echo is best accomplished by having the control reside as close as possible to the echo source. This suggests that the best placement for an acoustic-echo controller is between the PSTN and the TRAU function.

In practice, when the TRAU function is outside the MSC and inside the BSC, the network operator is faced with a choice of either placing the AEC between the MSC and the BSC on the A interface of the MSC, between the MSC and the PSTN, on the PSTN interface of the MSC, or as an adjunct to the MSC (see Figure 4.27). When the TRAU function is inside the MSC,[17] then a stand-alone system would have to reside either on the PSTN interface or as an adjunct to the MSC.

When placing the AEC equipment either on the A interface or as an MSC accessory, it supports all mobile-to-mobile applications including mobiles connected to the same MSC (see Figure 4.28). When placing the AEC equipment on the PSTN interface,

---

[16]   ITU-T Recommendation G.711 *Pulse Code Modulation (PCM) of Voice Frequencies (1988).*
[17]   Some CDMA MSC systems incorporate the TRAU function within the MSC. Global-systems mobile systems place the TRAU function beyond the A interface, opening the possibility for echo cancelers to be placed on the A interface.

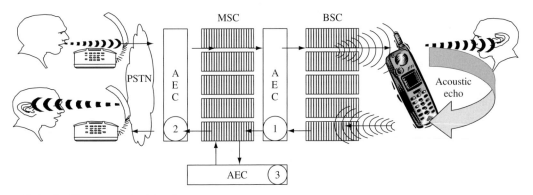

Option 1 – Placement on the A interface
Option 2 – Placement on the PSTN interface
Option 3 – Placement as an adjunct (or internal) to the MSC

Figure 4.27 AEC placement options.

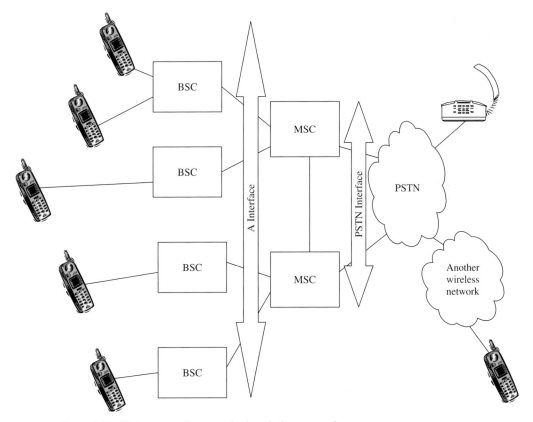

Figure 4.28 AEC system placement in the wireless network.

Acoustic echo control only

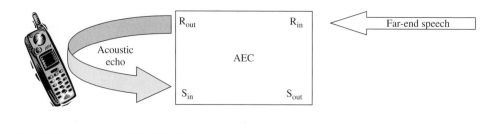

Hybrid-echo canceler with acoustic-echo control

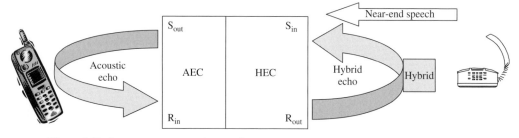

Figure 4.29 System port convention naming.

it serves mobile to PSTN connections or mobile-to-mobile calls transported over the PSTN[18] only.

When the AEC equipment is a stand-alone unit designed for the sole purpose of acoustic-echo suppression, the send and the receive ports are designated as described in Figure 4.29. However, when the AEC is an add-on to a network hybrid-echo canceler, the send and receive ports of the equipment take their designation from the hybrid EC (HEC). Figure 4.29 depicts the different port, near-end, and far-end designations, as they relate to the particular scope of the echo-control system.

Many wireless-network configurations include echo systems that control both hybrid and acoustic echo at the same time. In most configurations, the two applications inhabit the same system as they operate in opposite directions. The hybrid EC cancels echo that originates on the PSTN, while the AEC suppresses echo that originates on the mobile end. When adding an AEC application to an existing wireless system that includes hybrid EC on the PSTN interface as part of the MSC infrastructure, the AEC may not be built into the HEC system. In fact, it may even be appended to the TRAU system, and technically positioned on the A interface.

Positioning an AEC application on the A interface rather than the PSTN interface provides for a better acoustic-echo control coverage at the cost of having more AEC systems overall.

---

[18]   Calls between two different wireless networks may be transported over the PSTN.

4.10   Conclusions

Acoustic echo is one of the remaining, yet to be fully eradicated, frontiers reported as an existing impairment in wireless communications. Solutions and remedies do exist, but the better ones are not quite abundant yet.

Acoustic-echo control, if not done well, could cause more harm than good. Side effects stemming from a poor implementation in either mobile handsets or network VQS may trigger speech or noise clipping, or both. Either clipping experience is highly annoying and may be viewed as more impairing than the acoustic echo.

On the other hand, when a network AEC performs properly, it may enhance the communication experience, and may even put out of sight impairments (triggered by mobile handsets) that could have reached the scrutinizing ears of paying customers. A proper implementation of AEC (with an emphasis on the word *proper*) is vital medicine that could bring a notable comfort to users of mobile communications.

5 Noisy ambience, mobility, and noise reduction

Chapter 5 is devoted to the subject of noise reduction. Noise reduction is the most complicated feature among the voice-quality-assurance class of applications. It also requires a higher-level understanding of mathematics. This discussion, however, substitutes numerous mathematical expressions for intuition, ordinary analogies, and logical reasoning, supplemented by graphical and audio illustrations.

The analysis gets underway with the definition of noise, a definition consistent with the principles and characterization employed by a typical noise-reduction algorithm. It then introduces and explains the mathematical concept of time and frequency domains and the transformation process between the two. Once the reader is armed with the understanding of time- and frequency-domain representations, the analysis proceeds to a discussion of the noise-estimation process. The presentation then moves ahead to examine the suppression algorithm, which employs the noise-estimation results in its frequency-band attenuation procedures. The next segment contains a presentation covering the final algorithmic steps, which involve scaling and inverse transformation from frequency to time domains.

The next section in Chapter 5 reflects on key potential side effects associated with noise-reduction algorithms including treatment of non-voice signals. It points to key trade-offs and adverse-feature interactions that may occur in various GSM and CDMA networks – a subject that is covered much more thoroughly in Part V – *Managing the network*. The final section offers an examination of the network topology and placement of the noise-reduction application within it.

5.1 Noise in wireless networks

Background acoustic noise is a major voice-quality irritant that is, unfortunately, abundant in wireless communications. Many wireless calls are conducted from noisy environments that not only accompany the transmitted voice but also contribute to distortion of information used and manipulated by wireless codecs.

Although human beings are able to decipher audio content and separate it mentally from irrelevant or noisy interferences, there are many instances where the definition of what constitutes noisy ambiance is left for the particular caller's subjective interpretation, and even that is inconsistent and subject to changes over time due to context variations. For example, background music, or background voices, may or may not be deemed noisy interferences, while a car engine and wind, office fans or street traffic, would for most times be classified as noise.

Just like the ERL measure that signifies the relative[1] potential echo impairment to voice quality, the degree by which noise affects listening and comprehension difficulty is measured relative to the voice or intended transmission. The metric employed for the task is the signal-to-noise ratio (SNR).

The phrase signal-to-noise ratio, often abbreviated SNR or S/N, is a measure of signal (meaningful information) strength relative to an accompanying background noise. Because many signals have a very wide dynamic range, SNR is expressed in terms of the logarithmic decibel (dB) scale.

When the voltage reading of the source signal is denoted as P_S, and the voltage reading of the noise signal is denoted as P_N, the signal-to-noise ratio, SNR, in decibels, is stated as:

$$\text{SNR} = 20 \, \log_{10}(P_S/P_N).$$

Due to the definition of decibel the SNR is the same regardless of the metric (power, current, voltage, etc.) used for measuring the signal strength.

In situations where the signal borders on being masked entirely by the noise, the reading may approach $\text{SNR} = 0$ because $P_S = P_N$.

In most cases, P_S is greater than P_N, so SNR is positive. As an example, when $P_S = 0.01$ volts and $P_N = 0.001$ volts, the $\text{SNR} = 20 \, \log_{10}(10.0) = 20.0 \, \text{dB}$. In general, a higher SNR makes listening more comfortable, and comprehension more precise.

Reed and Bilger[2] studied thresholds of auditory detection. They concluded that over a wide range of levels and frequencies the relevant signal energy must be 5–15 dB more intense than the noise level for the signal to be detected. They contended that as the signal frequency becomes higher the SNR must also approach the upper point (15 dB) of the auditory detection threshold, and at low frequencies the SNR may be lower (still above 5 dB) for the signal to be detected.

Poor SNR may set off impairments over and above masking of the speech signal. Since most wireless codecs make use of voice-activity detection (VAD) as an integral part of their DTX (in GSM) or variable-rate application (in CDMA), extensive speech clipping may result when the VAD mistakes signal for noise. Since one key criterion for discriminating between the two is energy level,[3] poor SNR conditions on the wireless end of a call may have the VAD treat some speech segments as if they were noise and some noisy segments as if they were speech.

A single noise source may cause difficulties on both receiving ends of a communications channel. As illustrated in Figure 5.1, the acoustic noise surrounding the speaker on the mobile end is transmitted together with the voice signal to the PSTN receiving end. The receiving end on the PSTN may also be subject to speech clipping due to VAD errors

---

[1] Relative to the source signal.
[2] C. M. Reed and R. C. Bilger, A comparative study of S/N and E/No., *Journal of the Acoustic Society of America*. **53** (1973).
[3] When noise energy is much lower than speech energy, detection of speech becomes significantly more reliable, since signals containing energy higher than the "noise floor" are far less likely to be noise.

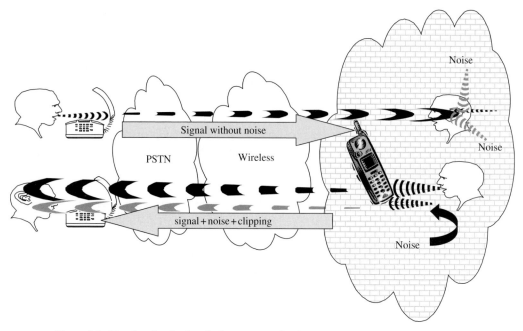

Figure 5.1 Signal and noise in wireless communications.

caused by poor SNR. At the same time, the clear speech transmission from the PSTN to the mobile end blends in the background noise inside the receiving ear of the wireless listener.

High noise and poor SNR may trigger erroneous behavior of hybrid-echo cancelation, acoustic-echo control, and automatic level control, not counting all the impairments already listed.

Discussions and analysis of these noise-triggered errors that may offer inappropriate treatments, dispensing more harm than good, are also presented in Chapters 3, 4, 6, and 12. These chapters cover the subjects of voice-quality-enhancement algorithms and the course of action they should apply so as to minimize mistaking noise input for hybrid echo, acoustic echo, low-level speech, or intended comfort noise.

5.2   Introduction to noise reduction

The topic of noise reduction in wireless communications burst into the scene in the late 1990s[4] as the application was built into CDMA codecs[5] and echo-cancelation systems. The major factors leading to the transformation of echo-cancelation technology into

---

[4]   Gillian M. Davis (editor), *Noise Reduction in Speech Applications*, CRC Press (2002).

S. Gustafsson, Enhancement of audio signals by combined acoustic echo cancelation and noise reduction, Ph.D. dissertation, ABDN Band 11, P. Vary, Aachen, Hrsg. Verlag der Augustinus Buchhandlung (1999) www.ind.rwth-aachen.de/publications/sg_diss99.html.

S. V. Vaseghi, *Advanced Signal Processing and Digital Noise Reduction*, John Wiley and Teubner (1996).

E. Beranek and L. Leo, Noise Reduction, McGraw Hill (1960).

[5]   The EVRC in CDMA was first to introduce noise reduction as an integral part of the codec.

voice-quality-enhancement systems including noise reduction were rooted in the improvements in cellular coverage and codec performance that have identified issues previously considered insignificant.

In addition, advancements in digital-signal processing (DSP) and memory technology have allowed computationally intensive algorithms, such as NR, to be implemented more economically, thereby expanding the feature content of products. At the same time, echo-canceler equipment vendors were eager to enhance the appeal of their products and differentiate them from the growing number of "switch integrated solutions" being offered in the marketplace.[6] Rightfully, noise-reduction (NR) applications became popular and were broadly promoted as the next fad in voice-quality enhancement.

The NR application reduces noise that accompanies speech transmitted from a noisy environment (e.g. airport, restaurant, factory, moving vehicle, etc.), where the surrounding noise level entering the telephone's microphone (i.e. the handset's mouthpiece) can degrade the speaker's voice quality. In most comparable situations, a high level of background noise can make it difficult for the subscriber on the other end of the call to understand the person located in a noisy environment.

Ironically, NR benefits the person located at the far end (not necessarily the noisy end) of a call connection, such that most potential beneficiaries of a NR application, paid for by a wireless-service provider, are non-subscribers. They are either PSTN or far-end (with a potentially different service) mobile listeners.

Ideally, the noise-reduction (NR) function should isolate the speaker's voice from the surrounding noise, as it should automatically attenuate the background-noise level that is sent as part of the call. This method of suppressing the background noise only (by selective signal attenuation), allows the subscriber at the other end of a call to hear clearly the voice signal of the subscriber located in a noisy environment.

The NR function is essential in wireless, public-telephone and conference-call applications, where surrounding noise often interferes with communications.

This chapter addresses the main principles behind the various noise-reduction algorithms including their key trade-offs and side effects.

5.3   The noise-reduction process

Unlike the echo-cancelation process, which bases its algorithm on a feedback mechanism that checks for consistency and error magnitude through a two-pass process, the NR course of action is confined to a single pass with no opportunity for feedback. The echo-cancelation process observes the pure speech signal before examining, correlating, and operating on the pure echo signal, while NR must perform its magic with a lesser amount of information, lacking the luxury of observing pure, uncontaminated speech, or taking advantage of (non-existent) correlation between noise and speech.

---

[6]   Vendors of "switch integrated EC solutions" typically concentrate on electrical (hybrid) echo-cancelation capabilities, and were slower to incorporate other voice-enhancement features (e.g. acoustic-echo control, noise reduction, automatic gain control, etc).

The task of separating out the noise from the speech and suppressing it without effecting speech distortion must be preceded by the task of identifying noise components embedded within the speech signal.

There are two different modes, speech and speech pauses, where noise segments may be captured and characterized.

Noise during speech pauses is the purer among the two, since noise, if it exists, is presented without being "distorted" by an accompanying speech. For this mode to be effective, the NR algorithm must include a reliable voice-activity detector (VAD) that can tell with high accuracy whether a particular signal capture is speech, pure noise, or silence. The accuracy of the VAD has a large impact on the performance of algorithms that depend on it.

A second consideration determining the effectiveness of this approach is that there is a noise type that exhibits stationary behavior over an entire speech period. In view of the fact that the main objective of an NR algorithm is enhancing speech performance during noisy conditions, an NR algorithm that characterizes noise during speech pauses only is highly dependent on the very stationary behavior of the noise signal. When noise during speech pauses is unvarying throughout a conversation segment, then information can be obtained with little errors during pauses and applied throughout. However, when noise exhibits frequent changes in character, amplitude, and color, then employing an estimate that is based on speech pauses alone may prove ineffective.[7]

The second approach, estimating noise during speech and speech pauses without differentiating between the two states, is more effective when noise is not stationary over an entire speech-activity period. This approach employs a noise estimator that is capable of tracking slow changes in noise characteristics. Consequently, this approach is more general and less prone to errors when noise characteristics vary in time.

The second approach is based on the main characteristic of noise that separates it away from speech. Noise can be characterized as a sequence of short-time stationary intervals. In short-time stationary intervals, the parameters of the noise signal remain constant. At the same time, speech is characterized as a signal with high periodicity. Obviously, there are potential overlaps between speech and noise. These occur when particular speech segments have a lower periodicity or when the noise character changes more rapidly. As an example, the maximum duration of a phoneme in normally spoken speech is roughly 200 ms,[8] and a typical duration of a sequence of phonemes spoken without a pause is roughly 2000 ms.[9]

---

[7]   M. Omologo, P. Svaizer, and M. Matassoni, Environmental conditions and acoustic transduction in hands-free speech recognition, *Speech Communication*, **25** (1998), 75–95.
   Walter Etter, Restoration of a discrete-time signal segment by interpolation based on the left-sided and right-sided autoregressive parameters, *IEEE Transactions on Signal Processing*, **44**: 5 (1996), 1124–1135.
   Walter Etter and George S. Moschytz, Noise reduction by noise adaptive spectral magnitude expansion, *Journal of the Audio Engineering Society*, **42**: 5 (1994), 341–349.

[8]   T. H. Crystal and A. S. House, Segmental durations in connected speech signals: preliminary results, *Journal of the Acoustical Society of America*, **72**: 3 (1982), 705.

[9]   D. H. Klatt, Linguistic uses of segmental duration in English: acoustic and perceptual evidence, *Journal of the Acoustical Society of America*, **59**: 5 (1976), 1208.

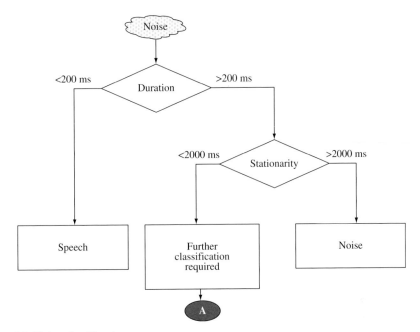

Figure 5.2 Noise-classification process.

A NR algorithm may classify a signal as noise if its stationary duration exceeds 2000 ms. When a stationary duration of a signal is greater than 200 ms but less than 2000 ms, the signal requires further classification since it might also be a sequence of phonemes spoken without a pause. The process is depicted in Figure 5.2.

Figures 5.3 and 5.4 accentuate the striking difference between speech and noise signals. Both figures present three different views of the same 1500 ms capture. The first view portrays the power of the signal as it progresses through time. The second view exposes the same information after the signal capture has been processed by an FFT. The power maintained by each one of the frequency bands (on the horizontal axis) is represented by its color. The brighter the color, the higher the energy level. A very bright color represents a very high energy level. A dark color or black represents low or no energy respectively. The vertical axis depicts the frequency band and the horizontal axis represents time. Accordingly, the second view depicts the power distribution per frequency band as the capture progresses through time. The third view gives a picture of the average power (over the entire 1500 ms range) per frequency band. The vertical axis represents the power and the horizontal axis corresponds to the frequency band.

As seen through the figures, particularly through the second view, the car-noise signal exhibits an ultra-stationary picture while the signal carries most of its power in the lower frequency bands, and its dynamic range is quite narrow. On the other hand, the speech and its echo are not stationary. Their dynamic range is extensive and their power is distributed across the spectrum. The echo signal is a delayed weak equivalent of the speech, where the upper bands present a better ERL in comparison to the lower bands.

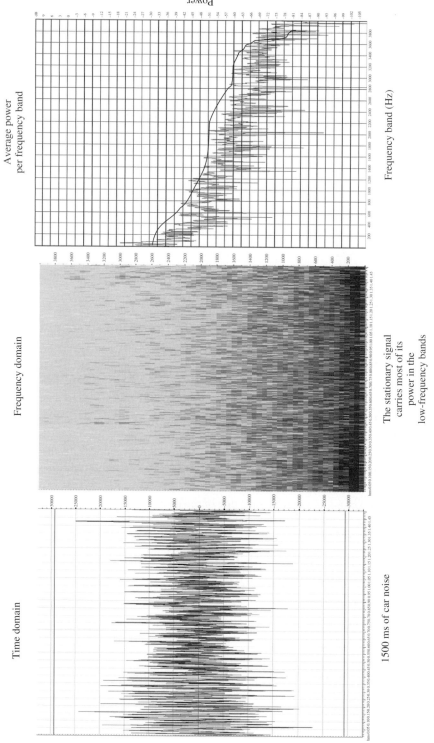

Figure 5.3 Time and frequency domain depiction of car noise.

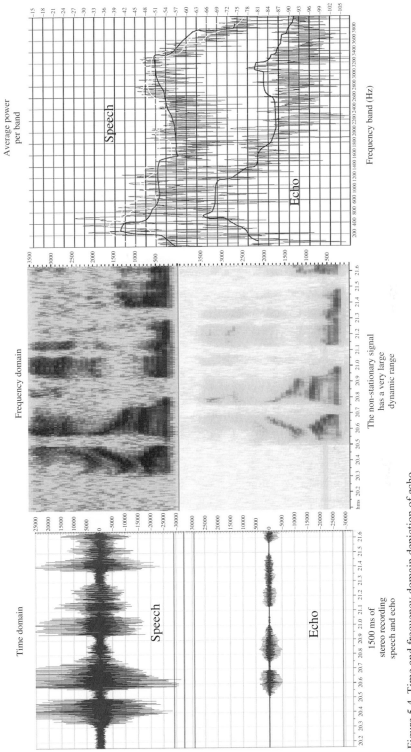

Figure 5.4 Time and frequency domain depiction of echo.

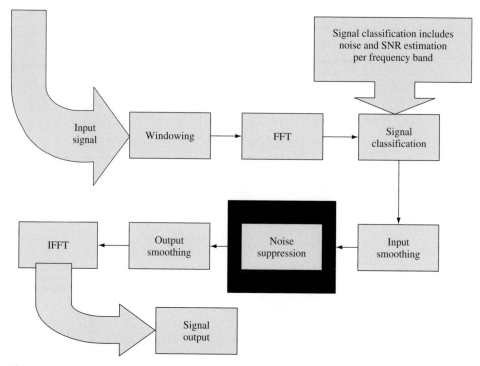

Figure 5.5　Noise-reduction process.

Figures 5.3 and 5.4 make it clear that characterizing and classifying an audio signal is best accomplished in the frequency domain. The first step of the process involves the capturing and buffering of a short time slice comprising a few milliseconds of signal information; this is referred to as windowing. The second phase employs a fast Fourier transform (FFT) that converts the representation of the particular segment from the time domain into the frequency domain (see Figure 5.5). Signal classification, which includes noise and SNR estimation, is enabled in the frequency domain, and it consumes the following phase. The fourth and the sixth phases involve smoothing the inputs and outputs going in and out of the actual noise-reduction procedure, whereas the actual noise reduction is carried out in between the two, during the fifth phase. After noise is characterized and attenuated the process continues by converting the output back into the time domain, where it is sent through the system to its destination. The conversion from the frequency to time domain is accomplished in the seventh phase by the inverse FFT (IFFT).[10] Figure 5.5 depicts the high level process.

10　Y. Ephraim and D. Malah, Speech enhancement using a minimum mean square error short time spectral amplitude estimator, *IEEE Transactions on Acoustic, Speech, and Signal Processing*, **ASSP-32** (1984), 1109–1121.
　　Y. Ephraim and N. Merhav, Hidden Markov processes, *IEEE Transactions on Information Theory*, **48** (2002), 1518–1569.

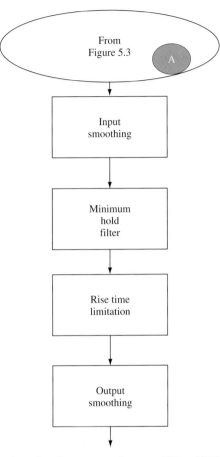

Figure 5.6 Signal classification of stationary states between 200 and 2000 ms. From W. Etter and G. S. Moschytz, Noise reduction by noise adaptive spectral magnitude expansion, *Journal of the Audio Engineering Society*, **42**: 5 (1994).

Signal classification

Signal classification is performed in the frequency domain, where noise and SNR estimates are carried out for each frequency band. In addition to scrutinizing the stationary nature of the signal and exploiting it as the main characteristic distinguishing between speech and noise, some NR algorithms add further screening procedures in the process of telling noise apart from speech.

The key to classifying a signal in a particular frequency band as noise is its stationary behavior. As discussed earlier, the more stationary the signal is, the higher the likelihood that it is noise. The stationary time trigger is usually a parameter threshold greater than 200 ms. When the stationary behavior is identified as a potential noise signal, then the input and output smoothing effects provide additional screening and filtering tools that solidify the inference (see Figure 5.6).

Input smoothing

The dynamic-range contrast between noise and speech can be accentuated through the smoothing of the input signal. The main reason for this result is the fact that speech contains long-lasting low-level signals caused by pauses as well as valleys between two formant poles derived from unvoiced consonants such as p, t, and k. These distinctive pauses are required for building up air pressure used for the formation of these consonants. As it turns out, noise may contain similar valleys, but with one key exception these valleys are conspicuously shorter in duration. Consequently, smoothing the input may eliminate the noise valleys from the scene and may paint the stationary character of the noise in sharper colors. This property that tells speech apart from noise is brought into play as a minimum-hold filter.[11]

Some NR algorithms employ a rise-time limitation[12] as another important parameter controlling the signal characterization. Rise time is the rate by which noise power increases over time. Abrupt changes in noise levels are smoothed by the filter, and are presented to the noise-reduction function as a more gradual increase in signal level. This procedure prevents a noise estimate from jumping up to unrealistic levels. Figure 5.6 illustrates the signal-classification steps designed to separate speech from noise.

Output smoothing

One popular screening device that differentiates unwanted noise from the sought-after speech is the dynamic range. As seen in Figure 5.3, noise tends to concentrate in lower frequency bands. No signal attenuation is prompted at higher bands when placing a bound on the highest possible frequency to be scrutinized for noise.

In trying to smooth transitions and minimize the risk of excessive distortion, most NR algorithms consider history by limiting the extent of a change to the attenuation amounts. The NR loss is smoothed in both time and frequency. In many algorithms, the time smoothing is performed first followed by a smoothing in the frequency domain.

Noise reduction

Noise reduction is carried out on each frequency bin by applying a measured attenuation on each band that is designated as noisy. It should be noted, however, that many of the frequency bins that contain noise might also contain some speech information. The attenuation of these frequency bands reduces the noise content as well as the speech component contained within the same bin. Since noise is not present in every frequency band, many remain unaffected, and the resulting impact on speech is an uneven attenuation of speech frequencies that changes the color and causes distortion.

Evidently, some speech distortion is inevitable when reducing noise. Nonetheless, NR is still a viable application due to careful implementation. Design trade-offs, which reduce noise where it counts while minimizing attenuation of frequencies that contain

---

[11]   Crystal and House, Segmental durations in connected speech signals, p. 705.
[12]   Etter and Moschytz, Noise reduction by noise adaptive spectral magnitude expansion.

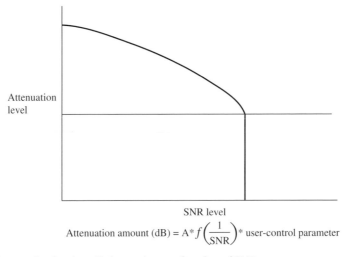

$$\text{Attenuation amount (dB)} = A * f\left(\frac{1}{\text{SNR}}\right) * \text{user-control parameter}$$

Figure 5.7 Attenuation level applied as an inverse function of SNR.

considerable speech contents, contribute to minor speech distortions that may hardly be perceived by a listener, and to highly noticeable gains in noise reduction.

The main strategy guiding proper implementations of NR is employment of the SNR as a gage, steering the amount of attenuation applied to particular frequency bands. A rather small SNR points to a large noise component relative to the speech contents in the appraised frequency band, and a large SNR points to a relatively small noise component. At the same time, a large SNR may not interfere with listening efforts and the value derived from reducing noise, under these conditions, may be marginal at best. These observations yield implementation rules that reduce, minimize, or avoid attenuation of frequency bands with a large SNR, while increasing the attenuation amount for bands with a small SNR as illustrated in Figure 5.7. The attenuation amount (in dB) per frequency band becomes an inverse function of SNR, a fixed parameter (A), and a parameter applied by the network administrator for controlling how aggressive the attenuation is. When the SNR grows beyond a certain level,[13] the attenuation level is reduced to 0 dB.

Many noise-reduction implementations provide users (a.k.a. network administrators) with the flexibility of controlling the level of aggressive working by the algorithm. When NR is provisioned to be aggressive, it attenuates noisy frequency bands by more dB than when the aggression level is set to a lower point. Aggressive behavior tends to take more of the noise out, trading it off for a higher level of distorted speech.

When transforming signals from the time domain to the frequency domain, the number of frequency bands used is finite. Some algorithms employ a 32, 64, 128, or, in general, 2^n point FFT.[14] A larger point FFT provides for better resolution and finer, more accurate, administration of measured attenuation at the cost of increased processing. Still, quantization is unavoidable, and smoothing must be administered among neighboring frequency

---

[13] Different implementations apply different values. As a rule, the value is greater than 9 dB.
[14] Where n can be arbitrarily large. Many popular algorithms use $n = 8$. The enhanced variable rate codec uses $n = 9$ (128 point FFT) for its NR algorithm.

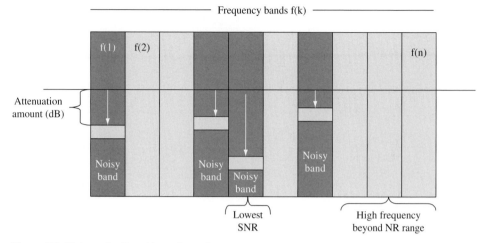

Figure 5.8 Noise-reduction attenuation scheme.

bands to avoid taking large drastic steps (between neighbors) that accentuate the bending effect of quantization.

The NR process may be likened to an application of a graphic equalizer containing 2^n levers. A more authentic sound is always attained when the equalizer is configured to provide smooth transitions between neighboring frequency bands.

Figure 5.8 illustrates an NR attenuation scheme. The illustration employs the graphic equalizer analogy, since it provides an intuitive understanding of the signal color-changing process.

5.4   Noise reduction and GSM DTX

Discontinuous transmission (DTX)[15] is a method (used mostly in GSM) of powering down, or muting, a mobile or portable telephone set when there is no voice input to the set. This procedure optimizes the overall efficiency of a wireless voice-communications system. A DTX circuit operates by using voice-activity detection (VAD). Sophisticated engineering is necessary to ensure that circuits of this type operate properly. In wireless transmitters, VAD is sometimes called voice-operated transmission (VOX).

In a typical two-way conversation, each individual speaks slightly less than half of the time. If the transmitter signal is switched on only during periods of voice input, and can be cut to less than 50% utilization, then the handset conserves battery power and the network infrastructure eases the workload of the components in the transmitter. The action of replacing speech with comfort noise frees the channel so that time-division multiple access (TDMA) can take advantage of the available bandwidth by sharing the channel with other signals.

When the communications channel employs DTX, it replaces the effected silence with comfort-noise matching. The transition from speech to comfort noise is smoothed by

---

[15]   GSM 06.31 *Digital Cellular Telecommunications System (Phase 2+); Full Rate Speech; Discontinuous Transmission (DTX) for Full Rate Speech Traffic Channels.*

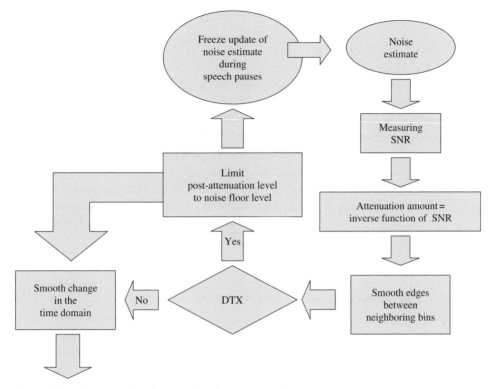

Figure 5.9 Reducing noise of a particular frequency band.

having the comfort noise and the background noise (during speech periods) ensue at a noticeable level in order to mask potential bumpiness during the transition.

The noise-reduction algorithm reduces the level of the background acoustic noise. Consequently, it may unmask the imperfections hiding under the untreated noise level. A rectification to the NR algorithm running under the DTX condition must ensure that any attenuation carried out by the NR algorithm does not cross the masking threshold.

The adjustment works as follows:

if   (signal attenuation is about to reduce the signal level below the DTX threshold)
then   (reduce the intended attenuation amount and halt its progress before reaching the threshold point)

Some NR algorithms include voice-activity detectors that control the algorithm activities as they detect a change in state between speech and speech pauses. The DTX application brings about a change in noise color when states shift between the two conditions as it lets the actual background noise through during speech and a different self-generated noise during speech pauses. An NR algorithm studying the noise input would obtain erroneous information during speech pauses that would result in an erroneous noise estimate during state-transition periods. For that reason, freezing the noise estimate during speech pauses is one possible remedy for that particular case. Figure 5.9 illustrates the key considerations applied to a well-designed NR algorithm.

5.5   Noise reduction and CDMA EVRC and SMV codecs

Both EVRC and SMV incorporate noise-reduction procedures within their codec operations.[16] Thus, by the time a signal arrives at the mobile-switching center many frequency bands have already improved their particular SNR. A separate NR application residing within the MSC boundaries takes hold of the noise-reduced signal. It attenuates frequency bands, which exhibit an SNR still low enough to justify further attenuation.

Obviously, applying a noise-reduction procedure to a signal that has already been processed by a different noise-reduction algorithm yields a lesser marginal change in the overall color of the speech in comparison with the first application. There is a risk, however, that the double enhancement may result in an overly aggressive treatment, which may not yield the same trade-off balance (between noise reduction and speech distortion) intended by the network administrator.

5.6   Noise-reduction level compensation

The process of noise reduction tends to attenuate noisy frequency bands that embrace SNR values less than some provisioned rate.[17] Attenuation is maxed at a very low SNR,[18] and it grows slighter as the SNR increases. Since many of the attenuated frequencies contain speech components, the overall attenuation impacts the overall speech level as well. In other words, NR tends to attenuate speech in addition to attenuating noise. Since the NR process concentrates on noisy frequency bands, NR contributes to a notable improvement in the overall SNR for the complete spectrum when the complete spectrum contains noisy frequency bands.

Noise attenuation produces an undesired side effect in the form of speech attenuation. A properly designed NR algorithm may be able to readjust the speech level by lifting the signal back to its former average level. There is a side effect, however. The process of signal amplification restores the average speech level back to its original plane, while at the same time it raises the noise level as well. A carefully designed algorithm may sustain the post NR SNR through the level-compensation process as illustrated in Figure 5.10.

Figure 5.10 illustrates the ideal operation of a level compensation built into a noise-reduction algorithm. In the above example, the SNR before the noise reduction is measured at 5 dB. After the noise reduction, the SNR grows to 12 dB, and after the NRLC it is restored to 12 dB. At the same time, the noise level is reduced by an average of 9 dB before

---

[16]   EIA/TIA Recommendation IS-127, *Enhanced Variable Rate Codec (EVRC)* (1998).
        Selectable Mode Vocoder Service Option 56 for Wideband Spread Spectrum Communication Systems, 3GPP2
        C.S0030-0 Version 2.0 (December 2001).
[17]   The popular threshold is 9 dB.
[18]   The popular threshold is 3 dB.

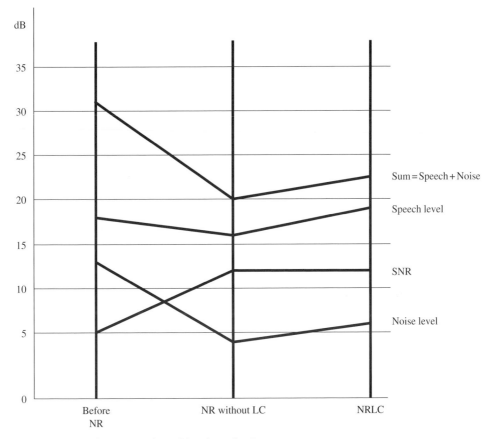

Figure 5.10 Level compensation with noise reduction.

applying the level compensation, but ends up with a 7 dB average attenuation after the LC, while speech remains at the same average level before and after NR in consequence of the LC.

It should be noted that NRLC adjusts the average level; even so, the attenuated frequencies remain attenuated relative to the ones that have been unaffected. Figure 5.11 illustrates the effect of NRLC on the individual frequency bands. The graph illustrates the change in band power relative to its original state. The average level is restored to its pre-NR level while the individual frequency bands maintain identical levels, relative to each other, before and after NRLC.

Noise reduction with level compensation is a useful tool that contributes to a notable reduction in speech distortion caused by attenuation. There is still, however, a question mark concerning the usefulness of NRLC provided that the VQ system includes an automatic level control (ALC) that appears to duplicate the effort. The issue is clarified in Chapter 6, which gives details of ALC. At this point, it is suitable to point out that the two features, NRLC and ALC, are different in many ways, and each serves a different objective.

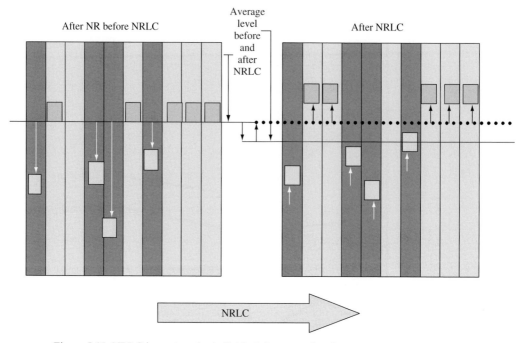

Figure 5.11 NRLC impact on the individual frequency bands.

5.7   Noise reduction and signals other than voice

Noise-reduction algorithms tend to manipulate signals crossing their path in an attempt to reduce suspected noisy components. Many of the signals that do are ones that accompany speech communications on a regular basis. Many non-speech signals carry stationary characteristics that would make them look like noise unless the algorithm is equipped with special detectors that disable the NR and let the signal through unscathed.

The most prevalent examples of non-voice signals using voice-band transmission that require special treatment are: ring-back tone, DTMF, background music, fax, and modems that do not send a disabling tone.

Ring-back tone

In the United States, carriers use a de-facto ring-back standard during call initialization. This ring-back tone is the sum of a 440 Hz and a 480 Hz signal for 2 seconds duration. Each tone pulse has 4 seconds of silence between pulses. Research has shown that most ring-back tones exist at levels from –20 dBm0 to –15 dBm0.

Unless programmed to detect the ring-back tone, an NR algorithm might interpret the stationary signal to be noise. It would take proper action and attenuate it over its duration. The attenuation would slow down due to the 4 second pauses, which would update the noise estimate and lessen the attenuation.

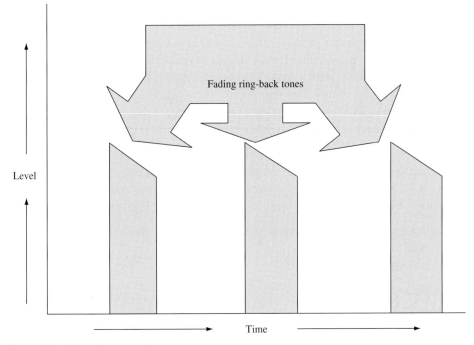

Figure 5.12 NR without ring-back tone detector.

In the absence of proper detection and disabling, the NR would affect fading of the ring-back tone during each of the 2 second ring durations. It would then ring again, starting at a normal level, only to fade again (see Figure 5.12).

DTMF

DTMF, a.k.a. touch-tone, refers to the Bellcore GR-506[19] and the ITU-T Q.23[20] standards for in-band digits. Each dialing digit has a specified sum of two frequencies as well as power and duration requirements. Voice-quality systems must be sensitive to preserve DTMF signals, otherwise DTMF receivers elsewhere in the network would fail to detect such digits.

Unless it is equipped with a DTMF detector, the NR may interpret a DTMF signal to be noise because of its stationary nature. When this happens, the NR would distort the original DTMF and mobile-phone users might not be able to communicate with voice mail and other interactive systems requiring the user's digits as a response.

Background music

Noise reduction may interfere with mobile phones that send low-level background music accompanying the phone call. It may clip notes, distort, and attenuate phrases. A potential

---

[19] Bellcore, Call processing, *GR-505-CORE*, Issue 1 (December 1997).
[20] ITU-T Recommendation Q.23, *Technical Features of Push-Button Telephone Sets* (1988).

remedy for this specific application is to limit the potency of the NR by raising the floor on both SNR and background ambience. During active speech periods, the noise reduction need not attenuate a particular frequency band as long as the SNR is above a set threshold value. With music played from mobile phones, the threshold may be lowered on a permanent basis, so that music may be little affected during active speech. At the same time, if music volume does not exceed a maximum threshold level, the signal need not be considered for attenuation during speech pauses.

The rules can be summarized as follows:

- Set SNR threshold to a new threshold assuming a lower value.
- If (SNR > new SNR threshold) then do not attenuate.
- Set new maximum noise threshold level to a higher value.
- If (noise level < new maximum noise threshold level) then do not attenuate.

By softening NR aggressive application, background music played off mobile phones may pass unharmed.

Fax

Both ITU-T G.164 and G.165 specified a 2100 Hz tone without and with phase reversal respectively. The reason for the G.165 phase reversal was evidence that fax machines running at 9.6 kbps benefited from line-echo cancelers. Fax machines were designed to send a 2100 Hz tone without a reversed phase. Consequently, echo cancelers provisioned to detect a phase-reversed 2100 Hz tone were not about to disable themselves. They stayed enabled in the face of fax transmission.

The case is different with NR. When detecting a 2100 Hz tone, the noise reducer must disable itself and let the fax or voice-band data signal pass through without any noise reduction processing, regardless of whether the tone is or is not phase reversed.

5.8   Network topology of noise reduction

Noise reduction, when implemented in wireless networks, is intended to suppress acoustic noise that surrounds the mobile terminal. When a mobile subscriber makes a call from noisy surroundings, the far-end listener is the one who benefits from the reduced noise. The far end may be a PSTN or another mobile subscriber who may even subscribe to a different service provider. In short, noise reduction has no direct impact on perceived voice quality for those who pay to implement it in their network unless both end points subscribe to the same service. The service providers are faced with a tricky situation. By benefiting the other end, the session, as a whole, turns out for the better, but the service providers' subscribers do not realize the benefit directly, and have no appreciation of the feature.

One proposed remedy to the service provider's noise-reduction dilemma is a bi-directional implementation. Under this type of architecture, the service provider's own subscribers are always covered by the application. And when the same provider

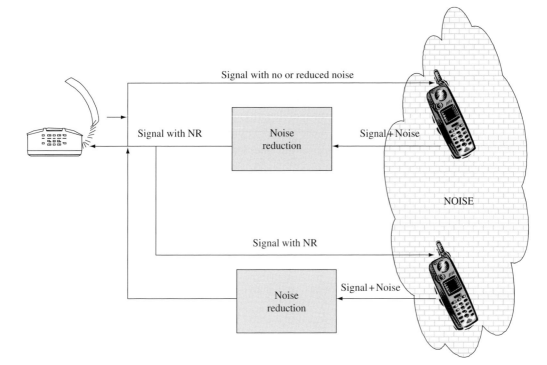

Figure 5.13 One-way noise-reduction topology.

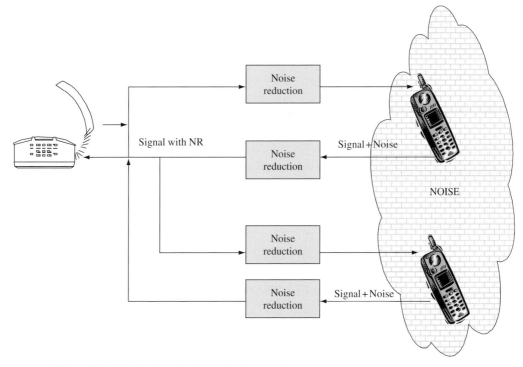

Figure 5.14 Bi-directional noise-reduction topology.

services the two ends to the call, each would be granted double enhancement (see Chapter 9 for further details).

As stated earlier in the chapter, double enhancement is not necessarily a positive development, since it is equivalent to an overly aggressive noise reduction, leading to a more ostensible speech distortion. An architecture consisting of double enhancement must beware of mitigating the parts that make up the sum of the total NR enhancement so that the double treatment in the mobile-to-mobile (within the same network) case does not become too aggressive.

Figures 5.13 and 5.14 describe the one-way and the bi-directional topologies that support the NR application.

6 Speech-level control

6.1 Introduction

Chapter 6 is dedicated to the subject of level-control optimization. The presentation is divided into three parts and an introduction. The chapter starts the ball rolling in the introduction by defining standard methodologies for measuring and quantifying signal levels. The first part deals with automatic level control (ALC), how it works, and its placement within the network. The second part describes the adaptive level control, a.k.a. noise compensation (NC), how it works under different codecs, and where it is placed in the network. The third part describes the high-level compensation procedure along the same outline.

6.2 Basic signal-level measurements and definitions

One of the most critical speech-quality attributes is the perceived-speech level. When it is too loud, it may either overload or hurt the eardrum.[1] When it is too soft, the listener or even the codec may have difficulties picking up words. There are several different measures and metrics used to measure speech levels. The common thread linking them together to a standardized scale is the unit of measurement defined as decibel and abbreviated as dB.

The human auditory system has a dynamic range of 100 000 000 000 000 (10^{14}) intensity units.[2] This dynamic range is better represented by a logarithmic scale, as a ratio of two intensities, P_1 and P_2.[3] The expression "$\log(P_1/P_2)$" is labeled a bel. The resulting number is still too large so, instead, the measure employed is one tenth of a bel or a decibel (dB).

Representing the dynamic range of human auditory system in dB requires a scale ranging from 0 to 140 dB[4] as opposed to a scale with a range of 10^{14} intensity units.

The dB is used to describe, logarithmically, a difference (gain or loss) between two measures of speech intensity, power, or voltage. It does not describe an absolute level or a

---

[1] When an audio system overloads it exceeds the maximum power or amplitude that can be stored or transmitted by the system. The peak amplitude is then clipped and the signal sounds cracked or broken.

[2] Sound intensity is defined in units of power/area = watts/cm^2.

[3] A ratio of two intensities is equal to a ratio of two power measures, or energy measures. When dealing with voltage measures the formula is dB = 20 log(V_1/V_2), since power depends on the square of voltage.

[4] 10 log(10^{14}) = 10 × 14 = 140 dB.

level relative to a reference. The metrics used for these purposes are denoted as "dB" with one or more characters appended to the end of dB. The first metric to be introduced is dBm.

The term dBm is used in telecommunications as a measure of an absolute power value, where 0 dBm is equal to 1 milliwatt (1 mW). A negative number means less than 1 mW and a positive number means more than 1 mW. For example: a signal level of -10 dBm means that the signal is 10 dB less than that at 1 mW. The same concept can be expressed mathematically as

$$\text{absolute speech level} = 10 \log(P_L/P_0) \, \text{dBm},$$

where P_L is a particular power level and P_0 is the reference value, i.e., 1 mW.

Throughout this book, I use relative rather than absolute levels. Relative levels are practical when specifying loss or gain, or even when characterizing communication equipment. The unit dBr is used to characterize relative levels. Once a specific signal level in a specific medium or equipment is designated as 0 dBr, then signal levels in other parts of the system may be expressed in relative terms to the reference level.

For example: the signal level of a digital 64 kbits/s PCM bit stream is designated as 0 dBr, provided that there are no digital loss or gain pads in its path. When a digital loss or gain pad is included in the digital bit stream, 0 dBr is assigned to the signal level at the input side of the pad. If the pad provides 6 dB loss at the output side, then the signal level at the output side is -6 dBr. If the pad provides 6 dB gain at the output side then the signal level is 0 dBr at the output side and -6 dBr at the input.

The most useful measure brought into play throughout the book is the metric dBm0. It is the absolute power in dBm referred to or measured at a 0 dBm transmission-level point. This metric relates the dBm and the dBr by having the relationships expressed as:

$$\text{dBm0} = \text{dBm} - \text{dBr}.$$

For example: a signal with a relative level (-6) dBr is a signal level that is 6 dB lower than its reference. If the absolute power requirement for the signal is -20 dBm, then the absolute power measured at a 0 dBm transmission point is: $(-20) - (-6) = (-14)$ dBm0.

Example 2: if a plain old telephone service (POTS) line is supposed to have a receive level of -9 dBm, then the reference signal is -9 dBr. If an ideal voice-signal power level is (-18) dBm0, then the ideal absolute power of a voice signal within a POTS line is $(-9) + (-18) = (-27)$ dBm.

The implication of the above examples is that two different lines or two different pieces of equipment may run a signal at the same dBm absolute power levels while at the same time they may produce different perceived loudness as measured in dBm0. Also, two different lines or two different pieces of equipment may run a signal at different dBm power levels while the perceived loudness is the same as measured in dBm0.

All signal levels possessing the same dBm0 power level deliver identical perceived-signal loudness if the dBr is set to take account of the specific equipment the signal is running through. Accordingly, dBm0 is used throughout the book as the metric measuring target and actual signal (loudness) level in telecommunications networks.

The sound-pressure level (SPL) and the sensation level (SL) are two other important metrics used in expressing perceived loudness. In the 1930s it was determined through experiments that 20 micropascals (20 µPa) was the minimum sound pressure needed for average healthy human beings to detect the presence of 1000–4000 Hz sinusoid. When using 20 µPa as a reference signal pressure, units of pressure can be measured as dB SPL. For example, when the intensity is 50 dB above 20 µPa, then it can be expressed as 50 dB SPL.

The sensation level (SL) is a measure used in specific experiments where a 50 dB SL means that the signal detected is 50 dB above a minimum intensity threshold required before human beings can detect it in a particular circumstance.

The sending-loudness rating (SLR) is the ratio in dB of the sound pressure produced by a speaker to the voltage produced by a telephone and loop as defined by ITU-T Recommendation P.79.[5]

The receive-loudness rating (RLR) is the ratio in dB of the voltage entering a loop and telephone to the sound pressure produced by the telephone's receiver.

According to ITU-T Recommendation G.223,[6] the mean channel-load capacity should be −15 dBm0, with speech pauses included and consideration taken of some extraneous signals. This translates into −11 dBm0 for the actual speech periods. For more definitions see ITU-T Recommendation G.109.[7]

6.3   Automatic level control (ALC)

Signal levels in networks are, in general, unbalanced. Signals originating from the wireless end of a network connection tend to be "spicier" and louder in comparison with signals originating from the public switched-telephone network (PSTN), as shown in Figure 6.1. Unbalanced signals are one principal cause for reduced speech-signal quality because they increase the complexity of the echo-cancelation task, reduce codec performance by increasing error rates, and make listening a grueling activity that requires additional subscriber effort.

One of the most unintended and unadvertised speech-level drawbacks in telecom networks involving sessions between mobile and PSTN is the mobile relative loudness in comparison with the PSTN side. Mobile speech is typically louder than PSTN speech by 6 dB or more.

The ALC feature is molded to adjust and drive the (receive and transmit) signal levels to a pre-set optimal point (under normal, quiet environment conditions). This has a corresponding effect of steering the listening effort to a minimum, which rewards the subscriber with a pleasant experience. The ALC applies a negative or a positive adjustment to either hot or faint signals respectively. The function applies a computed gain or loss to the incoming signal without modifying its frequency content. In many instances, traversing signals build up either excessive gain or undue loss. Under these circumstances, it is highly desirable to remedy the situation and mold the signal into a fitting level. Automatic level

5   ITU-T Recommendation P.79, *Calculation of Loudness Ratings for Telephone Sets* (1993).
6   ITU-T Recommendation G.223, *Assumptions for the Calculation of Noise on Hypothetical Reference Circuits for Telephony* (1988).
7   ITU-T Recommendation G.109, *Definition of Categories of Speech Transmission Quality* (1999).

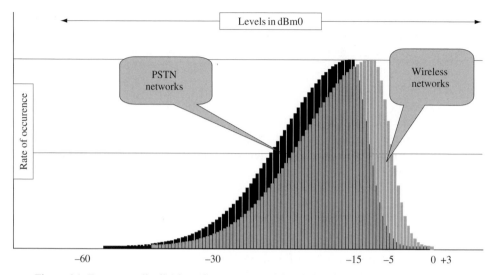

Figure 6.1 Frequency distribution of average speech levels in wireless and PSTN networks.

control functions in a variety of network environments including mobile and landline configurations.

One of the key ingredients of ALC is its ability to preserve dynamic speech artifacts while converging on a preset target level. Consequently, a properly designed ALC feature must perform adjustments smoothly and gradually rather than abruptly. This means that the ALC algorithm developer must establish acceptable trade-offs between the speed of adjustment and the preservation of accurate speech characteristics. The ITU-T Recommendation G.169[8] specifies many of the requirements and constraints governing proper implementation of an ALC mechanism. Among the most notable aspects of G.169 is the guidance provided for algorithm designers, ensuring against devastating side effects that an overly aggressive ALC might wreak on network operations and subscribers (see Chapters 9 and 12).

A properly designed ALC must have the following characteristics:

- It can be configured separately and independently in the receive and send directions.
- It can function appropriately in any serene or noisy environment including mobile-to-mobile, mobile-to-land, and land-to-land.
- It includes an adjustable maximum gain.
- It conforms to the ITU Recommendation G.169 standard.
- Its adaptation rates are adjustable.
- It includes a wide range of adjustable target values.
- It preserves speech dynamics, such as yelling and whispering.
- It does not interfere with signaling tones – DTMF, MF, 2100 Hz, and pure tones.
- It preserves frequency content.
- It preserves loop stability by minimizing saturation of voice signals.

---

[8]   ITU-T Recommendation G.169, *Automatic Level Control Devices* (1999).

A well-designed ALC lets users provision several key parameters in their effort to optimize the listening comfort of their customers. These parameters include target values in the receive and send directions, applied gain, attack and release rates, and noise floor definition.

Target value

Automatic level control may apply gain or loss to achieve this predetermined value. This parameter may be configured separately in each direction. It usually ranges from -23 dBm0 to -6 dBm0.

Applied gain

This gain is applied to the incoming signal to achieve the target value and is automatically computed by the ALC function. The ALC function may apply a maximum loss of -12 dB and a maximum gain of 12 dB. The maximum gain should stay within the range 3–12 dB.

Attack and release rates

The release rate is the rate at which ALC increases the applied gain. The attack rate is the rate at which ALC decreases the applied gain. These may be configured in the range of 1–9 dB/s of sufficient voice signal. Caution should be placed on setting high values, since this action may cause the loss of speech dynamics preservation.

Noise floor

Automatic level control is designed to operate in noisy conditions such as mobile environments. It should attempt to adjust the gain based on the voice signal only, without adapting on background noise. This parameter represents the noise-floor threshold where ALC should cease its functioning. It may be configured between -30 and -40 dBm0.

Logical flow depicting a constrained approach

The ALC logic flowchart is presented in Figure 6.2. The key elements making up the algorithm are the constrained mechanisms that control and limit the amount and rate of gain or loss applied to an existing signal. As the signal level approaches the target level, the gain adjustments are slowed down.

Placement of the ALC within the voice-quality system

Ideally, the ALC application on the send side should be placed in sequence after the echo canceler has completed its task, as illustrated in Figure 6.3. On the receive side, the ALC should be placed after the noise reduction and the acoustic-echo control have completed

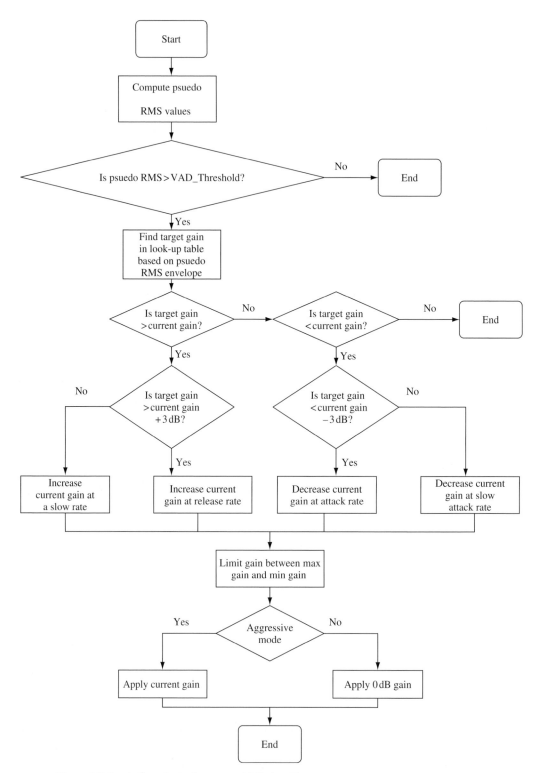

Figure 6.2 Logic flowchart of a proper ALC algorithm.

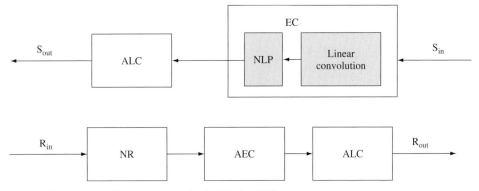

Figure 6.3 Proper application sequencing inside the VQS.

their tasks. When introduced as an afterthought, it may be placed either before or in between any of the above applications. There are risks associated with placements preceding any other voice quality functionality. In the following sub-sections, I will analyze these particular architectures (see Figure 6.3 for proper placement of the ALC with the VQS applications).

ALC applied prior to S_{in}

When an ALC function is placed upstream of S_{in} (see Figure 6.4), it may change the ERL value perceived by the EC. If it is a fixed and consistent gain, then the ERL threshold may be set to a lower value to compensate for the louder echo. If the gain were adaptive and time variant, then the EC would have to re-converge at every gain adaptation (see Figure 6.5).

Since echo is usually quieter than the source triggering it, an ALC may amplify it whenever the echo signal is sufficiently hot and above the VAD threshold of the EC. In some cases, the amplification may even result in an echo signal louder than the source; the EC may treat it as double talk, and it may fail to converge on it.

In many EC implementations, the ALC dynamic setting inside its core cannot be affected or reset on a call boundary. The echo return loss enhancement average convergence rate for most commercial echo cancelers is about 5–10 dB/s in the relevant part of the adaptation curve. Accordingly, ALC settings could be carried over from one call to the next. When a particular call experiences a relatively loud S_{in} signal (including loud echo) due to network characteristics, the ALC may try to lower the applied gain, but in cases where a previous call had its gain set to a high value, it would take 5–10 s before the correct gain is applied. Consequently, it may take longer for the NLP to operate and eliminate the residual echo during the beginning part of the call.

When its adjacent switch controls the EC on a per call basis, the ERL threshold setting could become adaptive; it may reset itself at the beginning of every call. However, adaptive ALC operating before the signal moves through the S_{in} port may amplify the echo over time. At the very beginning of the call the echo signal may not reach its final level; it may appear to be of a lower volume than it is 30 seconds to a minute after the call is well into its session. At the beginning of the call when the ERL setting is established the EC may

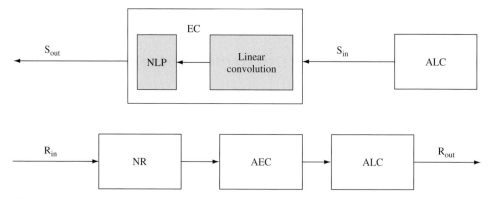

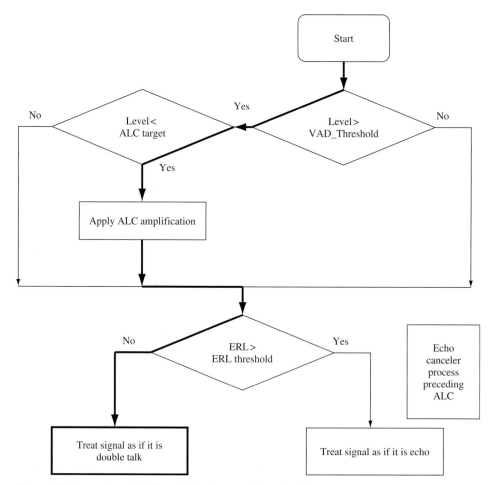

Figure 6.4 Improper sequence of VQ applications, placing ALC ahead of EC.

Figure 6.5 Issues when ALC is applied prior to EC functionality. The heavy line follows a high-probability path.

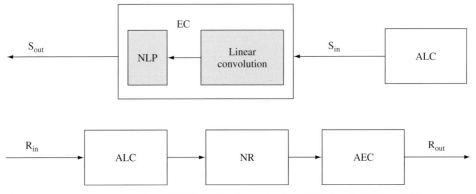

Figure 6.6 Improper sequencing of VQS applications, placing ALC ahead of NR and EC.

overestimate its value.[9] Under these conditions, at about 30 seconds to a minute later into the call the echo signal may become louder and may be treated as double talk. The EC may freeze its convergence state and cease adaptation.

ALC applied prior to NR and AEC on the receive side

The implications of placing an ALC upstream of the R_{in} port are equivalent to having it upstream of the S_{in}, only the logic, in this case, is applied to acoustic rather than electrical echo. There is one additional situation, however, and that is the noise-reduction application. It must also operate ahead of the ALC. The logic flow presented in Figure 6.2 suits the acoustic-echo situation as well as it describes the electrical one. Placing the ALC ahead of NR may bring about additional issues worth discussing.

When placing an ALC beyond the NR application (see Figure 6.6), it may amplify both speech and noise by the same amount, but would preserve the SNR arrived at after the NR completed its task. When the background noise is higher than the provisioned noise floor, placing an ALC ahead of the NR application may amplify the noise component during speech pauses, mistaking it for low-level speech. Consequently, it might reduce the true SNR, and would make the noise-reduction task more difficult and more prone to speech distortions.

6.4 Noise compensation (NC)

The preset ALC target level may not be optimal when a listener (especially a mobile subscriber) is exposed to varying levels of background noise. When background noise is taken into account, the optimal listening level becomes a moving target. That is, the optimal target-signal level must be repositioned with respect to surrounding noise conditions. The louder the noise, the louder the optimal target level.

---

[9] When the echo gets louder relative to its source signal, the associated ERL becomes smaller. The ERL is the level difference between the source signal and its corresponding echo.

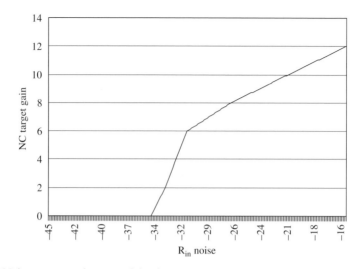

Figure 6.7 Noise-compensation potential-gain map.

To accommodate this type of performance, the noise compensation (NC), (a.k.a. adaptive level control) application effectively adjusts the optimal signal level by means of real-time adaptation to the noise level surrounding the listener. It compensates for the noise. The NC adjustment is not instantaneous. It is delayed by two factors – round-trip and adaptation delays. The round-trip delay is a physical unavoidable fact in cellular communications. The adaptation delay is programmed into the feature to allow for smooth unnoticeable rises in speech level while preserving its natural dynamics. When the noisy channel becomes quiet, the NC feature does nothing in response, and leaves the incoming signal to its natural state.

Target gain is determined by measuring the instantaneous R_{in} noise. The R_{in} noise is computed by taking the average noise estimate across all frequency bins. The illustrated gain map in Figure 6.7 illustrates the relationship between R_{in} noise and NC target gain. The gain has a faster ramp up at lower noise levels because the human ear does not perceive small levels of gain. For most implementations, the applied gain is limited to a range between 0 dB and 12 dB under all circumstances. The gain adaptation only occurs when the input signal at R_{in} is larger than a noise floor (typically set at −30 dBm0). When the R_{in} signal level drops below this value, the gain may be leaked down slowly at a rate of <0.5 dB/s. This helps alleviate the improper boosting of small signals, such as echo and noise, at R_{in}.

However, to complicate things a bit, the NC must bend the rules associated with the purely linear relationships to introduce a non-linear saturation-avoidance constraint. In so doing, it effectively "puts on the brakes"; decelerating its gain-boosting function (if necessary all the way down to a complete stop) as the signal level approaches saturation. In typical implementations, the noise-compensation gain is limited such that amplification is ceased when the signal level reaches about −6 dBm0 (see Figure 6.8 for level and gain relationships).

The placement of the NC application within the VQS is illustrated in Figure 6.9. The noise-level input at R_{in} is used by the NC application as the key parameter controlling the

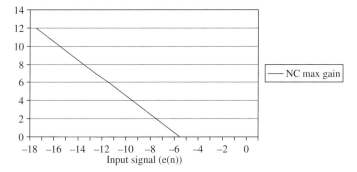

Figure 6.8 Noise compensation saturation prevention gain map.

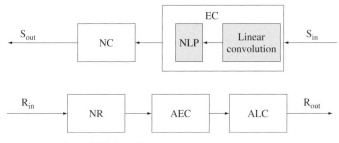

Figure 6.9 Proper sequencing of VQS applications.

amount of gain it appends to the incoming signal on the send side. An ALC operating on the receive side prior to entering the R_{in} port (as illustrated in Figure 6.6) may distort the noise level information, and may instigate an overly aggressive gain treatment.

6.5 Combined NC and ALC

Noise compensation and automatic level control may be combined into a more intelligent level control. The combination may utilize most of the ALC logic with one exception – the optimal target level, rather than being fixed and provisioned by the network engineer, may become adaptive to noise level at R_{in}. If the gain rates for ALC and NC are the same, then the combination of the two applications becomes an ALC with a dynamic target level.

Many applications, however, prefer separate controls for setting gain rates. A faster reaction by NC is required when noise levels vary rapidly throughout a session due to fast-varying street noises. Under this philosophy, the combination is confined to adjustments in the gain amount applied under the NC application; the combination of ALC (send side) gain and NC gain may not be allowed to exceed MAX_GAIN dB (12 dB for most implementations). Accordingly, when computing the NC gain, the ALC (send side) gain is taken into account by ensuring that the sum of the two (ALC + NC) does not exceed the MAX_GAIN. The resulting gain is then applied to the signal.

6.6   High-level compensation

Some network or mobile terminal conditions produce very hot signals that bring about severe impairments to the voice quality. In addition to giving rise to saturated speech by clipping its peaks, hot signals present a problem for echo cancelers as well. When peaks are clipped, the difference in level between speech and its corresponding echo is significantly reduced, and echo cancelers may have difficulty distinguishing between the two. Some echo cancelers may interpret the high-level echo to be double talk, and the erroneous treatment may result in very loud echo returned to the wireless speaker.

The high-level compensation (HLC) feature is intended to ensure that hot signals are cooled off before being processed by the other voice-quality enhancement applications. The HLC consists of two key features; far-end level regulator and near-end hybrid equalizer.

The far-end level regulator (FLR) is designed to maintain a comfortable listening (power) level by attenuating far-end (receive side) hot signals characteristic of numerous mobile systems.[10] The feature is capable of attenuating hot signals by 6 to 12 dB. It is recommended that the feature be turned on when the average signal-level difference between the hot mobile end and the PSTN end presents an average disparity greater than 5 dB. The PSTN listener would benefit from a more comfortable (not as loud) listening level while the mobile listener would benefit from an improved echo-canceler performance caused by from a significant drop in clipping, which is, more often than not, a by-product of asymmetrical levels.

One of the major parameters employed in distinguishing between echo and speech signals is a level difference. Network hybrids returning echo signals are supposed to provide a minimum of 6 dB attenuation to the returning echo signal. Every echo canceler is designed to differentiate between speech and echo by bringing into play that specific characterization. In some networks, there is an overload of substandard hybrids, each failing to attenuate echo signals by a minimum of 6 dB.

The near-end hybrid-equalizer (NHE) feature is designed to remedy and compensate for substandard hybrids by attenuating and equalizing the S_{in} signal prior to its all-inclusive treatment by the echo canceler.[11] The feature is enabled when many hybrids in a specific portion of the network serviced by a specific trunk show evidence of unsatisfactory performance.

---

[10]   The mobile side is generally hotter than the PSTN side. In this example, the receive side is the mobile side.
[11]   Attenuating the S_{in} signal would attenuate the echo signal as well as the near end speech. The trade-off may be worthwhile when echo is generally too loud and is not canceled because of its high level.

Part III

Wireless architectures

Part III

Wireless architectures

7 Mobile-to-mobile stand-alone VQS architectures and their implications for data communications

7.1 Introduction

Chapter 7 reviews the optional placements of the VQS functions relative to the mobile-switching center and the base-station controller, since placement impacts voice performance, applications, deployment cost, and data-detection algorithms. The first section of this chapter covers wireless-network architectures that provide comprehensive signal processing coverage for mobile-to-mobile call applications. The topic of economics and architectural trade-off associated with voice-enhancement systems is also addressed. The second part of the chapter presents an analysis of the techniques employed by a voice-quality system when coping with data communications without interfering or blocking its error-free transmission. The analysis includes descriptions of data-detection algorithms based on bit-pattern recognitions. The scope encompasses circuit-switched and high-speed circuit-switched data (CSD and HSCSD respectively) services. Finally, the third section describes tandem-free operation (TFO),[1] its potential impact on speech transmission and data communication, and potential features and architectures.

The chapter characterizes two major themes and their joint interplay: (1) mobile-to-mobile network architectures with voice-quality enhancements, and (2) mobile data communications. It elaborates on various applications, technical challenges, and potential solutions. It is also intended to impart a sharper awareness of where technology is heading, and what constitutes winning features in the race to provide products that deliver superior voice quality in the wireless-communications arena.

The surge in data communications has spilled its fervor into wireless applications. Demand for higher-speed data access and the remarkable growth of internet applications have fueled the growth of this industry. As a result, traditional voice-quality technology has been transformed into a more intelligent resource capable of interworking with a set of enhanced data services.

---

[1] ETSI TS 128 062 (2002–03), *Inband Tandem Free Operation (TFO) of Speech Codecs, Stage 3 – Service Description* (2002).

7.2   Wireless-network architectures

Signal processing in mobile-to-mobile call applications

Early implementations of wireless networks were developed with the premise that electrical (hybrid) echo cancelation was the only signal-processing feature required to insure voice quality in the path between a mobile subscriber and the public switched-telephone network (PSTN). This theory was based on an incorrect assumption that acoustic echo and noise originating at the mobile end of a call were insignificant annoyances easily controlled at the source (i.e., eliminated by the design of the mobile phone itself). This led to the implementation of network architectures with echo cancellers (ECs) positioned between the PSTN switch and the mobile-switching center (MSC), with the sole intention of controlling the electrical (hybrid) echo originating in the PSTN, as illustrated in Figure 7.1.

   Although the arrangement shown in Figure 7.1 eliminated electrical (hybrid) echo originating in the PSTN, subscribers began experiencing degraded voice quality caused by acoustic echo, noise, and unbalanced speech levels. The primary cause of these problems was deficient wireless-phone design (i.e. the original assumption that acoustic echo and background noise would be eliminated by wireless phones was invalid). Moreover, wireless-network operators had very limited influence regarding the type or quality of mobile phones being used by their subscribers. In fact, many subscribers demanded high-quality voice service, regardless of the type of mobile phones they were using. Wireless carriers were expected to fix the problems caused by cheap mobile phones. In response to

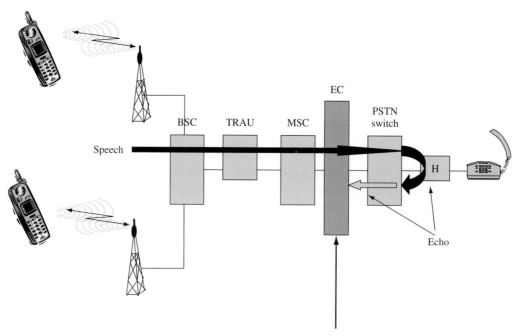

Echo canceler positioned between MSC and PSTN

Figure 7.1 Electrical (hybrid) echo in mobile-to-PSTN applications.

this situation, most wireless-network operators requested new features and enhancements for existing echo-canceler equipment – including acoustic-echo control, noise reduction, and automatic level control.

Figure 7.2 illustrates the operating sequence of voice-enhancement applications within a generic VQS. These applications include noise reduction (NR), acoustic-echo control (AEC), automatic level control (ALC), and adaptive level control, a.k.a. noise compensation (NC). All of these applications were appended to the existing echo-cancelation (EC) function in the mobile-to-PSTN signal path to enhance the quality of communications.

It is important to note that the sequence of these features is crucial to proper operation. That is, NR must precede AEC, AEC must precede ALC, and EC must precede ALC and NC. These constraints become obvious when various system architectures are discussed in later sections of this chapter.

The deployment of voice-quality applications in addition to echo cancelation was initially limited to connections between mobile and PSTN subscribers. However, the rapid growth in wireless communications increased the number of mobile-to-mobile calls, which created additional delay and noise in the signal path. As a result, it soon became apparent that even though electrical (hybrid) echo was not present in two-way wireless calls, the remedies for mobile-to-PSTN connections (i.e., NR, NC, ALC, AEC) were even more important for ensuring voice quality in genuine wireless applications.

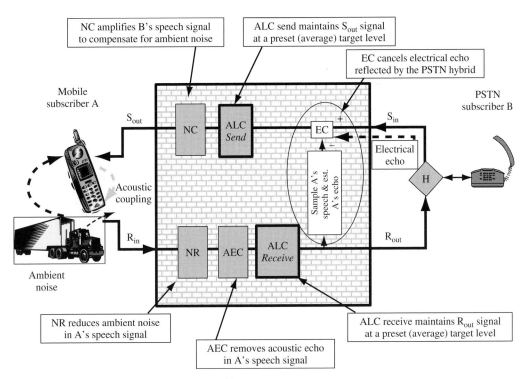

Figure 7.2 Voice-quality system architecture.

Back-to-back echo-canceler arrangement

Back-to-back echo canceler deployment is used to support mobile-to-mobile communications on the same MSC. Figure 7.3 depicts a typical back-to-back EC deployment for a wireless application, which is a natural evolution of the MSC–EC–PSTN architecture also shown in this diagram. The back-to-back EC arrangement is added on, rather than a change involving replacement or modification of existing connections. It is, however, relatively expensive with respect to indirect MSC opportunity cost.

The PSTN to MSC VQ application architecture typically requires two ports out of the MSC (see Figure 7.3). In comparison, the back-to-back echo-canceler arrangement for mobile-to-mobile applications is doubled (i.e., it requires four ports), as shown in Figure 7.3. Moreover, network planners must allocate the available MSC ports and pre-assign a fixed number of ports for each application. Pre-assignment yields inefficiencies and increased cost due to splitting (rather than sharing) of ports among different applications (i.e., mobile-to-PSTN vs. mobile-to-mobile).

If the number of mobile-to-mobile calls on the same MSC is relatively small, the inefficiencies associated with a back-to-back VQS configuration are inconsequential.

Voice-quality systems installed in a back-to-back arrangement to enhance mobile-to-mobile voice quality are not required to provide electrical (hybrid)-echo cancelation (i.e., the hybrid is absent in mobile-to-mobile applications, since electrical echo does not exist).

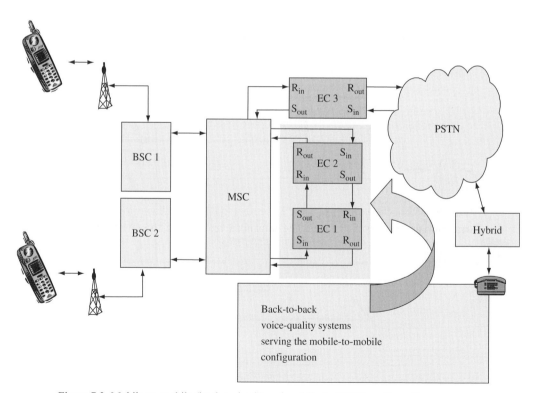

Figure 7.3 Mobile-to-mobile (back-to-back) and mobile-to-PSTN configurations.

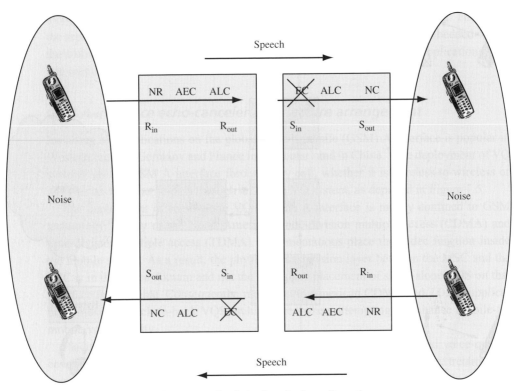

Figure 7.4 VQ application sequencing in back-to-back configuration.

Consequently, the electrical (hybrid) echo-cancelation function may be permanently disabled in these VQS units, as shown in Figure 7.4. Disabling the electrical (hybrid)-echo cancelation function reduces the possibility of speech clipping, which can result from erroneous operation of the non-linear processor (NLP) in the EC module.

Combined PSTN and A-interface echo-canceler architecture arrangements

One approach for enhancing the voice quality of mobile-to-mobile calls in existing (traditional) networks[2] is to include advanced features on the (A-law or μ-law – pulse code modulation [PCM]) mobile switching center (MSC) to the TRAU link (i.e., the A-interface).[3] This architecture allows for a relatively graceful evolution to the superior voice quality platform that embraces the entire suite of voice quality enhancements, without removing (or replacing) the existing electrical (hybrid) echo cancellers from their position on the MSC to PSTN connection, as shown in Figure 7.5.

---

[2]   Traditional networks with electrical (hybrid)-echo cancelers deployed on PSTN-to-MSC connections without
       enhancements such as NR, ALC, AEC, etc.
[3]   This A-interface architecture is standard in GSM. It is not always feasible under CDMA topologies.

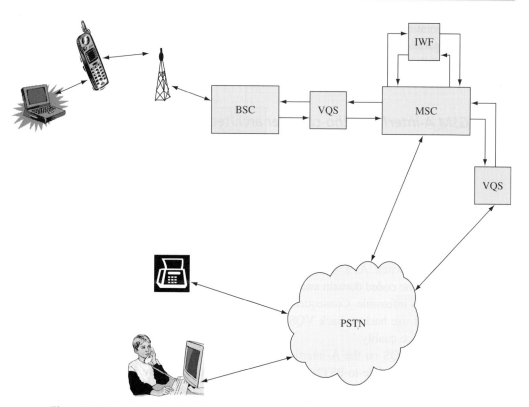

Figure 7.6 Voiceband data in 2G wireless communications.

standard modem (located on the PSTN side of the call). The 16 kbps signaling channel between the MSC and the BSC is used to control the routing of data traffic to the IWF before it is sent to its final destination in the PSTN, as shown in Figure 7.6.

As seen in Figure 7.6, the VQS is placed ahead of the IWF. Consequently, the signal is not yet modulated as voice-band when it moves through the VQS, and the bit sequence assumes a unique pattern (discussed in the following section) that a well-designed VQS may be able to detect. The VQS must disable its signal processing on detection of a non-voice signal and let the data signal move on to the IWF before being shipped to its final destination.

Circuit-switched data communications with VQ systems on the MSC to PSTN interface

Voice-quality systems on the PSTN interface must be disabled to allow circuit-switched data communications. Similarly, VQ applications on the PSTN interface are also usually disabled for fax communications.[7] There are two methodologies for disabling VQ applications.

---

[7] It is not necessary to disable echo cancelation for some types of fax transmission (e.g., fast fax); however, the standards (i.e., ITU-T G.164 and G.165) specify that the straight tone (e.g., 2100 Hz) generated by fax terminals should not be used to disable the echo cancelation function, but should be used to disable echo suppressors and VQ applications.

Voice-band data The data terminal on the PSTN or the IWF (on the MSC) generates a 2100 Hz tone with phase reversal. The echo canceler detects this tone and automatically disables voice-quality applications for the respective channel (i.e., voice-quality applications remain disabled as long as there is energy on the line). If a fax machine generates the 2100 Hz tone, then it is likely to be without phase reversal. Under these conditions, with the exception of the EC, all other VQ applications must disable themselves.

Digital data The MSC issues a message (commonly called per call control [PCC] or signal processing control [SPC] commands) that places the VQ application in the clear channel mode. While in the clear channel mode, voice-quality enhancement is disabled and typically up to 64 kbps bandwidth is available for the respective channel (i.e., VQ enhancement remains disabled until a subsequent PCC/SPC command re-enables it). In this arrangement, both the VQ system and MSC equipment designs must support the physical interface, protocol, language, and syntax used for the PCC/SPC messages. This task requires considerable customization and coordination between the vendors of the MSC and EC products.

Data communications with VQ systems on the GSM A-interface[8]

The A-interface between a mobile terminal and the BSC supports a 16 kbps data channel. Most GSM data communications are conducted at approximately 9.6 kbps. A more comprehensive service, high-speed circuit-switched data (HSCSD) specifies mobile data-communication speeds at rates of up to 56 kbps. This higher-speed service may be implemented by merging several (up to four) 16 kbps channels (on the IWF to mobile side) into a single higher-bandwidth channel (on the IWF to PSTN side) as illustrated in Figure 7.6.[9]

Voice-quality systems deployed on the A-interface must be disabled to allow data and fax communications at any speed. In this application, the data signal is digital, therefore the integrity of every bit is essential. Voice-quality systems must be equipped with detection mechanisms[10] that allow operation on the A-interface without disrupting data communications.

Ensuring transparency by the VQ system requires the understanding of all possible HSCSD (high-speed circuit-switched data) configurations.[11]

---

[8] Although the GSM 2.5G supports packet switched data – GPRS and EDGE – neither one of these architectures has any impact on VQS, as explained in detail in Chapter 8. Accordingly, the discussion in this section is confined to circuit-switched data.

[9] GSM 05.02, *Digital Cellular Telecommunications System (Phase 2+); Multiplexing and Multiple Access on the Radio Path*, version 8.5.1 (1999).

[10] There is no 2100 Hz tone on the A-interface. Detection of data communications on the GSM A-interface is performed by monitoring the channel and identifying a bit pattern characteristic of data.

[11] GSM 03.34, *Digital Cellular Telecommunications System (Phase 2+): High Speed Circuit Switched Data (HSCSD) – Stage 2*, version 7.0.0 (1998).

General terminology

The air-interface user rate (AIUR) is the rate seen by the mobile terminal (e.g., modem). This data stream can be broken into sub-streams before transmission over one or more radio channels. This is analogous to the fixed-network user rate (FNUR) on the PSTN side. Possible values of the AIUR are ≤2.4, 4.8, 9.6, 14.4, 19.2, 28.8, 38.4, 48.0, and 56.0 kbps. Depending on the vendor, not all rates may be supported by HSCSD.

The following logical **radio traffic channels** are defined to carry user circuit-switched data:[12]

- full-rate traffic channel for 9.6 kb/s user data (TCH/F9.6);
- full-rate traffic channel for 4.8 kb/s user data (TCH/F4.8);
- half-rate traffic channel for 4.8 kb/s user data (TCH/H4.8);
- half-rate traffic channel for ≤2.4 kb/s user data (TCH/H2.4);
- full-rate traffic channel for ≤2.4 kb/s user data (TCH/F2.4);
- full-rate traffic channel for 14.4 kb/s user data (TCH/F14.4);
- enhanced circuit-switched full-rate traffic channel for 28.8 kb/s user data (E-TCH/F28.8);
- enhanced circuit-switched full-rate traffic channel for 32.0 kb/s user data (E-TCH/F32.0);
- enhanced circuit-switched full-rate traffic channel for 43.2 kb/s user data (E-TCH/F43.2).

Only **TCH/F4.8** and **TCH/F9.6** are supported by HSCSD.[13]

The radio interface rate is the rate that is transmitted over each TCH, including any overhead relating to the data protocol. This is higher than the designation in the traffic channel because it contains overhead.[14] Possible radio-interface rates are 3.6, 6.0, 12.0, and 14.5 kbps.

Each sub-stream is rate adapted to an **intermediate rate** before transmitting over the A-interface. Possible intermediate rates are 8, 16, 32, and 64 kbps. Only rates of **8 kbps, 16 kbps, and 64 kbps** are supported by HSCSD.

In single-slot operation, each intermediate sub-stream is transmitted over a single timeslot (octet) of an A-interface 64 kbps circuit.

In multi-slot operation, two to four intermediate sub-streams are multiplexed over a single timeslot (octet) of an A-interface 64 kbps circuit.

High-speed circuit-switched data (HSCSD)

Without HSCSD, the AIUR is limited to 9.6 kbps using the 12.0 kbps radio-interface rate. The HSCSD system combines multiple full-rate traffic channels (TCH/F) in a single physical layer structure for the purpose of increasing the AIUR.

---

[12]   Although the GSM 2.5G supports packet-switched data – GPRS and EDGE – neither one of these architectures has any impact on VQS, as explained in detail in Chapter 8. Accordingly, the discussion in this section is confined to circuit-switched data.

[13]   There is no 2100 Hz tone on the A-interface. Detection of data communications on the GSM A-interface is performed by monitoring the channel and identifying a bit-pattern characteristic of data.

[14]   GSM 03.10, *Digital Cellular Telecommunications System (Phase 2+); GSM Public Land Mobile Network (PLMN) Connection Types*, version 7.0.1 (1998).

Table 7.1. *HSCSD supported channel configurations*

User rates (kbps)	Sub-streams × Traffic channels	Intermediate rates (kbps)
≤2.4	TCH/F4.8	8
4.8	TCH/F4.8	8
9.6	2 × TCH/F4.8	2 × 8
9.6	TCH/F9.6	16
14.4	3 × TCH/F4.8	3 × 8
14.4	2 × TCH/F9.6	2 × 16
19.2	4 × TCH/F4.8	4 × 8
19.2	2 × TCH/F9.6	2 × 16
28.8	3 × TCH/F9.6	3 × 16
38.4	4 × TCH/F9.6	4 × 16
48.0	5 × TCH/F9.6	64[a]
56.0	5 × TCH/F9.6	64
64.0	6 × TCH/F9.6	64

Note:
[a] For AIUR over 38.4 kbps, the sub-streams are combined to form either 32 bit or 64 bit V.110 frames. Then they are transmitted over the A-interface as a 64 kbps stream in a single-slot operation.

Source: This table was created by combining Tables 8 and 11 in GSM 08.20 *Digital Cellular Telecommunications System (Phase 2+); Rate Adaptation on the Base Station System – Mobile Services Switching Center (BSS-MSC) Interface* version 8.1.0 (1999).

The data stream is split into multiple sub-streams, and each sub-stream is transmitted over a single TCH/F. Each sub-stream is then rate adapted to an intermediate rate before transmission over the A-interface. Depending on the AIUR and channel configuration, these sub-streams may be multiplexed onto a single 64 kbps circuit (timeslot.) The reverse process occurs in the opposite direction.

Table 7.1 shows the relationship between AIUR, number of sub-streams, channel coding, and intermediate rates for HSCSD.

VQ system transparency

HSCSD requires that multiple sub-streams be multiplexed onto a single A-interface circuit (octet). Recommendations GSM 08.20 and ITU I.460[15] list the possible A-interface octet patterns, as shown in Table 7.2, where an X indicates that there are data or status transmitted in the bit. In either case, the VQS may not be able to foretell the value ahead of time.

Each letter X, Y, Z, W designates data for a particular sub-stream.

The first five cases can be detected by disabling upon detection of a XXXXXX11 pattern. Detection of the 4 × 8 kbps case is realized by monitoring for an additional pattern, namely X1X1X1X1.

For the last two cases, there are no single-bit patterns, and the transparency algorithm must go a step farther and examine a multi-octet pattern.

---

[15] ITU-T Recommendation I.460, *Multiplexing, Rate Adaptation and Support of Existing Interfaces.*

Table 7.2. *Possible A-interface patterns*

Sub-streams × intermediate rates (kbps)	A-interface octet
8 kbps	X1111111
2 × 8 kbps	X1Y11111
16 kbps	XX111111
3 × 8 kbps	X1Y1Z111
2 × 16 kbps	XXYY1111
4 × 8 kbps	X1Y1Z1W1
4 × 16 kbps	XXYYZZWW
64 kbps*	XXXXXXXX

An intermediate rate of 64 kbps can only be formed using either 32-bit or 64-bit ITU V.110 frames.[16]

The 32-bit V.110 frame exhibits the following pattern of octets:

1XXXXXXX
0XXXXXXX
1XXXXXXX
1XXXXXXX

The 64-bit V.110 frame exhibits the following pattern of octets:

XXXXXXX1
XXXXXXX1
XXXXXXX1
XXXXXXX1
XXXXXXX1
XXXXXXX1
XXXXXXX1
XXXXXXX1

On detection of this pattern, the VQS must enter 64-clear transparency to allow data to pass until the pattern is removed.

The 4 × 16 kbps case relies on the fact that each 16 kbps sub-stream carries 80-bit modified V.110 frames. These frames exhibit the following pattern:

00000000
1XXXXXXX
1XXXXXXX
1XXXXXXX
1XXXXXXX
1XXXXXXX

---

[16] GSM 03.10, *Digital Cellular Telecommunications System (Phase 2+)*.
ITU-T Recommendation V.110, *Support of Data Terminal Equipment's (DTEs) with V-Series Interfaces by an Integrated Services Digital Network*.

1XXXXXXX
1XXXXXXX
1XXXXXXX
1XXXXXXX

The first two bits need to be extracted and examined for the following pattern:

00 00 00 00 1X XX XX XX 1X XX XX XX 1X XX XX XX 1X XX XX XX
1X XX XX XX 1X XX XX XX 1X XX XX XX 1X XX XX XX 1X XX XX XX

An extra benefit of having A-interface-capable VQ systems installed on the IWF to BSC link is the elimination of PCC/SPC messages (i.e., a customized control interface between the VQS and MSC is not required). Since A-interface-capable VQS can detect the presence of digital data on the A-interface, there is no need for the MSC to issue a PCC/SPC message to invoke the Clear Channel mode of operation because this is done automatically by the VQS.

7.4   TFO (a.k.a. vocoder bypass)

Tandem-free operations (TFO)[17] is a voice-quality standard intended to eliminate unnecessary signal-transcoding operations (i.e., conversion of signal formats) for mobile-to-mobile calls. TFO signals traverse the network without going through additional steps of transcoding performed by wireless codecs (see Figure 7.7). All transcoding takes place at the mobile terminals, as shown in Figure 7.8. Tandem-free operations is a voice standard only, and does not address data communications because signal transcoding is not used in data-transport applications.

The objective of TFO is to improve (or eliminate further degradation of) voice quality in mobile-to-mobile calls. Using this technique, the transported signal skips over (at least) one extra step of transcoding. Transcoding introduces distortion caused by quantization effects (i.e., the process of assigning discrete values when converting continuous signals into digital formats). Accordingly, quantization errors may be magnified by multiple stages of transcoding, resulting in notable signal deterioration that could cross the threshold of acceptable voice quality. By applying the TFO technique, the transported speech signal skips over (at least) one extra step of transcoding, thereby eliminating unnecessary signal processing (and voice quality degradation).

Tandem-free operations is also known as "vocoder bypass," because this approach is based on taking no notice of coded information in the voice-signal path. This is accomplished by adding TFO control messages as in-band information (i.e., the control messages are embedded in the voice-signal bit stream) that is used to decode the signal and output the same stream of data by using the two lower bits (the lowest bit for HR), replacing them with the coded stream (the input of the codec). In TFO operation mode, the remote TRAU then uses the two (one) lower bits and forwards it to the decoder.

---

[17]   ITU-T Recommendation G.711, *Pulse Code Modulation (PCM) of Voice Frequencies* (1988).

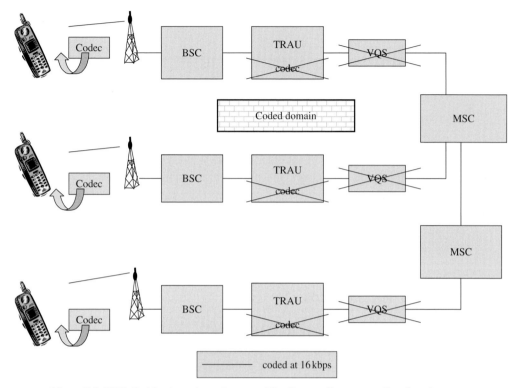

Figure 7.7 TFO disables in-path equipment while disregarding transcoding functions.

In addition, the present TFO requirements specify that any type of signal-processing function (including echo cancellers and VQS) located in the voice path must detect TFO messages and disable all VQ functionality to prevent disruption of the coded voice signal (i.e., voice signal encoding and echo cancelation or other voice-enhancement functions are incompatible).

Voice-quality systems could detect the presence of a TFO coded signal automatically and then initiate appropriate actions. However, merely disabling VQS in the mobile-to-mobile path would also remove the benefits of acoustic-echo control, level control, and noise reduction. These voice-enhancement features are important for ensuring high-quality voice transmission, especially considering the wide range of environmental conditions that surround mobile subscribers (e.g., noisy public areas, moving vehicles, outdoor locations, etc.) and the corresponding negative effects on voice quality. Therefore, the trade-off between disabling voice-quality applications and eliminating extra transcoding may not always yield optimal performance in the presence of high background noise, less than optimal level, or acoustic echo, unless voice-quality applications such as AEC, NR, and ALC are performed in the mobile handset before being coded.

Mobile-to-mobile voice communication in a digital-wireless network is usually implemented with voice-operated codec equipment located both at the origin and destination

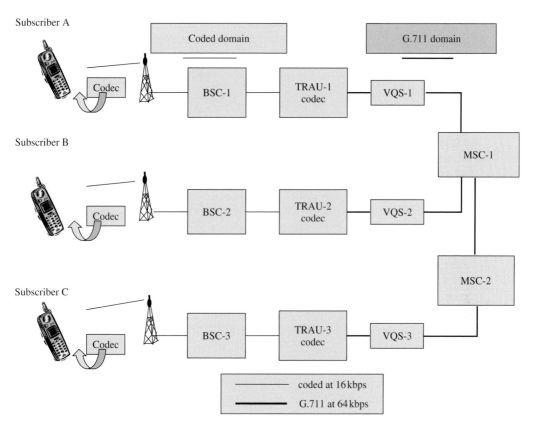

Figure 7.8 Pre-TFO mobile-to-mobile configuration.

ends of the communications path. A typical mobile-to-mobile voice-signal path is illustrated in Figure 7.8 where the VQS are placed on the A-interface, the coding and decoding between two mobile stations are done both at the handset and at the TRAU. Conversions from coded domain to G.711[18] and back are carried out in the TRAU. Altogether there are four conversions[19] on any mobile-to-mobile connection.

When subscriber A calls subscriber B the first codec located in the mobile unit (e.g., handset) encodes subscriber A's audible sounds (e.g., speech), and generates a digital-coded-speech signal for transmission over the wireless air interface (A-interface), typically using code-division multiple access (CDMA), time-division multiple access (TDMA) or global-system mobile (GSM) transmission technology. The second codec, located in the base-station controller (BSC-1), decodes subscriber A's coded speech signal before it is sent to the mobile-switching center (MSC-1). This is done to provide a signal format that is compatible with the public switched-telephone network (PSTN). The third codec located at the BSC-2 converts subscriber A's PSTN compatible signal (G.711) to a wireless coded signal. The fourth codec is located at subscriber B's handset. It decodes the signal and converts it back to speech. However, because this is a mobile-to-mobile call, subscriber A's voice signal encounters additional stages of coding and decoding as it traverses the wireless network. In this example, the decoded-speech signal is applied to VQS-1 and VQS-2, which may provide voice-enhancement features (e.g., acoustic-echo control [AEC], automatic gain control [ALC], and noise reduction [NR]) to improve the voice quality of the call.

A similar coding and decoding process follows a communications session between subscriber A and subscriber C. Only, in this case, the signal moves through another MSC, and possibly through other VQS (not shown in the figure) between the MSCs.

Each time a voice signal passes through a codec, additional distortion is added, degrading the quality of the voice. Accordingly, tandem-free operation (TFO) was developed to alleviate this potential problem.

A total TFO network voice service requires interworking of mobile-phone codecs, BSC codecs, MSC connections, in-path transmission equipment such as VQS, and in-path backbone-network equipment. Coordination of network operations on that scale of magnitude is cumbersome, especially when the voice path traverses multiple networks managed by an assortment of different service providers that purchase equipment from a variety of vendors. Because of these complexities, a "total TFO network" is likely to be a future realization. Most present TFO implementations are confined to mobile-to-mobile transmission within the same network, within the same region.

8 The VQS evolution to 3G

8.1 Introduction

This chapter portrays the 2G, and 3G network topologies, and their impact on VQA feasibility and architecture. It provides an evolutionary examination of the process leading from the 2G to 3G wireless architecture, and it presents a parallel progression of placement and applicability of the VQS that supports the evolving infrastructure.

3G promotions and promises

Anyone following the latest developments in wireless telecommunications has certainly noticed the hype concerning the 3G wireless era. "It sets in motion high-speed web-cruising via your cellphone," the exuberant technologists exclaim. "It lets you watch a movie on your cellphone," the ecstatic entrepreneurs predict, waiting for you to part your lips in awe. "It represents an amazing political transformation," the sociological gurus gasp. "Imagine – a single standard, with the GSM and TDMA (IS-54) advocates surrendering to a competing religion – CDMA," the 3G zealots chant in unison. Although these are exciting and enticing statements, they represent rumour and, at best, are actually only half-truths.

Considering the complexities, it may take quite some time beyond launching the early pieces of 3G networks for wireless-data rates to match existing wireline digital subscriber line (DSL) and cable-modem speeds. Presently, the 3G theoretical speed limits are severely constrained by implementation intricacies.

The 3G IMT-2000,[1] UMTS,[2] and cdma2000[3] capabilities are founded on the basis of foresight and planned service capabilities that require wide-band virtual channels, which enable full-motion video transmission and very high-speed data-transfer options. Considering bandwidth limitations, the only practical implementation of wide-band virtual channels is through packet technology, where resource sharing can maximize utilization of the scarce spectrum. Likewise, packet communications is ideal for handling "bursty" traffic containing calls that require a large measure of spectrum for a relatively brief duration.

---

[1] See www.imt-2000.com.
[2] See www.umtsworld.com/technology/overview.htm.
[3] *International Mobile Telecommunications – 2000 (IMT-2000, Universal Mobile Telephone System (UMTS), and Code Division Multiple Access – 2000 (cdma-2000) Wireless Standards.*

Dense population centers typically offer bandwidth-rich infrastructures capable of supporting a large number of users simultaneously, while at the same time allowing individual subscribers to access wide-band packet bursts for short intervals. Urban areas tend to have increasing numbers of users concentrated in relatively small geographic quarters, and investing in wide-band infrastructures to support high-population areas makes economic sense. Conversely, deploying wide-band capacity in rural or sparsely populated areas may prove wasteful and economically unattractive. In anticipation of relatively low levels of activity, there is a high probability that a significant portion of a rural wide-band infrastructure's capacity would remain (partially or even fully) idle for considerable periods. Since bandwidth is a limiting factor for transporting wide-band packet bursts, sparsely populated rural settings that are equipped with lower bandwidth infrastructures may not be capable of offering this type of service. As a result, rural subscribers may not be able to realize the benefits of high-capacity wide-band packet-burst services available in large population centers.

The wide-band designation, for example W-CDMA,[4] indicates a channel bandwidth of 5 MHz. This is four times the bandwidth of cdmaOne and 25 times that of GSM. The bandwidth of the wide-band channel service was chosen to allow high data-rate transport. Unlike cdmaOne, which sends every bit 64 times, W-CDMA (for example) adjusts gain depending upon signal strength. That is, every bit is sent between 4 and 128 times, with lower rates being used in situations when the signal is strong. This means that greater bandwidth is available in areas with strong signal characteristics. Distance between the mobile terminal and the base transmitter station (BTS) is the prime determinant of signal strength. Therefore, in regions with higher population density, the signal strength tends to be stronger because of the relatively short distance between the mobile subscriber and the BTS.

All things considered, data communication rates may vary widely between rural and urban areas. Whereas urban centers benefit from a richer spectrum, rural boroughs may not profit (even under a relatively mature 3G network) from the potential increase in bandwidth. The implication is apparent; the 3G infrastructures contain a montage of 2G and 2.5G components. Therefore, mobile-cellphone users may roam through pure 3G, enhanced 2.5G, and standard 2G backbones during a single voice or data session. Given this situation, the quality of communications and uniformity of service can only be maintained if transitions between various infrastructure segments are semi-transparent.

3G applications

With all due respect to the promotional hype, watching movies on a tiny cellphone display may not be as popular as viewing them on a big screen. It may be deemed a fascinating novelty by some hobbyists, but is likely to be tiring for most wireless subscribers. Cellphones are designed for mobility, convenience, and relatively short periods of intensive interactive use. In contrast, watching a full-length motion picture requires a span of about two tranquil hours of concentrated attention. Perhaps 3G cellphones have a more

---

[4]   See www.umtsworld.com/technology/wcdma.htm.

practical use as a wireless camera; sending video images to a distant desktop (or laptop)-computer screen to be viewed by another subscriber. This type of feature would seem to have greater appeal for commercial and business applications where the immediate display of an image is of value (e.g., security, news, ads, real estate, manufacturing).

The 3G standard is designed to support backward compatibility. As such, 3G cell phones must be capable of operating across the full range of 2G, 2.5G and 3G environments. Moreover, spectrum allocation for 3G communications is not uniform around the world, and chip rates of earlier standards (cdmaOne, GSM, and TDMA equipment) are dissimilar. Consequently, 3G cell phones must be designed as multi-band, and multi-standard terminals, if the underlying structure of the 3G air-interface is to remain consistent with its predecessors. The 3G standards have grown into two major CDMA versions. The first version, W-CDMA, is designated as the standard succeeding GSM and TDMA. Similarly, cdma2000 is the anticipated successor to cdmaOne.

The 3G data- and video-communications features are designed for packet transmission. However, voice transmission over 3G networks utilizes both packet and circuit switching. Circuit switching is the only medium that 2G and 2.5G networks use to transmit voice signals, and it is essential that 3G be compatible with the embedded technology. Therefore, evolution to 3G must dictate a mixture of technologies characterized by a slow and partial migration from voice-over-circuit to voice-over-packet. It is evident that the path to 3G traverses a combined 2G and 2.5G landscape. The 2.5G standard is intended to increase data-handling capabilities by merging the two diverse air interfaces of GSM and cdmaOne respectively. The 2.5G standards are general packet-radio service (GPRS), designed to augment GSM, and cdmaTwo, which is planned as the next step up from cdmaOne.

3G deployment

Implementation schedules for 3G infrastructures have descended to a new reality. Two major restrictions have surfaced, and they have instigated a slowdown of forward movement along with reassessment of implementation plans. First, competitive bidding (particularly in Europe) for 3G spectra has raised the trophy price to an altitude equivalent to the likes of Mount Everest. The frenzy accompanying the bidding process subsequently brought about a severe financial drain for potential service providers. Furthermore, an objective business analysis that takes into account the first round pay-out for spectrum finds it tricky, if not unattainable, to demonstrate profitability. At the same time, it is apparent that 2.5G "green applications" (i.e., GPRS[5] and cdmaTwo[6]) can provide a significant piece of 3G functionality. It could do so for a relatively inconsequential portion of the total investment cost of an all-inclusive 3G deployment. Accordingly, 2.5G became a major force dragging down the urgency of 3G deployments. Moreover, it is widely recognized as a stepping stone on the path to a pure 3G infrastructure. Therefore, 2.5G is

---

[5]   See www.gsmworld.com/technology/gprs/intro.shtml.
      GSM 03.60 *GPRS Service Description Stage 2* (Phase 2+), version 7.0.0 (1998).
[6]   The cdmaTwo standard is defined in IS-95b.

positioned as an attractive form of progression that also provides a comfortable rest area where service providers can be revitalized while they take a temporary siesta.

Second, the United States' wireless market is confronted with a fundamental impediment as service providers move toward full deployment of a 3G infrastructure. Unlike other world regions, the US spectrum has already been fully allocated to 2G and other radio applications. Since the USA has no spectrum reserved exclusively for 3G communications, 3G can only be realized by swapping bandwidth with existing 2G allocations. This situation may cause severe service interruptions if acted upon hurriedly by simultaneously converting sizeable chunks of 2G spectrum. Consequently, the move to 3G in the USA is carried out slowly by resorting to a "dietary" course of action; converting small slices of 2G into 3G infrastructures while maintaining a sufficient level of 2G to sustain service throughout the transformation process.

8.2   The 2.5G network voice architecture

2.5G voice and data network topology

The 2.5G architecture does not impact fundamental voice transmission or VQ concepts. The primary improvement offered by 2.5G is its higher-speed data communications capability. The wireless transmission path (shared voice and data) originating from a mobile terminal is segregated into voice and data at the base-station controller (BSC), as shown in Figure 8.1. After segregation, data is transported over a separate link to a packet control unit (PCU) that performs the signal conversion from wireless protocol to packet (IP or ATM) protocol. From that point the data signal continues its journey through the internet into its final destination.

GSM data – GPRS

The 2.5G architecture is best defined as GSM, where data applications are dominated by the GPRS (Global Packet Radio Service).[7] A higher communication speed is realized by sharing the 14.4 kbps GSM channel bandwidth among many users. Since GSM divides a given frequency-division multiplex (FDM) channel into eight TDMA timeslots, the theoretical bandwidth afforded by GPRS transmission is 115.2 kbps (8×14.4 kbps). Reality dictates that the average data speeds are no better than existing circuit-switched services such as HSCSD (high-speed circuit-switched data), which has a theoretical speed of 57.6 kbps. These theoretical speed limits have been elusive for some time, because battery technology limitations caused severe overheating problems at speeds higher than 14.4 kbps.

---

[7]   GSM 03.60 *GPRS Service Description Stage 2.*

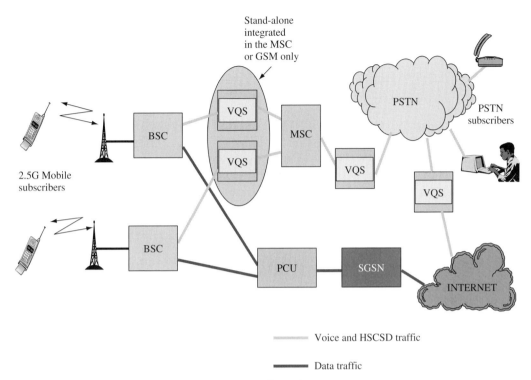

Figure 8.1 2.5G wireless-network VQS configuration.

CDMA data – cdmaTwo

Mobile phones conforming to cdmaOne standards can receive and transmit data at 16 kbps. Although most of this data is currently redundant information, it does not have to be. The cdmaTwo standard[8] uses four Walsh codes to yield the maximum rate of 64 kbps (instead of 16 kbps delivered by cdmaOne using a single Walsh code). The cdmaTwo standard supports CDMA 2.5G data capability (equivalent to GSM GPRS). The voice-service architecture is unchanged from cdmaOne; hence voice-quality systems are positioned using the same topology as previously illustrated in Figure 8.1.

8.3   The 3G network architecture

Evolution to 3G

Figure 8.2 shows the leading milestones that were encountered during the transition from 2G to 3G wireless networks. As previously explained, the existing 2G starting point dictates the overall shape of the endgame. The essence of the 3G story is its plurality.

---

[8]   Also known as IS-95b.

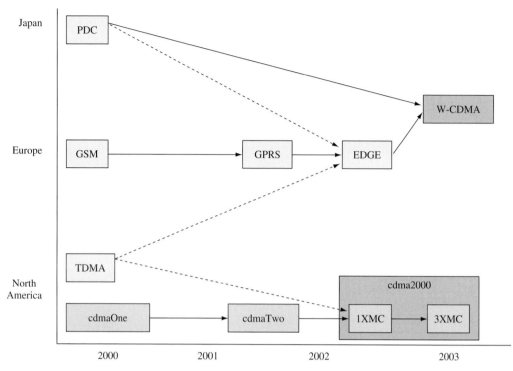

Figure 8.2 2G to 3G network evolution.

There are two 3G standards. Although both are technically CDMA, in reality (from an implementation perspective) they are different. As shown in Figure 8.2, cdma2000 (1XMC followed by 3XMC) is the successor to cdmaOne and cdmaTwo. Similarly, W-CDMA is the successor of GSM. Each standard is backward compatible with its predecessor, and each architecture provides established interfaces into the other.

The evolution from GSM via GPRS into W-CDMA moves through another phase referred to as EDGE (enhanced data rates for GSM evolution).[9] EDGE was designed to provide a smooth evolution for the digital – advanced mobile phone service (D-AMPS, TDMA) into W-CDMA. EDGE employs a different modulation scheme (8-PSK; every bit has eight states instead of two) allowing it to increase (triple) the raw-data rate from 21.4 kbps to 64.2 kbps. Voice traffic is also a potential payload for EDGE, and in some technical literature EDGE is identified as a 3G methodology (equivalent to W-CDMA and cdma2000).

The evolution of cdmaOne into cdma2000 has occurred via cdmaTwo, 1XMC, and then 3XMC. The 1XMC technology doubles the capacity of cdmaTwo by using each of the IS-95's Walsh codes twice. Completing the transition, 3XMC is part of the cdma2000 IMT-2000 proposal, which encompasses the wide-ranging 3G standard.

[9]   Hari Shankar, *EDGE in Wireless Data*, www.commsdesign.com/main/2000/01/0001feat2.htm (August 2005).

Voice services and VQS as 3G killer applications

The dominant factors guiding the evolution of voice communications in 3G networks are based on a parallel progression of business considerations, which are partially independent of packet-technology details, that make up the 3G architecture. Global-communications service providers have been steering through a process of consolidation, resulting in a smaller number of larger companies. Similarly, mobile carriers have been merging, and mergers have led to establishing long (international) wireline links as part of large wireless carrier service offerings.

This process has given rise to changes in MSC architectures and creation of Gateway MSCs (GMSCs). A GMSC is essentially a hub that interconnects multiple MSCs. The GMSC also assumes the home-location register (HLR)[10] or visitor-location register (VLR)[11] function, thereby eliminating unnecessary billing when routing a local call via an international HLR. The phenomenon we refer to as the "Tripper's trap,"[12] caused by customer billing based on rigid interpretations of network topologies, can be better controlled when the HLR/VLR is located in the GMSC.

In situations with high traffic concentration, the GMSC has become a broadband hub. As a result, VQS and EC equipment has been moving from the MSC-to-PSTN link[13] to the GMSC-to-PSTN link to accommodate the increased echo tail lengths associated with long-distance connections. As part of this evolution, the integration of EC capabilities with packet handler or codec equipment within the MSC is less critical and may even become redundant.

Packet-switched voice (VoIP and VoATM) is a major 3G attraction. The 3G VOP (voice over packet) signal travels along the same route assigned for data and video applications. In most cases, it travels by way of the internet to the PSTN, another VOP terminal, or a wireless terminal (refer to Figure 8.3). At the same time, circuit-switched voice continues to be the dominant voice application (compared with packet-switched voice) and may stay that way for some time. The main reasons for this are the requirement for backward compatibility with 2G, and the fact that 3G bandwidth is predominately concentrated in heavily populated urban centers. This leaves sparsely inhabited suburbs and rural regions with older infrastructures (2G or 2.5G).[14] In general, 3G networks must coexist with 2G infrastructures and mobile subscribers must traverse different network types as they move through various geographic regions.

---

[10] T. Haug, Overview of GSM: philosophy and results, *International Journal of Wireless Information Networks*, **1**:1 (1994), 7–16.

[11] GSM 09.18, *Visitors Location Register (VLR), Gs Interface, Layer 3 Specification*, version 1.2.0 (1997).

[12] Tripper's trap – when a Frenchman travels to the USA with a friend and then makes a wireless call to the friend's (French) cellphone number, the call is routed over two international links. One to France, paid by the caller, and one back to the USA, paid by the called party (the friend), even though the friend may be next door in the same hotel.

[13] Voice-quality systems installed on the GSM A-interface may not be moved. They continues to provide VQ functionality on mobile-to-mobile calls within the same MSC system.

[14] Broadband voice, data, or video communications are transported in packets. This approach is attractive because thick (virtual) pipes are shared among many users. Resource sharing becomes economically viable when many users drive the proliferation of large resource pools. Hence, dense population centers are the most economical breeding grounds for 3G broadband traffic, while rural and sparsely populated regions retain 2G circuit-switching mode operations.

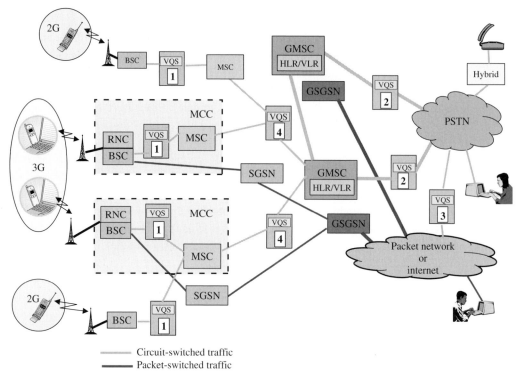

Figure 8.3 3G voice and data architecture.

2G/3G network topology

Figure 8.3 depicts the 3G network voice topology with respect to the placement of VQS and EC equipment. This is an integrated view that blends 2G and 3G kernels. The two main building blocks, BSCs and MSCs, of this architecture do and will coexist for a long time. The 3G BSCs have expanded their functionality and become RNCs (radio network controllers) supporting both packet and circuit-switched voice traffic, while MSCs continue to sustain their role of managing and routing circuit-switched traffic. The MSCs perform many of the functions they are responsible for in pure 2G applications. In addition, the circuit-switching layer provided in the GMSCs (in many cases) reduces the load presently carried by traditional MSCs.

In this arrangement, packet traffic is separated from circuit traffic at the RNC, and then it navigates towards its final destination (e.g., a standard two-wire telephone, a desktop computer, or a wireless phone) via the internet. The introduction of the HLR/VLR function in the GMSC eliminates (reduces) the previously described "Tripper's trap" phenomenon.

Traditional VQS and EC equipment continue to inhabit the circuit switched and 2G segments of the 3G infrastructure. Four different VQS elements, designated types ⬛1, ⬛2, ⬛3, and ⬛4, are illustrated in Figure 8.3. Each type of VQS element is briefly described in the following paragraphs.

Type ☐1 VQS is typically used to support the GSM 2G infrastructure. This equipment is deployed on the GSM A-interface. In the 3G architecture, the Type ☐1 VQS are designed to provide acoustic-echo control, noise reduction and level optimization for mobile-to-mobile calls within the same planetary system.[15] Type ☐1 VQS is not required to support hybrid-echo cancelation. Under 3G the BSC functionality moves closer to the MSC and together are referred to as the MCC (mobile-communication control center) as shown in Figure 8.3. Type ☐1 VQS could be positioned within the MCC.

Type ☐2 VQS provides electrical-hybrid-echo cancelation (EC). As previously indicated, the HLR/VLR and EC functions gradually move away from the MSC to be repositioned (closer to the PSTN) in the GMSC. There are several reasons for rearranging these elements in the evolving wireless infrastructure. The most important advantages are reduced cost and increased density. The availability of a broadband VQS (i.e., having STS-1, OC-3, or STM-1 facility interfaces) offers both these benefits. Lower cost is realized as direct savings attributed to higher density VQS, and (much greater) indirect savings are brought about by reducing space, concentrating cables, and eliminating unnecessary multiplexer equipment. Furthermore, the closer proximity of the voice-quality function to the hybrid (in a centralized position) virtually guarantees that every PSTN-to-mobile call traverses the voice-quality element. The new position of broadband Type ☐2 VQS may not cover mobile-to-mobile calls (either within the same MSC or even between MSCs.)[16] Therefore, it becomes apparent that an ideal architecture would position just the EC function (rather than the jam-packed VQS) at the GMSC location. Likewise, VQS without EC functionality could be positioned at the intra-MSC point for mobile-to-mobile connections (either on the A-interface for GSM configurations, or on the MSC-to-GMSC links for all other architectures [not excluding GSM]).

The Type ☐3 voice-quality element is designed for supporting VoIP or packet multimedia gateway applications. Although 3G voice service carries the promise of higher fidelity (due to expanded bandwidth) than 2G voice,[17] it may bear the same impairments (noise, level mismatch, and acoustic echo) that 2G circuits are renowned for. Furthermore, the universal translation of video signals requires a new gateway function capable of linking applications supported by diverse protocols. The packet multimedia gateway (PMG) and the VQ function could be combined into a single terminal denoted as a Type ☐3 VQ element (see Figure 8.3).

The Type ☐4 VQS is suitable for use in either the present GSM or non-GSM (cdmaOne) application. The Type ☐4 VQ function is identical to the Type ☐1 VQS installed on the A-interface, and is not expected to support hybrid EC.[18]

---

[15] A planetary system is defined as a sub-network consisting of MSCs and BSCs (the planets) that are all managed by the same HLR/VLR function in a given GMSC (the sun).

[16] In some applications MSCs may be connected directly without involving a GMSC.

[17] A portion of 3G voice traffic is transported over the packet stream allocated for all other media. This architecture would allow higher bandwidth on virtual voice circuits.

[18] The Type ☐1 and Type ☐4 VQS are mutually exclusive. Global-systems mobile may employ either Type ☐1 or Type ☐4 VQS (but not both), and cdmaOne may employ Type ☐4 elements only.

8.4   Conclusions

Voice-over-packet transmission over 3G is an alternative to the present 2G circuit-switched voice architecture. However, mobility requires multi-mode auto-switched cellphones that are compatible with the full range of 2G, 2.5G, and 3G applications. Mobile subscribers are roaming between 3G, 2.5G, and 2G regions; therefore their cellphones are designed to switch modes seamlessly on the fly. It is obvious that 2G, 2.5G, and circuit-switched voice applications must coexist with pure 3G, and compatibility is required to achieve subscriber acceptance. The infrastructure and VQ systems required to support mature architectures continue to experience growth characteristics matching the "newly baked technologies" used to implement infrastructures carrying packet-switched services. Equipment providing VQ functions under 3G consist of the four VQS types previously described. The main difference between 2G and 3G equipment rests with the application focus: mobile-to-mobile, mobile-to-PSTN, and internet-to-PSTN. Each application determines the functions embedded within a given VQS.

In summary, the market demand for VQ applications and EC systems continues to be robust. Voice-quality applications may be embedded within switching or other transmission products, but they are definitely sharing a piece of the 3G pie, whether as a 2G component or as a 3G application.

Part IV

A network operator's guide for selecting, appraising, and testing a VQS

9 A network operator's guide to testing and appraising voice-quality systems

9.1 Introduction

This chapter describes test and evaluation procedures of performance associated with the various voice-quality applications. The telecommunications equipment marketplace is filled with a variety of echo canceler (EC) and voice-quality systems (VQS) promoted by different vendors. Noticeably, the characteristics and performance of these products are not identical. In addition, the non-uniformity and arbitrary judgment that is often introduced into the EC and VQS product-selection process makes the network operator's final decision both risky and error prone. This chapter describes the criteria and standards that are available to facilitate methods for objectively analyzing the benefits of EC and VQA technology when confronted with multiple EC and VQS selection choices. The scope includes procedures for evaluating the performance of electrical (hybrid), acoustic-echo control, noise reduction, and level optimization via objective, subjective, laboratory, and field-testing methods.

This chapter brings to light a list of tools and standards designed to facilitate the voice-quality assessment process. It is intended to provide a methodology for objectively analyzing the benefits of voice-quality assurance technology in situations where network operators are confronted with multiple VQA system choices.

Voice-quality application features required by network operators typically include the following capabilities:

- network hybrid (electrical)-echo cancelation,
- acoustic-echo control,
- automatic volume control,
- adaptive level control (noise compensation) as a function of surrounding noise,
- noise reduction,
- non-voice applications.

Voice-quality application and performance assessment criteria encompass the following areas:

- application features and performance,
- maintenance features,
- environmental considerations,
- reliability thresholds,

- density and heat-dissipation characteristics,
- level of integration with key infrastructure elements.

9.2   Judging audio quality and performance of hybrid-echo cancelation

The literature and descriptive material distributed by many VQS vendors tends to concentrate on specific strengths, while glossing over the technical details that are vital in deriving a complete and accurate assessment of the VQS capabilities. Selecting a VQS (from an array of competitive vendor's products) is a comprehensive task that requires careful analysis. This can be a formidable challenge for network operators, who may not have the experience or knowledge needed to conduct an accurate VQS evaluation. Many different characteristics, including applications, performance, maintenance, environment, quality, reliability, and standards compliance, must be considered.

Although all echo cancelers are supposed to perform satisfactorily, many implementations, when running into the harsh realities of the imperfect world, exhibit severe weaknesses that may be unacceptable to service providers. Voice-quality systems must be put through a series of tests before placing them in live networks. Tests are divided into objective and subjective types, with some tests conducted in the laboratory and others in the field. Recordings of the inputs and outputs are essential for many of the subjective tests, since they let the examiner evaluate the contribution of the VQS operation by contrasting the two and inferring the VQS contribution from the difference.

The most important sequence of objective tests is prescribed in the ITU-T standard recommendation, G.168.[1] For historical reasons, the first test in the sequence is test 2. Test 1 was dropped from the list in the latest (2002) version of G.168 after test 2 was enhanced to cover the same goal of convergence speed and stability over time. Most of the G.168 tests deduce the results by measuring the signal power at the S_{out} port after the EC has operated on the signal that entered the S_{in} port. The tests in the G.168 recommendation employ special signals such as noise, tones, Group 3 facsimile signals, and a composite source signal (CSS). The CSS consists of a speech-like power-density spectrum (a pseudo-noise signal generated using 8192 point FFT) (see Annex C of G.168 and Recommendation P. 501)[2] with a random starting point. When running a specific test using the CCS it is recommended that the test be performed several times in a sequence. If the EC performance is dependent on call characteristics or on a shared resource, the performance of each run may look different.

The test equipment is set up as illustrated in Figure 9.1. The user interface must have enable and disable control over comfort-noise matching and NLP. During normal EC operation, comfort noise and NLP are never disabled. Nevertheless, since G.168 measures the leftover power at the S_{out} port after canceling echo that entered at the S_{in} port, it is crucial that the difference should represent the removed amount of the echo signal. An insertion of comfort noise to replace an echo signal prevents the G.168 test equipment from

---

[1]   ITU-T Recommendation G.168, *Digital Network Echo Cancellers* (1997, 2000, 2002).
[2]   ITU-T Recommendation P. 501, *Test Signals for Use in Telephonometry* (1996).

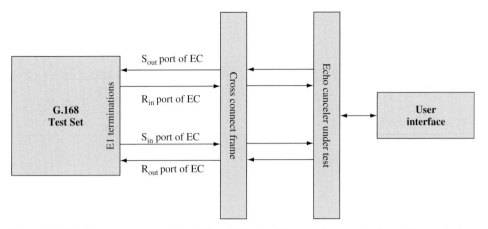

Figure 9.1 G.168 test set-up; provision EC to CCS signaling, synchronous timing to S_{in} terminal.

reading the true difference between the ports. Accordingly, comfort-noise matching must be disabled before running most of the G.168 tests. The NLP must be disabled for some of the test variants. A disabled NLP lets the test equipment assess the quality of the linear convolution processor (LCP), since the power measured at the S_{out} port does not include the NLP contribution to the echo cancelation, and is the result of the LCP alone.

The ITU-T prescribed tests carry varying levels of significance. The first two tests, test 2 and test 3, are more central than the others, and this chapter provides more detailed descriptions of these procedures. In many instances where test time is limited, tests 2 and 3 are the only ones performed. When they are, it is recommended that multiple runs should be carried out employing several different hybrids from the ITU-T G.168 prescribed list.

Network operators have the option of either performing the complete ITU-T G.168 test specifications independently, or requiring the vendor to provide a comprehensive documentation of echo-canceler performance. When requesting vendor documentation, the network operator should verify that the ITU-T G.168 tests were run properly. In ascertaining the validity of vendor supplied test results, the following points should be examined:

Was testing conducted on a single channel while all other channels were idle?

Some echo-canceler implementations may economize on resources by sharing them across several channels. When running the ITU-T G.168 tests on a single channel, EC performance may appear to be adequate. However, when the EC is subjected to the same tests while forcing simultaneous call starts on several channels, the performance of the EC may deteriorate (possibly indicating resource contention).

Do the test results only indicate "pass or fail" ratings?

The G.168 requirements specify minimum acceptable thresholds, and some implementations may simply indicate that EC performance meets these thresholds with a pass or fail rating. However, vendor-supplied data should reflect the extent by which the EC product

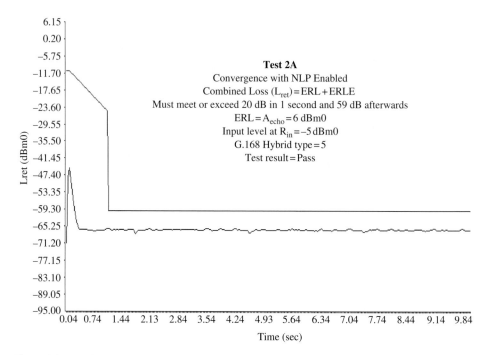

Figure 9.2 Test 2A – convergence with NLP enabled; combined loss = ERL + ERLE; must meet or exceed 20 dB in 1 second and 55 dB afterwards.

met each requirement (i.e., indicate margin with respect to the minimum threshold) to support accurate evaluations and comparisons with other EC products.

It should be noted that service providers could either acquire commercially available software[3] that runs all of the G.168 tests and insist on having it run in real time, or have it documented and guaranteed by the EC vendors. There are ten major categories of the ITU-T G.168 prescribed tests. Each category consists of sub-categories. A description of each test is outlined next.

Test 2 – convergence and steady-state residual and returned echo level

The object of test 2 is to assess the EC convergence speed. The test is divided into three parts. The first part is run with the NLP enabled where CSS is transmitted at the R_{in} port, and its corresponding echo is returned at the S_{in} port. The second part is run with the NLP disabled, and the third part with both NLP enabled and disabled, but with background noise entering the system at the S_{in} port. Examples of test runs are presented in Figures 9.2, and 9.3. Test 2A in Figure 9.2 is run[4] with the NLP enabled. The specific illustration exemplifies a successful test, where the EC exceeds the G.168 test requirement. Test 2B in Figure 9.3 illustrates a successful run, as the performance of the EC under test exceeds

---

[3]   *GL-Comm Echo Canceller Testing Solutions*, www.gl.com/echocan.html.
    DSPGenie Echo Canceler Tester, www.dspg.co.uk/.
[4]   Figures 9.2–9.6 are examples G.168 test runs of tests 2A, 2B, 3A, 3B and 3C. Bob Reeves, Rapporteur of the ITU Study Group 16, Question 6, performed the tests using the DSP Genie echo canceler tester and contributed the figures.

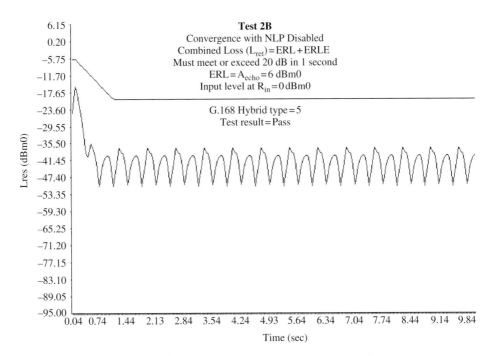

Figure 9.3 Test 2B – convergence with NLP disabled; combined loss = ERL + ERLE; must meet or exceed 20 dB in 1 second and 29 dB after 10 seconds.

the G.168 test requirements. The test conditions disable the NLP in addition to the comfort-noise injection. Test 2C repeats tests 2A and 2B only to add background noise. The simulated background noise is generated by a Hoth noise source (see ITU-T Recommendation P. 800)[5] and is added at the send side, opposite the CSS.

Test 3 – performance under conditions of double talk

The test consists of three parts. Each part tests the performance of the EC under different conditions of double talk. During conditions of double talk the echo canceler can give rise to unwanted artifacts, such as clipping, distortion, and noise contrast. The tests make the assumption that, on detection of double talk, measures are taken to prevent or slow adaptation in order to avoid excessive reduction in cancelation performance.

For this test, the R_{in} signal is CSS and the send-side (other than echo) signal is the double-talk CSS. While CSS is used for this test, it is recognized that it is only a statistical approximation of real speech. Double-talk tests performed with actual speech samples tend to produce results somewhat different than those shown in this test. It is possible that this test and its requirements may change as the correlation between CSS and real speech is better understood. The use of different languages in place of CSS has been shown to provide considerable variation in the results for tests 3A and 3B.[6]

5   ITU-T Recommendation P. 800, *Methods for Subjective Determination of Transmission Quality* (1996).
6   ITU-T G.168, *Digital Network Echo Cancellers*.

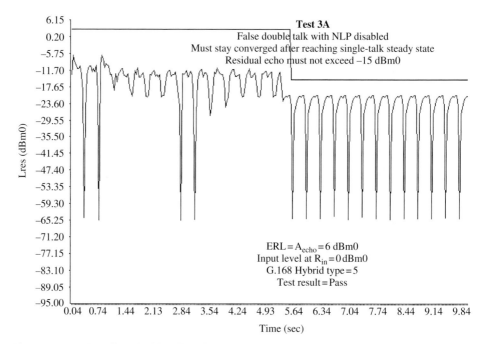

Figure 9.4 Test 3A – false double talk with NLP disabled; must stay converged after reaching single-talk steady state; residual echo must not exceed −30 dBm0.

Test 3A – false double talk

The objective behind test 3A is to ensure that the double-talk detection is not overly sensitive. An overly sensitive near-end speech detector may freeze the EC adaptation unnecessarily. It may falsely classify low sound (noise or background) as double-talk speech. The test procedure is to clear the h-register; then for some value of echo-path delay and ERL, a CSS signal is applied to R_{in}. Simultaneously, an interfering signal (double-talk CSS), which is sufficiently low in level, is applied at S_{in}. This signal level is low enough that it should allow adaptation and cancelation to occur. After the allowed convergence time the adaptation is inhibited and the residual echo measured. The NLP and the comfort noise insertion are disabled throughout the test. A successful test is illustrated in Figure 9.4.

Test 3B – true double talk

Test 3B (illustrated in Figure 9.5) is intended to measure the performance of an EC under double-talk conditions. It measures the reaction time to near-end-speech detection, and the amount of divergence caused by actual double talk.

The test procedure is to converge the echo canceler fully for a given echo path by applying CSS at R_{in}. After the canceler is converged on a single-talk signal, another signal is applied at S_{in}. The S_{in} signal matches the R_{in} signal strength. The EC is expected to freeze

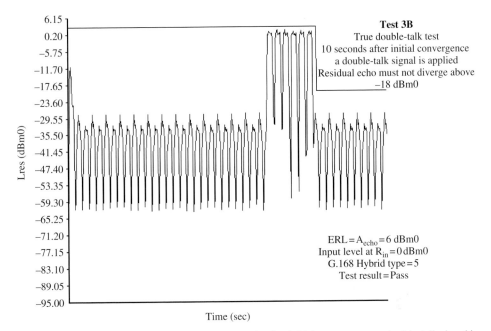

Figure 9.5 Test 3B – true double-talk test; 10 seconds after initial convergence a double-talk signal is applied; residual echo must not diverge above −29 dBm0.

its adaptation once it senses the double-talk condition. The NLP is disabled throughout the test.

Test 3C – double talk under simulated conversation

Test 3C is designed to assess whether the echo canceler produces undesirable artifacts during and after periods of double talk.

The procedure is to clear the h-register. Then a signal is applied at R_{in}.

Simultaneously, a signal matching the level of the R_{in} is applied at S_{in}. The signal level at S_{out} is measured afterwards. The main difference between tests 3C and 3B is the state of the NLP. The 3C test has the NLP enabled throughout. A successful test is shown in Figure 9.6. Note that the introduction of double talk creates a short diversion, which is quickly corrected by the EC.

Test 4 – leak-rate test

This test measures the rate by which the contents of the h-register leaks out. An excessively fast leak causes divergence when the EC is in steady state, while an excessively slow leak may not converge fast enough on a new call.

The test procedure is to converge the echo canceler fully using CSS for a given echo path and then to remove all signals from the echo canceler. Two minutes later, CSS is reapplied at R_{in}, and the residual echo is measured. The NLP is disabled throughout the test.

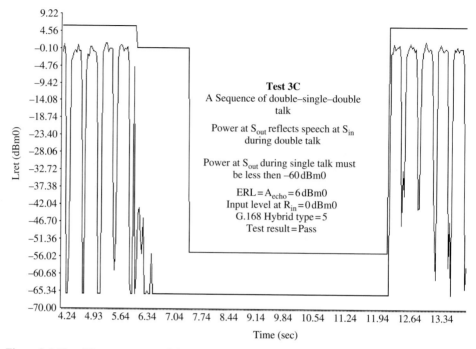

Figure 9.6 Test 3C – a sequence of double–single–double talk; power at S_{out} reflects speech at S_{in} during double talk; power at S_{out} during single talk must be less than -60 dBm0.

Test 5 – infinite-return-loss test with a change in echo path

This test measures the performance of an EC during a sudden change in the echo path.

The test procedure is to converge the echo canceler fully using CSS for a given echo path. The echo path is then interrupted while a CSS is applied at R_{in}, and the output at S_{out} is measured. The NLP is disabled throughout the test.

Test 6 – non-divergence on narrowband signals

This test measures the performance of an EC when the R_{in} signal changes from a wideband to a narrowband signal sequence after having converged on a wideband signal. The residual-echo level is measured before and after the application of a sinusoidal wave or a wave composed of two frequencies.

The procedure consists of completely converging the echo canceler. A sequence of tones is then applied at R_{in}. After completing the sequence, the adaptation is inhibited and the residual echo is measured. The NLP is disabled throughout the test.

Test 7 – stability under narrowband signals

The test measures how well an EC converges on a narrowband signal. Narrowband signals are different from speech since the ones under test are repetitive, and the periodicity may bring about divergence if the EC applies high gain or aggressive techniques that localize the gain application to the narrow region of a newly found impulse response. After resetting the h-register, the echo canceler is subject to a sinusoidal signal at the R_{in}. Two minutes later, the residual echo is measured using the applied signal. The NLP is disabled throughout the test.

Test 8 – compatibility with ITU-T signaling nos. 5, 6, and 7

A discussed in Chapter 3, signaling tones such as ITU-T Nos. 5, 6, and 7 must disable the EC before moving through it. When not disabled by the switch on a per-call basis, an EC must be able to recognize the specific signaling tones and disable itself on detection.

This test is intended to insure that echo cancelers do not remove or cancel a mono- or bi-frequency signal transmitted in a handshaking protocol in the send direction either before or after receiving an identical signal (except for amplitude and phase) in the receive direction. The NLP is enabled throughout the test.

Test 9 – comfort-noise level matching

This test measures the comfort-noise level under changing and steady noise conditions at the S_{out} port, relative to the actual noise that enters the S_{in} port during a period of near-end speech. White noise is used for all input signals for this test. The NLP and comfort-noise feature are enabled throughout the test.

Test 10 – fax performance for calling and called stations

This test is designed to confirm that the echo cancelers located at each end of a connection converge rapidly on the initial handshaking sequences of a facsimile call and have some means for preventing the unwanted generation of echo. The test is split into three parts. Test 10A looks at the performance of the echo canceler located on the calling-station side and Test 10B looks at the performance of the echo canceler on the called-station side. Test 10C looks at the performance of the echo canceler on the calling-station side during page transmission.

The test has been designed to run in a laboratory environment using an echo canceler and a fax simulator. Each test is run separately. When running the tests the G.165/G.168 tone disabler is switched on.[7]

9.3   Subjective testing

Unfortunately, the ITU-T objective tests do not capture many important performance features that differentiate echo cancelers from one another. This fact is particularly evident when evaluating the performance of the non-linear processing (NLP) portion of an echo-canceler product. Incorrect equipment operations related to timing, speed, loss insertion

---

[7] ITU-T Recommendation G.165, *Echo Cancellers* (1993).

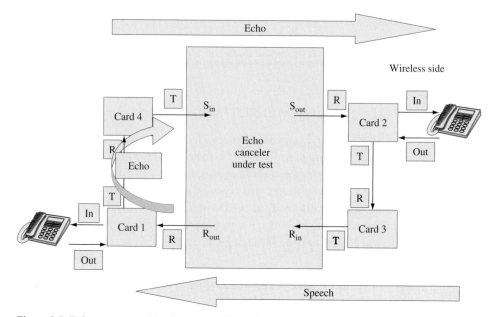

Figure 9.7 Echo-canceler subjective-test configuration.

level, etc., are significant points of disparity among competitive EC products, but are not revealed by ITU-T G.168 testing. Specifically, the amount of signal clipping during double talk, application of acoustic echo, continuity or discontinuity of background noise or audio signals during comfort-noise matching, quality of transmission when signal delay exceeds 500 ms, and signal (sound) quality or clarity in a noisy environment are all inherent EC functions that should be examined and appraised when comparing EC products.

The following sections describe test procedures designed to assist network operators in their efforts to record and analyze the performance of echo cancelers.

Subjective test configurations

A typical EC test configuration is depicted in Figure 9.7. This arrangement illustrates a PSTN-to-mobile application, however the mobile phone could be replaced with a standard PSTN phone when testing echo cancelers used in a long-distance (PSTN-to-PSTN) application.

A digital recording device is connected to the four ports (R_{in}, S_{out}, R_{out}, S_{in}) of the EC system to monitor facility signals (e.g., T1 or E1 signals) during testing. Temporary connection to the facility signals (via connector jacks) is usually available at a DSX (patch panel) or equivalent equipment. Cards are circuit packs in the test box that contain T1/E1 facility interfaces. Cards 3 and 4 are used for recordings. T and R are initials for transmit and receive respectively.

Table 9.1 lists six different types of tests to be conducted along with the corresponding port connections on the EC. A detailed description of each test is provided in the following sections.

Table 9.1. *Echo canceler subjective tests*

Test condition	S_{in}	S_{out}	R_{in}	R_{out}
Single talk echo cancelation – test 1a	Yes	Yes		
Single talk with background music – test 1b	Yes	Yes		
Double talk without clipping – test 2a	Yes	Yes		
Double talk with background noise – test 2b	Yes	Yes		
Double talk with background noise – test 2c			Yes	Yes
Convergence test – test 3	Yes	Yes		
Wireless data-transmission test	Yes	Yes	Yes	Yes

Single-talk echo cancelation – test 1a

The recording device is connected to the S_{in} and S_{out} ports of the EC. During the "single-talk echo cancelation – test 1a," the mobile subscriber speaks into the digital cellular phone (R_{in} port) while the PSTN subscriber (S_{in} port) remains silent. Echo from the mobile subscriber's speech is returned via the two-wire/four-wire hybrid and is recorded at the S_{in} port of the EC. After processing by the EC the echo signal should be canceled, and the result recorded at the S_{out} port. The echo signal should be canceled (removed) after processing by the EC. This is a typical situation, in which one subscriber is speaking while the other is listening. If the EC removes the echo signal completely at the S_{out} port, then the EC has passed this test.

Single talk with background music – test 1b

The recording device is connected to the S_{in} and S_{out} ports of the EC. During the "single talk with background music – test 1b," the mobile subscriber utters short phrases into the digital cellular phone (R_{in} port) while the PSTN subscriber (S_{in} port) plays background music at a level of about $-40\,dBm0$. The background music and echo from the mobile subscriber's speech is returned via the hybrid and recorded at the S_{in} port of the EC. After processing by the EC, the echo signal should be canceled (removed), but the background music should flow uninterrupted throughout the speech and the silent periods, as recorded at the S_{out} port. If the EC breaks the continuous flow of music at the S_{out} port when the mobile subscriber changes between speech and silence or allows echo to pass through the connection, then the EC has failed this test.

Double talk without clipping – test 2a

The recording device is connected to the S_{in} and S_{out} ports of the EC. During the "double talk without clipping – test 2a," both the mobile subscriber (R_{in} port) and the PSTN subscriber (S_{in} port) speak simultaneously, thereby creating a double-talk condition. The PSTN subscriber's speech along with the mobile subscriber's echo will be recorded at the S_{in} port of the EC. After processing by the EC, the echo signal should be canceled (removed) but the PSTN subscriber's speech should remain, as recorded at the S_{out} port. First verify that the mobile subscriber's echo signal was removed. In addition, listen for speech clipping. Clipping is a condition that can occur during double-talk when the EC uses "hard" non-linear processor (NLP) on/off switching, which causes short portions of speech to be clipped (inadvertently cut off). Echo cancelers with "soft" NLP on/off switching provide better continuity with minimal (or no) clipping while removing echo during double talk. Some EC products have inferior double-talk detection circuitry that can cause the EC to diverge (become unstable and possibly inject noise bursts) during double talk. This behavior indicates that the EC does *not* have an accurate near-end speech detector that is capable of dynamically measuring and adapting to the fluctuating signal strength of double-talk signals. If the EC injects echo, noise bursts or has objectionable speech-clipping characteristics at the S_{out} port, then the EC has failed this test.

Double talk with background noise – test 2b

The recording device is connected to the S_{in} and S_{out} ports of the EC/VQS. During the "double talk with background noise – test 2b," a loud acoustic noise is generated next to the cellular phone. The acoustic noise condition can be created by playing an audio recording (e.g., a loud whistle, siren, or similar sound) or by running a noisy electric fan or similar device next to the cellular phone. The acoustic noise should be periodically toggled on and off during this test. If the NC feature is on, then the results of this test should show signal-level changes recorded at the S_{in} and S_{out} ports that correlate with the acoustic noise. In the absence of acoustic noise the S_{out} signal level should be identical to the S_{in} signal level. However, when the acoustic noise is present (toggled on), the S_{out} signal level should be automatically increased to provide for an improved signal-to-noise ratio for the mobile subscriber. The level of amplification should be a direct function of the acoustic-noise level present in the mobile subscriber's environment. If the EC/VQS does not automatically adjust the volume of the signal at the S_{out} port to compensate for the acoustic-noise level surrounding the mobile subscriber, then the NC application does not work. More importantly, if speech is clipped at S_{out} during noisy intervals, then the EC has failed the test.

Double talk with background noise – test 2c

The recording device is connected to the R_{in} and R_{out} ports of the EC/VQS. The conditions for the "double talk with background noise – test 2c" are identical to "test 2b," but the EC port connections for recording results are different. During this test, the R_{in} port should contain the mobile subscriber's speech along with the acoustic noise signal that is being periodically toggled on and off. The results of this test should show signal-level changes recorded at the R_{in} and R_{out} ports that correlate with the acoustic-noise level. In the absence of acoustic noise the R_{out} signal should be identical to the R_{in} signal level. However if the NR feature is on, then, when acoustic noise is present (toggled on), the R_{out} signal should indicate that the acoustic noise has been attenuated while the speech signal remains constant. One objective is to verify that the EC/VQS can significantly reduce the magnitude of the background (acoustic) noise traveling towards the PSTN subscriber without degrading the mobile subscriber's speech signal. If the EC/VQS does not automatically reduce the level of acoustic background noise delivered to the PSTN subscriber while preserving the integrity of the mobile subscriber's speech signal, then the NR application does not work properly, or does not work at all. And, more importantly, if speech is clipped at R_{out} during noisy interrals, then the EC failed the test.

EC convergence test – test 3

Some echo cancelers, specifically those exhibiting very fast initial convergence characteristics that use a single rate during initial convergence and during adaptation periods, tend to diverge periodically and generate bursts of echo or noise. This phenomenon is especially evident when echo is reflected by a hybrid with a relatively wide dispersion and an echo-return loss (ERL) characteristically less than 9 dB. Wireless network operators have no control over the type of hybrid used at the far-end point in the PSTN network. Consequently, it is important to protect the network operations by deploying a stable EC, one that does not diverge in the presence of hybrids embracing poor dispersion characteristics.

A white noise (random Gaussian noise) signal is applied to the R_{in} port of the EC. After running this test for several minutes and monitoring the signal at the S_{out} port, it will be evident whether the EC remains in a stable converged state or exhibits unstable performance. If the EC is unstable or generates noise bursts, then the EC has failed this test.

9.4 Judging audio quality and performance of acoustic-echo control

The ITU-T G.167[8] standard recommendation presents a list of performance-test procedures intended for evaluating a particular acoustic-echo control system and its effect on voice quality. Although the standard recommendation is aimed at end terminals, some of the tests may be adapted to evaluating network equipment. The G.167 recommendation treats the AEC as a system devoted solely to this function. Accordingly, it defines the send and receive ports as in Figure 9.8. For the purpose of the discussion in this chapter, I will treat the AEC as a function that shares a system with the more dominant application of hybrid EC. Consequently, the port designation is reversed and it is designated as in Figure 9.9.

Since acoustic echo in wireless networks is time variant and its power is inconsistent, ranging from none to very loud, even within a particular call, it is essential that tests last for

Figure 9.8 G.167 definitions of acoustic-echo control interface ports.

---

8   ITU-T Recommendation G.167, *Acoustic Echo Controllers* (1993).

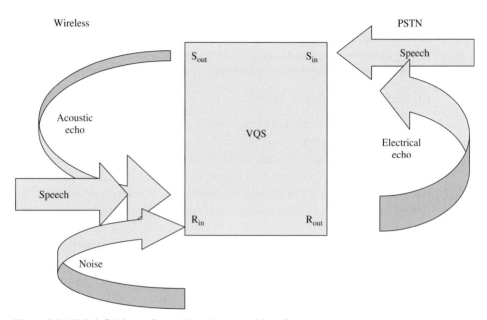

Figure 9.9 VQS definitions of acoustic-echo-control interface ports.

a relatively long duration.[9] All tests involve subjective listening and recordings made on all four ports. Acoustic echo is canceled on the receive side,[10] and single-talk speech is fed through the send side.

Acoustic echo is not always easy to generate, since only certain mobile phones produce it, and those that do tend to behave badly only under particular conditions. When running a test, an ideal set-up, which gives off acoustic echo, may prove difficult to assemble. The ideal configuration consists of speech signals that are transmitted through the system under test and onto a mobile network. The echo signal is supposed to be reflected from a mobile phone back into the test platform. A close emulation of that system could be constructed in a lab as depicted in Figure 9.10.

Test 1 – single-talk test

The AEC is disabled. A speech signal is transmitted through the send side for several seconds. Acoustic echo should be heard and recorded at the R_{in} and R_{out} ports. After a brief duration the AEC is enabled while speech continues to be transmitted on the send side. Acoustic echo should be canceled after entering the R_{in} port. The recording should reveal the existence of the echo at the R_{in} port, and its absence at the R_{out} port.

Test 2 – single talk with surrounding noise

The AEC is disabled. A speech signal is transmitted through the send side for several seconds. The receive side is surrounded by loud background noise measured at an average

---

[9]   A minimum of 30 seconds per test and possibly longer.
[10] As in Figure 9.9.

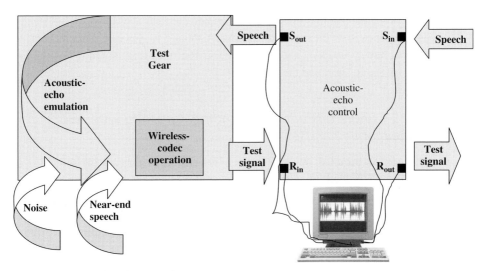

Figure 9.10 Acoustic-echo control test set-up.

of -20 dBm0. Acoustic echo and noise should be heard and recorded at the R_{in} and R_{out} ports. After a brief duration the AEC is enabled while speech and noise conditions persist. Acoustic echo should be canceled without clipping or without upsetting[11] the character of the background noise. The recording should reveal the existence of the echo and noise at the R_{in} port, the absence of echo, and the existence of equivalent noise at the R_{out} port.

Test 3 – Single talk with surrounding noise with no echo

The AEC is disabled. A speech signal is transmitted through the send side for several seconds. The receive side is surrounded by loud background noise measured at an average of -20 dBm0. No acoustic echo is heard at the R_{in} or R_{out} ports, while the background noise is. After a brief duration the AEC is enabled while speech and noise conditions persist. Noise should continue to advance without being clipped or without having its character being upset.[12] The recording should reveal the existence of noise at the R_{in} port, and the existence of equivalent noise at the R_{out} port.

Test 4 – single talk with music playing through the mobile phone with no acoustic echo

The AEC is disabled. A speech signal is transmitted through the send side, while music plays on the receive side for several seconds.[13] No acoustic echo is heard at the R_{in} or R_{out}

---

[11] When surrounding noise is loud and the comfort-noise injection is very different in level or color from the actual noise, the contrast may cause a distraction. The test administrator must decide (subjectively), whether the disparity is sufficiently distinct to cause a notable lessening of the overall voice quality.

[12] When surrounding noise is loud and the comfort-noise injection is very different in level or color from the actual noise, the contrast may cause a distraction. The test administrator must decide (subjectively), whether the disparity is sufficiently distinct to cause a notable lessening of the overall voice quality.

[13] Background music from a mobile phone is a common feature in parts of Asia. The music average power is about -30 dBm0.

ports, while the music is. After a brief duration the AEC is enabled while speech and music conditions persist. Ideally, music should continue to advance without being clipped or without having its character being upset.[14] The recording should reveal the existence of music at the R_{in} port and the R_{out} port.

Test 5 – single talk with surrounding noise and with AEC in the mobile handset

The AEC is disabled. A speech signal is transmitted through the send side for several seconds, while the AEC in the mobile handset reacts by suppressing the signal acoustic echo and noise on the receive side. The mobile handset does not inject matching noise so that noise entering the R_{in} port is clipped. No acoustic echo is heard at the R_{in} or R_{out} ports since it is suppressed by the mobile handset. After a brief duration the AEC is enabled while speech and clipped noise conditions persist. Ideally, the AEC should fill the clipped noise with matching noise and create noise continuity. The recording should show how the clipped noise at the R_{in} port is being repaired at the R_{out} port.

Test 6 – double talk

The AEC is disabled. A speech signal is transmitted through the send side for several seconds. Acoustic echo should be heard and recorded at the R_{in} and R_{out} ports. After a brief duration the AEC is enabled while speech continues to be transmitted on the send side, and a new speech signal is transmitted on the receive side to produce double talk. Acoustic echo should not be completely eliminated, and speech at the R_{out} port should not be clipped, but it should be attenuated. The recording should reflect the difference in levels between the R_{in} and the R_{out} ports while unclipped speech integrity is maintained in both.

Test 7 – double talk with surrounding noise

The AEC is disabled. A speech signal is transmitted through the send side for several seconds. Acoustic echo and noise should be heard and recorded at the R_{in} and R_{out} ports. After a brief duration the AEC is enabled while speech continues to be transmitted on the send side, and a new speech signal and noise are transmitted on the receive side to produce double talk. A residual acoustic echo may be retained, while speech and noise at the R_{out} should not be clipped but may be attenuated. The recording should reflect the difference in levels between the R_{in} and the R_{out} ports while unclipped noise and speech integrity are maintained in both ports.

---

[14] When surrounding noise is loud and the comfort-noise injection is very different in level or color from the actual noise, the contrast may cause a distraction. The test administrator must decide (subjectively), whether the disparity is sufficiently distinct to cause a notable lessening of the overall voice quality.

9.5   Judging audio quality and performance of acoustic-noise reduction

Just as the case with echo cancelers, performance testing of NR is divided into two separate phases – laboratory and field-testing. Laboratory testing is further divided into objective and subjective types. The objective testing is accomplished by way of computer procedures outlined in the ITU-T G.160[15] recommendation and specialized computer algorithms developed by SwissQual[16] utilizing their particular equipment.

ITU-T G.160 recommended procedure

The recent ITU draft recommendation G.160 proposed a set of three objective metrics designed to appraise the performance of a given NR algorithm.

- Signal-to-noise ratio improvement (SNRI): a test value measured during speech activity. It computes the difference in SNR between the pre-processed and post-processed signal. The present proposed minimum threshold is set at $SNRI \geq 4\,dB$ as an average over all test conditions.
- Total noise-level reduction (TNLR) measures the overall level of noise reduction experienced during both speech and speech pauses. The present proposed minimum threshold is set at $TLNR \leq -5\,dB$ as an average over all test conditions.
- Delta signal-to-noise (DSN): the delta measure between SNRI and noise power-level reduction (NPLR). The desired value for DSN is 0, which indicates neither attenuation nor amplification of the speech signal due to the NR algorithm. A $DSN < 0$ indicates speech attenuation as a side effect of noise reduction, and a $DSN > 0$ indicates speech amplification due to the NR algorithm. The present proposed thresholds are $-4\,dB \leq DSN \leq 3\,dB$ as an average over all test conditions.

One of the key NR objectives is to minimize changes in the power level of the speech signal. It is obvious that an improvement in the SNR of a noisy speech signal may be achieved by resorting to amplification of the signal during speech activity with no amplification during speech pauses when noise is the only measured signal. When amplification is applied only to the high-energy portions of the signal, i.e., to speech activity periods, it does not accomplish the true mission of noise reduction. In fact, it does not change the SNR when it really counts, i.e., during speech activity.

Changes in speech-signal level, devoid of changes to noise level during speech pauses, can be detected. The procedure applies an internal measure, noise power-level reduction (NPLR) – calculated during short speech pauses spread in between speech activities – contrasting it with SNRI. The SNRI-to-NPLR difference (DSN) provides an indication of possible speech attenuation or speech amplification produced by the NR algorithm under test.

[15]   ITU-T Recommendation G.160, *Voice Enhancement Devices*, draft 14 (2004).
[16]   ITU-T SG-15 *Question 6, Delayed Contribution On Objective Method for Evaluation of Noise-Reduction Systems*, Geneva, Switzerland, SwissQual, SQuad. (April 2002).

Test signals

The test signal employs a database containing two female and two male samples, mixed with car noise (two types) or street noise (two types), at 6, 12, and 18 dB SNR, for a total of 48 speech samples. The car interior noise is of a relatively constant power for a car moving at 120 km/h. The street noise has a slowly varying power level. The clean speech signals are of a relatively constant average power level within a particular sample, where "sample" refers to a file containing one or more utterances, and the noise signals are of a short-time stationary nature with no rapid changes in the power level and no speech-like components.

The samples are to be digitally filtered before NR processing by an appropriate filter[17] in an effort to represent a real cellular system frequency response. This filtering is carried out before the scaling of the samples.

The test samples abide by the following procedure.

- The clean speech material is scaled to the active speech level -26 dBov with the ITU-T Rec. P. 56 speech voltmeter, one file at a time, each file including a sequence of one (to four) utterance(s) from one speaker.
- A period of silence, lasting 2 seconds, is inserted at the beginning of each file.
- Augmented clean speech and noise files are then mixed to create the set of 48 noisy speech files with predefined SNR, sample rate, bits per sample, and file length.

Figure 9.11 presents a summary of a study of six laboratory simulations of NR algorithms (A through F, respectively), listing the average values for SNRI, TNLR, and DSN for each.

The objective of the ITU-T G.160 metric is to expose specific weaknesses in the NR algorithm performance as illustrated by the results above. To emphasise on various methodologies for reducing acoustic noise, I employed highly controlled computer simulations of various noise-reduction approaches. Some of these utilize the same engine with a different set of parameters. The various NR algorithms were designated as A, B, C, D, E, and F. They are arranged sequentially (and repeatedly) in three groups. Each group exhibits results of a particular test ranging from SNRI through TNLR to DSN respectively on every one of the A–F algorithms.

- Algorithm A accomplishes NR by focusing on overall signal attenuation. Consequently, its associated SNRI is poor and the amount of speech distortion (DSN) is high.
- Algorithm B implements NR but removes very little noise. Consequently, its SNRI and TLNR are insignificant. Nonetheless, it does not distort speech.
- Algorithms C and D accomplish NR by focusing on attenuation of the overall signal. Consequently, their notable TLNR is realized at the expense of a marginal SNRI and a very poor DSN, resulting in a high level of speech distortion through attenuation. The difference between C and D is the level of aggressiveness. D is more aggressive and, therefore, both its noise attenuation and speech distortion are more significant.

---

[17]   As specified in ITU-T G.160.

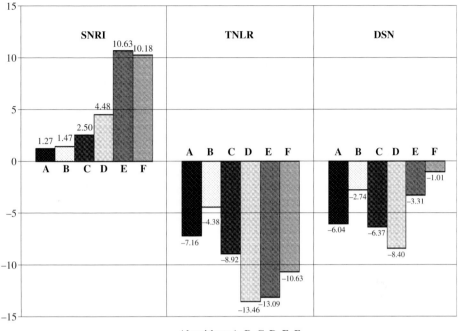

Figure 9.11 G.160 simulated results.

- Algorithms E and F provide the most robust NR. The difference between E and F is mainly in the level of aggression. The best SNRI and TLNR are accomplished at the expense of a slight reduction – within acceptable limits – in the DSN performance.

It should be noted that the test values of SNRI, TNLR, and DSN represent an average over all of the tested cases. A further investigation into the specific value for each of the subcategories[18] is a useful exercise, which might unearth certain performance details concerning the algorithm impact under different noise conditions.

As illustrated in Figure 9.12, when employing algorithms E and F above, segmenting and studying the G.160 results under the separate noise levels of 6 dB, 12 dB, and 18 dB, respectively, it becomes evident that the NR algorithm exhibits distinctive effectiveness under the different SNR conditions. The findings suggest that the NR algorithm performs better under the 12 dB (average) SNR than under the 6 and 18 dB conditions. The findings may be a bit surprising, since one might expect the case of 6 dB to present the best opportunity, owing to its highest noise level relative to speech.

The logical explanation for the 12 dB SNR pre-eminence over the 6 dB condition is the fact that although the 6 dB condition comprises the higher noise level and the better-seeming opportunity, it is also true that the 6 dB SNR condition contains more substantial

[18] Among the 48 cases, one could examine separately the results associated with the subcategories. A subcategory comprises a specific intersection of a person (4), noise type (4), and SNR level (3). A further inspection of the details behind the gross averages may shed more light on the performance of the NR algorithm.

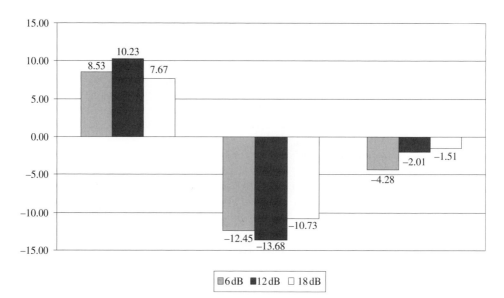

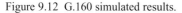

Figure 9.12 G.160 simulated results.

noise components spread out all through the various frequency bands. These bands contain significant speech parts, and aggressive attenuation of these many bands may bring about noticeable speech distortion. Accordingly, a well-balanced NR algorithm may restrain its appetite and maintain a reasonable equilibrium between the two opposite potential outcomes (noise reduction and speech distortion) that may be affected by aggressive action.

The logical explanation for the 12 dB pre-eminence over the 18 dB SNR condition is derived from the worth effect discussed in Chapter 2. Although noise (in the 18 dB SNR case) may be reduced with little speech distortion (because noise is likely to be concentrated in a small number of frequency bands), the initial condition (of 18 dB SNR) is not very debilitating. When observing the graph in Figure 9.12, it is not at all obvious that the case of 12 dB SNR is superior to the case of 18 dB SNR. In fact, it may seem that NR contributes to a notable improvement for the case of 18 dB SNR. Nevertheless, phone conversations at 18 dB SNR without NR may not be terribly impaired to start with, and hence improving an already reasonable quality carries less weight than elevating poorer quality to a reasonable one.

The ITU G.160 recommendation presents a sound picture of a particular NR algorithm performance. The three measurements provide insights into the approach adopted by a specific algorithm. The previous examples illustrate and highlight the trade-offs picked by each of the algorithms for realizing their final goal.

A more convincing picture of the 12 dB SNR pre-eminence is available through mean-opinion scoring. It is accomplished by appending one more step that computes a quality index, a value equivalent to the MOS-LQO, by treating the values of the SNRI, TNLR, and DSN measurements in a single score that weights the various trade-offs and merges them into a distinct bottom line.

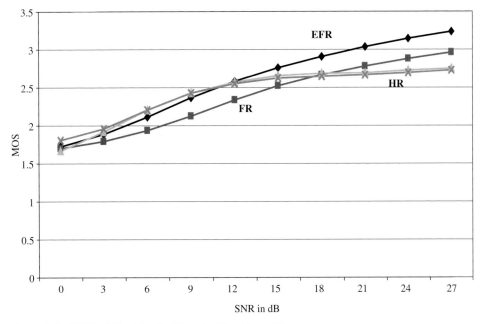

Figure 9.13 MOS – HR with NR, FR and EFR without NR.

There are two classes of procedures for generating objective MOS, intrusive and non-intrusive. The intrusive procedures most often employed for the task are PESQ (ITU-T P. 862)[19] and the – not yet standard, but popular among service providers – SQuad-NS.[20] The most recent non-intrusive procedure is based on the ITU-T P. 563[21] standard recommendation.

Intrusive procedures may be employed during laboratory evaluations of voice-quality algorithms. They take MOS measurements under different NR state conditions, and then compare the results to derive an assessment of improvement due to a change in the NR status.

In Chapter 2 we learned that voice quality drawn from GSM HR codec responds more negatively to acoustic noise than either the FR or the EFR codecs. It would, therefore, present a greater opportunity for a noise-reduction algorithm to lay bare its contribution to overall voice quality. Figure 9.13 presents results of a PESQ study contrasting HR enabling noise reduction with FR and EFR disabling NR. Under the various noise conditions, HR demonstrates its greatest improvement for SNR values less than 12 dB.

Figure 9.14 presents a PESQ study that my colleagues and I conducted in 2005, involving many noise types encompassing airport, babble, café, bus, subway, train, and factory. The figure presents an average score. Although the signal employed for the study

19 ITU-T Recommendation P. 862, *Perceptual Evaluation of Speech Quality (PESQ): An Objective Method For End-to-End Speech Quality Assessment of Narrow-Band Telephone Networks and Speech Codecs* (2001).
20 A product by Swissqual, www.swissqual.com.
21 ITU-T Recommendation P. 563, *Single Ended Method for Objective Speech Quality Assessment in Narrow-Band Telephony Applications* (2004).

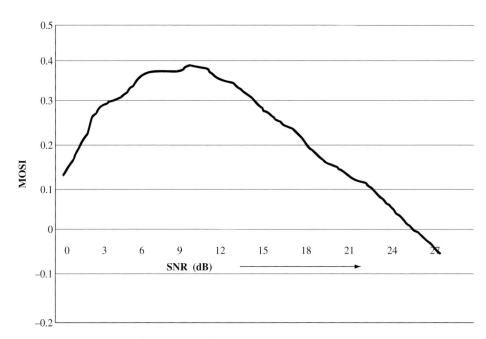

Figure 9.14 Average MOS improvement due to NR.

was not processed by a codec, the PESQ study is consistent with studies employing coded signals. It suggests that NR may become a negative contributor to the overall speech quality when applied to signals with SNR values greater than 24 dB. It also suggests that NR is most effective at noise conditions around 8–12 dB of SNR.

Bringing SQuad-NS into play yielded conclusions equivalent to the ones obtained from the PESQ studies. The data used for the study included algorithms E and F in Figure 9.11. The average MOS improvement generated by the algorithms is presented in Figure 9.15.

The conclusions are consistent with the PESQ results, as they all suggest that when appraising the quality of a particular NR algorithm, noise-level conditions should be controlled and results should be assessed as for specific noise levels.

The case of double enhancement

Noise reduction is an application paid for by the mobile service provider for the benefit of the other end, a PSTN subscriber, or possibly a subscriber from another wireless operator. This fact sets some frustration in motion, which spawns ideas such as, "we need a bi-directional NR to benefit our own subscriber base."

A bi-directional NR may be implemented within a VQS as an add-on to an existing NR application operating on the receive side. Having NR installed on the send side as well may not take as many DSP MIPS as the initial receive side application, due to its incremental nature. However, a bi-directional NR may carry some unintentional outcomes including performance inconsistencies that need to be understood prior to taking the plunge.

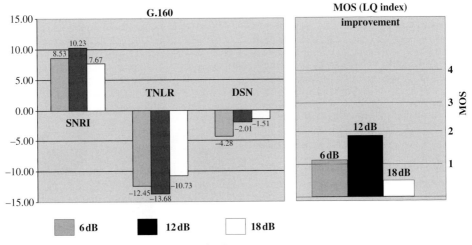

Figure 9.15 MOS improvements by SQuad-NS.

Having NR installed on both the receive and send sides of the VQS leads to double enhancement wherever the far end is a mobile subscriber of the same service provider, or another provider who employs NR on the receive side of their own VQS (see Figures 5.13 and 5.14).

One may argue that double enhancement is not a bad feature. After all, if no noise is left after the first enhancement, then the second round may do little to make any difference. Nevertheless, if there is still noise left after the first treatment, then the subsequent application may clean it up. It does sound good. But there is a counter-argument supported by simulation employing a sequence of G.160 runs. The study that my colleagues and I conducted shows that double enhancement is equivalent to a single run assuming a more aggressive setting. Double enhancement tends to take more noise out at the expense of more speech distortion.

Figure 9.16 exhibits the study results. As indicated in the figure, in both cases, A and B, the double enhancement yields less SNR improvement and less noise reduction than a single run, assuming a doubling of the aggressive level respectively. At the same time, the DSN measure indicates that speech distortion is greater when applying double enhance- ment sequentially in comparison with doubling the aggressive level in a single run. In general, sequential applications of NR carry more noise reduction than a single application but the speech distortion level rises correspondingly.

Caution in relying solely on automated MOS algorithms

Perceptual evaluation of speech quality and SQuad-NS are well-designed tools. Their equations and parameters were chosen to maximize their linear correlation to outcomes generated by subjective testing. It has been documented in P. 862 that correlation has been found to exceed 90%. Still, relying solely on these automated objective method- ologies carries a high risk since both PESQ and SQuad-NS are limited in scope. These

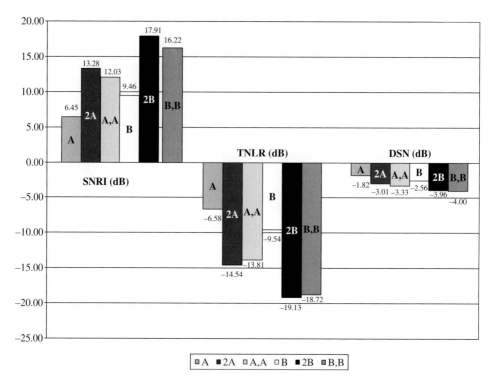

Figure 9.16 G.160 comparisons of single versus sequential applications of NR: A – single run, least aggressive; 2A – single run, double-A aggressive level; A, A – sequential runs using A level in both; B – single run, medium aggressive; 2B – single run, double-B aggressive level; B, B – sequential runs using B level in both.

procedures are useful in pointing to areas that may require further investigation through subjective discovery. They may be able to narrow down the investigation and focus it on particular aspects of noise reduction, but they should never be used as the final decision tools.

In its very introduction to PESQ, the P. 862 standard issues a word of caution: "It should be noted that the PESQ algorithm does not provide a comprehensive evaluation of transmission quality. It only measures the effects of one-way speech distortion and noise on speech quality. The effects of loudness loss, delay, side-tone, echo, and other impairments related to two-way interaction (e.g., center clipper) are not reflected in the PESQ scores. Therefore, it is possible to have high PESQ scores, yet poor quality of the connection overall."

Consequently, PESQ becomes an insufficient tool for evaluating noise-reduction algorithms. Noise-reduction tools employ selective and uneven attenuation of frequency bands within a given signal. This action may attenuate frequency bands containing speech, affecting speech distortion as a result. Since PESQ is not designed to evaluate loudness loss it may not be ideal for evaluating performance of noise-reduction algorithms.

The SQuad-NS procedure may be viewed as an extension to the ITU G.160 standard recommendation. It provides equivalent measurements, as described in Section 9.4.

SwissQual's added contribution to the ITU G.160[22] approach is the supplementary MOS scoring. The results obtained from the measurements above are then processed through a human psychoacoustics model to emulate a subjective voice-quality score that would or could be obtained by subjecting humans to the output of the noise reduction algorithm.

The inherent bias embedded in the SQuad-NS procedure is its knowledge of the signal quality before and after applying the NR algorithm. It bases its quality-scoring mechanism on the knowledge of the improvement due to the NR. Personal experiences (my own) through technical field trials suggest that when people listen to a noisy signal while possessing the ability to turn NR on and off in real time, they develop a stronger appreciation for the application, especially when it is aggressive. They pay attention and can clearly hear the noise reduced, because they know what it sounds like when noise is excessive. On the other hand, when people listen to a noise-reduced signal with no reference and no knowledge of what it sounds prior to the NR application, they listen more carefully to the speech quality. And they may pay more attention to speech distortions rather than noise reduction (they do not know how much noise was reduced, since they have no knowledge of the signal prior to the NR application).

The high correlation between subjective studies and SQuad-NS is derived from subjective tests that expose listeners to the same information used by the SQuad-NS. This information includes the clear speech and the signal quality prior to the NR application.

In real life, listeners are not aware of what the signal would sound like if... Consequently, subjective studies that emulate the real life experience, i.e., lack the additional (pre-NR) input, may not correlate as highly with SQuad-NS as those that include the additional input.

For that reason, the SQuad-NS scoring mechanism tends to favor a slightly aggressive provisioning of the algorithm, whereas subjective MOS studies that involve real calls through a network exhibit a notable bias towards moderate aggression, i.e., less noise suppression and less speech distortion.

Both PESQ and SQuad-NS MOS are of a very narrow scope. They do not cover the following situations:

- Communications in an interactive (bi-directional) session. Both PESQ and SQuad-NS evaluate a one-way link. When a signal causes impairment in the opposite direction, PESQ and SQuad-NS would miss it and ignore it completely. Neither PESQ nor SQuad-NS would relate cause and effect and might completely miss potential problems seen in some networks. Consequently, they may err and propose a high MOS where in fact there may exist severe voice-quality problems.

- For example, a poor AEC treatment inside mobile phones may cause severe noise clipping when the mobile phone transmits from noisy surroundings. The noise clipping may be triggered by signals originating at the quiet end of the call (the other end). At the same time, the SQuad-NS may miss the noise clipping. It would not run a signal that

---

22   ITU-T SG-15 Question 6, *Delayed Contribution on Objective Method for Evaluation of Noise-Reduction Systems.*
     ITU-T SG-12 Question 9, Contribution 32, *Cause Analysis of Objective Speech Quality Degradations Used in SQuad*, SwissQual (August 2001).

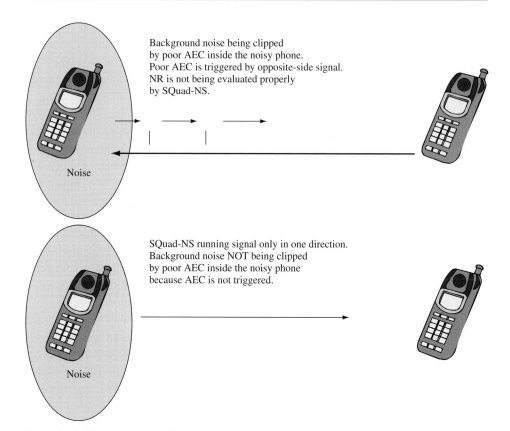

Figure 9.17 Distortions caused by conversational speech rather than a one-way communication.

triggers the AEC inside the phone. Consequently, it would rate the call as being high quality because it would look only at the clean signal (on the other side of the call), while missing its effect on the opposite side. Figure 9.17 illustrates this particular example.

- The G.160 test used as input for the MOS computed by the SQuad-NS is a step in the right direction, but, at present, it is an immature test. It requires significant improvements before it can be used to replace human listening effectively. It does not measure real speech distortion if the NR algorithm triggers it as a side effect. The G.160 recommendation measures overall attenuation by relating a total signal to noise ratio improvement (SNRI) to a total change in noise level during speech pauses. Although this is a first step in detecting distortion it is a long way from depicting a complete picture. The DSN measure detects attenuation level differences between speech and speech pauses. A negative value suggests that the NR algorithm is more aggressive during speech pauses. A positive value suggests that the method used by NR resorts to amplification during active speech with no action or minor attenuation during speech pauses.

Speech distortion consists of different relative changes among frequency bands in addition to an overall attenuation. When the overall attenuation is unchanged, the DSN would indicate a distortion-free algorithm while, in fact, there may be notable changes

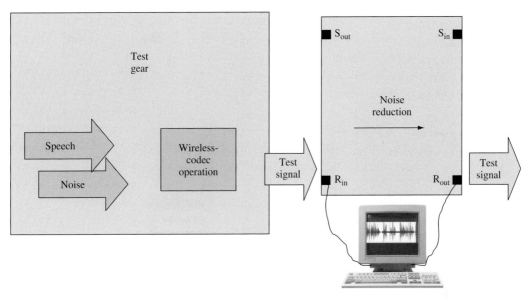

Figure 9.18 Noise reduction test set-up.

among various frequency bands, which tend to move in opposite direction and cancel each other's impacts when added together. Evaluating distortion caused by an NR algorithm requires a frequency-domain analysis, where a distortion measure is applied to each of the frequency bands.

9.6   Equipment set-up for lab testing

Figure 9.18 shows the key components and their interrelationships. The test gear includes a digital player with recordings of the various speech and noise segments. The recordings may be processed by a wireless codec for a more authentic representation before being met by the NR algorithm. The only ports require monitoring and recording are the R_{in} and the R_{out} ports.[23]

In case there is a need to evaluate the performance of double enhancement, the signal captured at the R_{out} port of the test in Figure 9.18 is fed again through the VQS R_{in} port, as illustrated in Figure 9.19.

9.7   Automating MOS improvement estimate due to NR

Objective MOS tools are best implemented in a lab environment where the scoring device has the full knowledge of the noise free, clear speech, before it is mixed with the noise. The

---

[23]   Note that the designation of the ports assume that the NR algorithm is part of a larger system including a hybrid EC.

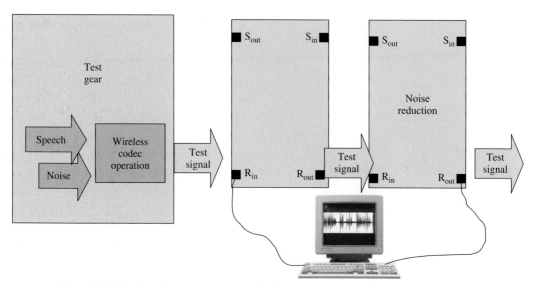

Figure 9.19 Double-enhancement noise-reduction test set-up.

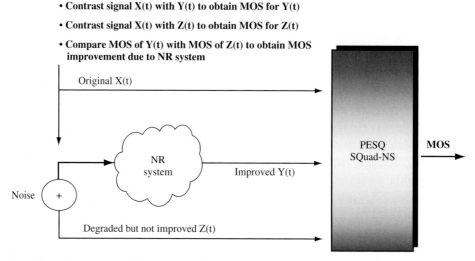

Figure 9.20 Intrusive lab tools using a reference signal.

speech-quality assessment procedure computes a MOS estimate for the degraded signals by contrasting them with the clear speech. The procedure may be used to assess the contribution of the NR algorithm by taking the score of the NR-improved signal and subtracting the score of the unimproved signal from it. The difference between the two scores is the contribution due to NR. When employing a speech-quality assessment procedure for the task of estimating the NR contribution, testers must run it twice – once with NR disabled, and then with NR enabled.

Figure 9.20 illustrates the process applied for the task.

9.8 Judging audio quality and performance of automatic and adaptive level control

Objective and intrusive laboratory tests

The key assumption behind the guidelines in this section is that the ALC and NC applications are part of a complete VQS, and they are tested for conformance in digital wireless networks by the wireless-service provider who plans on installing it as part of a voice-quality infrastructure. Automatic level control (ALC) and adaptive level control, a.k.a. noise compensation (NC), may be verified for adequate performance by resorting to parts of the ITU-T G.169[24] standard recommendation that apply to the wireless environment. Throughout this presentation, the G.169 test suite is not taken at face value but rather modified and augmented to suit the wireless-service provider in the decision-making and evaluation process.

Wireless calls are often launched from noisy environments. When applying ALC in the receive side of the VQ system, there is a risk that noise surrounding the wireless caller might be amplified by the ALC device and shipped over to the far-end caller. Acoustic echo may be reflected from a wireless handset. When applying ALC in the receive side there is a risk that the acoustic echo might be amplified and returned to the far-end caller at an elevated level. Wireless infrastructures include echo cancelers protecting their subscribers from hybrid echo. Residual echo left over from an echo-cancelation process (particularly during double talk) may be amplified by an ALC or an NC device installed in the send path and bother the wireless subscriber.

Figure 9.21 depicts the test set-up.

These risks may be reduced or eliminated by properly designed ALC and NC devices. The following procedures outline tests that focus on ensuring that amplification of unwanted signal components does not accompany any positive contribution of level control applications.

Test 1 – testing for ALC (send side) stability when operating post an echo canceler

(1) Employ suitable test signals from a set of sinusoids (e.g., 204 Hz, 304 Hz, 504 Hz, 1004 Hz, 1804 Hz, and 2804 Hz), on/off modulated, if necessary, at the rate of 1 second and on 1 second off. Emulate a 6 dB ERL hybrid and an active echo canceler that removes echo from the send side.[25]

(2) Enable the ALC device on the send side and apply the test signal at a level of -2 dBm0 through the S_{in} port. A proper ALC algorithm may not amplify a signal as hot as the one applied.

(3) Take measurements of the signal level at the S_{out} port.

[24] ITU-T Recommendation G.169, *Automatic Level Control Devices* (1999).
[25] Since speech dynamics make it difficult to assess level changes due to an ALC function, the better choice is a stable tone transmitted through the ALC input port. Any changes to the tone level would be caused by signal processing rather than the tone's own level instability.

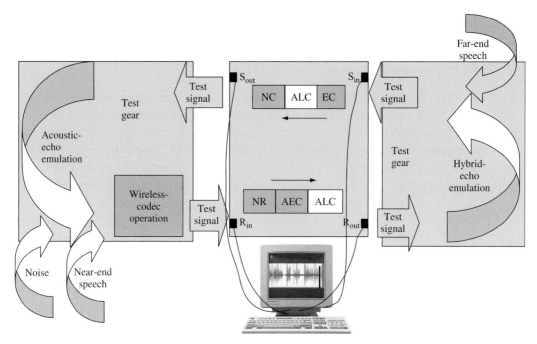

Figure 9.21 Level control test set-up.

(4) Remove the signal from the S_{in} port and apply it through the R_{in} port at the same level. The test signal at the R_{in} port should be frequently interrupted (or on/off modulated at approximately 0.5 Hz).
(5) After 1 minute, remove the test signal from the R_{in} port, and re-apply it to the S_{in} port. Take measurements of the signal level at the instant it arrives at the S_{out} port.
(6) A proper ALC application should not exceed the level noted in step 3 by more than 12% (or the level measurement should be greater by more than 1 dB).

Test 2 – noise tolerance

One of the major risks that an ALC function may bring to bear is the risk of amplifying noise, whether it is circuit noise or acoustic background noise. Test 2 is specifically designed to expose this aspect of an ALC.

(1) Employ suitable test signals from a set of sinusoids (e.g., 204 Hz, 304 Hz, 504 Hz, 1004 Hz, 1804 Hz, and 2804 Hz), on/off modulated, if necessary, at the rate of 1 second on and 1 second off.[26]
(2) Enable the ALC (receive side) and disable the NR and the AEC functions if they are part of the system.

---

[26] Since speech dynamics make it difficult to assess level changes due to an ALC function, the better choice is a stable tone transmitted through the ALC input port. Any changes to the tone level would be caused by signal processing rather than the tone's own level instability.

(3) Apply a test signal (on/off modulated if necessary; see clause 6 of the recommendation) to the R_{in} port, at a level of -5 dBm0. This level is sufficiently high to ensure that the ALC function may not amplify the signal any further.

(4) Measure the level of the signal R_{out} port of the ALC system, and infer the value of the ALC device gain.

(5) Replace the test signal at the R_{in} port with a continuous Gaussian, white-noise signal (band limited 0–3.4 kHz) at a level of -40 dBm0.

(6) After about 20 seconds, recheck the ALC device gain by measuring the level of the noise signal at the R_{out} port.

(7) A properly operating ALC function may not amplify the noise signal by more than 1 dB.

(8) Repeat step 5–7 but with acoustic background noise at -28 dBm0.

Test 3 – high-level compensation

(1) Employ suitable test signals from a set of sinusoids (e.g., 204 Hz, 304 Hz, 504 Hz, 1004 Hz, 1804 Hz, and 2804 Hz), on/off modulated, if necessary, at the rate of 1 second on and off 1 second.[27]

(2) Enable ALC and NC (send side).

(3) Insert acoustic noise at -28 dBm0 at the R_{in} port.

(4) Apply a test signal (on/off modulated if necessary; see clause 6 of the recommendation) to the S_{in} port, at a level of -2 dBm0. This level is sufficiently high to ensure that the NC function does not amplify the signal any further, and the ALC function will actually attenuate the signal.

(5) After about 2 seconds, measure the level at the S_{out} port.

(6) A properly operating NC function will not amplify the noise signal while a properly operating ALC function would attempt to attenuate the signal. Any amplification due to NC overpowering the ALC attempt to attenuate is to be considered improper.

(7) Disable NC and repeat steps 4–5.

(8) A properly operating ALC would attenuate the level at the S_{out} port by about 5 dB.

Test 4 – maximum gain

(1) Employ suitable test signals from a set of sinusoids (e.g., 204 Hz, 304 Hz, 504 Hz, 1004 Hz, 1804 Hz, and 2804 Hz), on/off modulated, if necessary, at the rate of 1 second on and 1 second off.[28]

(2) Enable ALC and NC (send side).

---

[27] Since speech dynamics make it difficult to assess level changes due to an ALC function, the better choice is a stable tone transmitted through the ALC input port. Any changes to the tone level would be caused by signal processing rather than the tone's own level instability.

[28] Since speech dynamics make it difficult to assess level changes due to an ALC function, the better choice is a stable tone transmitted through the ALC input port. Any changes to the tone level would be caused by signal processing rather than the tone's own level instability.

(3) Insert acoustic noise at −28 dBm0 at the R_{in} port.

(4) Apply a test signal (on/off modulated if necessary; see clause 6 of the recommendation) to the S_{in} port, at a level of −25 dBm0. This level is sufficiently high to ensure that the combination of NC and ALC functions would amplify the signal. After about 10 seconds, measure the level at the S_{out} port.

(5) A properly operating combination of NC and ALC functions may not amplify the test signal beyond −5 dBm0 at the S_{out} port.

(6) Disable NC and repeat steps 3–4.

(7) A properly operating ALC may not amplify the test signal by more than 15 dB.

Test 5 – rate of change in level

(1) Employ suitable test signals from a set of sinusoids (e.g., 204 Hz, 304 Hz, 504 Hz, 1004 Hz, 1804 Hz, and 2804 Hz), on and off modulated, if necessary, at the rate of 1 second on and off 1 second.[29]

(2) Enable ALC and NC (send side).

(3) Insert acoustic noise at −28 dBm0 at the R_{in} port.

(4) Apply a test signal (on/off modulated if necessary; see clause 6 of the recommendation) to the S_{in} port, at a level of −25 dBm0. This level is sufficiently high to ensure that the combination of NC and ALC functions would amplify the signal. Immediately and then after every second for about 10 seconds, measure the levels at the S_{out} port.

(5) A properly operating combination of NC and ALC functions may not amplify the test signal at the S_{out} port by more than 15 dB per second.

(6) Disable NC and repeat steps 3–4.

(7) A properly operating ALC may not amplify the test signal by more than 10 dB per second.

(8) Apply a test signal (on/off modulated if necessary; see clause 6 of the recommendation) to the S_{in} port, at a level of −1 dBm0. This level is sufficiently high to ensure that the ALC function would attenuate the signal.

(9) Measure the elapsed time it takes the ALC to drive the signal down to −4 dBm0.

(10) A properly operating ALC may not impose an attenuation rate greater than 10 dB per second.

9.9   Testing for data compatibility on a fully featured system

Almost all of the voice-quality enhancement algorithms interfere with data integrity. These applications are designed to alter signal characteristics (NR), change level (ALC and NC), and suppress (what they consider to be) undesired bytes (AEC and EC). While these operations help improve speech quality they turn poisonous when operating on data.

---

[29] Since speech dynamics make it difficult to assess level changes due to an ALC function, the better choice is a stable tone transmitted through the ALC input port. Any changes to the tone level would be caused by signal processing rather than the tone's own level instability.

Consequently, properly designed systems must detect any non-voice signal that comes across, and disable themselves for the duration of the transmission, or until the channel returns to carrying speech.

In particular, noise-reduction algorithms tend to manipulate signals crossing their path in an attempt to reduce suspected noisy components. Many of the signals that do are ones that accompany speech communications on a regular basis. Many non-speech signals carry stationary characteristics that would make them look like noise unless the algorithm is equipped with special detectors that disable the NR and let the signal through unscathed.

The ability to transmit and receive voice-band data signals reliably, either facsimile (fax) or computer-file transfers, must be verified by network operators as part of the VQS evaluation process. It has been observed that VQ equipment having inferior performance characteristics can prevent successful call completion or cause signal corruption during data transfers, especially for applications that involve an A-interface (i.e., wireless data transfer).

The data signals are monitored at the R_{out} and S_{out} ports of the VQS. During this test, the modem data call should establish properly and remain active throughout the test session, and there should be no instances of data corruption (e.g., illegal characters, missing data, etc.) at either the PSTN or wireless data/fax terminal. After running this test for several minutes, it will be evident whether the VQS is operating properly. If the modem data call cannot be established, is unstable, or data corruption occurs, then the VQS is unable to pass this test.

The following sections discuss testing of various signals that must not be harmed by the VQ system when plotting a course through its path. The most prevalent examples of non-voice signals using voice band transmission that require special treatment are: DTMF, ring-back tone, background music, fax, and modems that do not send a disabling tone.

Dual-tone multi-frequency (DTMF)

Dual-tone multi-frequency, a.k.a. touch tone, refers to the Bellcore GR-506[30] and the ITU-T Q.23[31] standards for in-band digits. Each dialing digit has a specified sum of two frequencies as well as power and duration requirements. Voice-quality systems must be sensitive so as to preserve DTMF signals otherwise DTMF receivers elsewhere in the network would fail to detect such digits.

Unless it is equipped with a DTMF detector, the NR may interpret a DTMF signal as noise because of its stationary nature. When this happens, the NR algorithm would distort the original DTMF and mobile-phone users may not be able to communicate with voice mail and other interactive systems requiring the user's digits as a response.

The DTMF method is widely used for interactive communication between human beings and machines. Ordinary examples include voice mail, phone-directory services, various voice menus, and many more DTMF control interface applications. The wireless service provider may be most concerned with having uncorrupted digits leaving the S_{out} port of the voice-quality system after being inserted via the wireless caller's mobile handset.

[30]   Bellcore, *LSSGR: Signaling for Analog Interfaces*, GR-506-CORE, Issue 1 (June 1996).
[31]   ITU-T Recommendation Q.23, *Technical Features of Push-Button Telephone Sets* (1988).

Table 9.2 *DTMF test sequences*

	Frequency offset (%)		Level (dBm0)	
Set no.	(low frequency)	(high frequency)	(low frequency)	(high frequency)
1	0	0	0	0
2	0	0	−18	−18
3	1.5	1.5	−10	−10
4	−1.5	−1.5	−10	−10
5	0	0	−12	−18
6	0	0	−14	−10
7	1.5	1.5	−14	−20
8	−1.5	1.5	−14	−20
9	1.5	−1.5	−6	−12
10	−1.5	−1.5	−6	−12
11	0	0	−12	−18
12	0	0	−14	−10
13	1.5	1.5	−10	−6
14	−1.5	1.5	−10	−6
15	1.5	−1.5	−18	−14
16	−1.5	−1.5	−18	−14

Source: Specified in ITU-T G.169 and G.160.

Touch-tone detectors may vary with respect to their tolerance levels. Some may detect correctly in the face of minor distortions, but others may fail to discern when VQ systems alter particular aspects (such as small level changes or minor frequency shifts) of the signal.

Testing for compatibility with DTMF transmission requires a means of generating and sending a sequence of DTMF characters to the test circuit and means for detecting the output after it has passed through the system under test.

The DTMF test sequence consists of 256 different DTMF signals, arranged as 16 sets of the 16 DTMF characters (0 ... 9, #, *, A, B, C, D). Each set has the frequency offset specified in Table 9.2. The replayed levels of the discrete frequencies making up the tone pairs are also specified in Table 9.2. Each DTMF character should last for 50 ms, with an interval between characters of 100 ms. The interval between each set is to be set to 1 second, and the 16 sets of DTMF characters are to be replayed as a continuous sequence of 256 characters.

Test sequence

(1) Disable all of the voice-quality applications.
(2) Replay a test signal consisting of DTMF test sequences including added noise over the test circuit. Record the percentage of detection failures and detection errors.
(3) Repeat step 2 a sufficient number of times to satisfactorily establish the standard of DTMF detection performance.
(4) If DTMF detection is not reliable, lower the noise level and repeat steps 2 and 3.

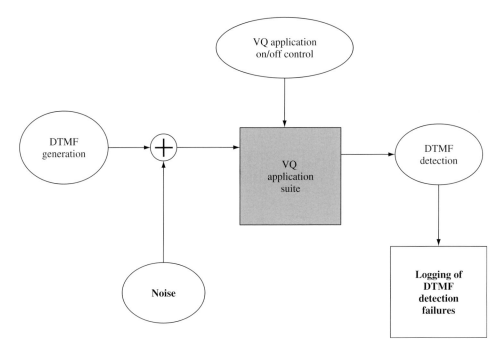

Figure 9.22 DTMF detection test set-up.

(5) Enable all of the VQ applications and repeat steps 2 and 3. If DTMF detection performance is noticeably worse with the applications enabled, than the VQ system interferes with the transmission of DTMF signals.

(6) If the VQ system, as a whole, failed the test, then isolate the particular problem application by enabling one application at a time while repeating steps 2–3.[32]

(7) Stop the DTMF and insert a speech signal. Check to verify that all of the VQ applications are enabled properly.

Figure 9.22 illustrates the test set-up for the DTMF detection and disabling.

Ring-back tone

In the United States, carriers use a de-facto ring-back standard during call initialization. This ring-back tone is the sum of a 440 Hz and 480 Hz signal for 2 seconds duration. Each tone pulse has 4 seconds of silence between pulses. Most ring-back tones exist at levels from −20 dBm0 to −15 dBm0.

Unless programmed to detect the ring-back tone, an NR algorithm would interpret the stationary signal to be noise. It would take proper action and attenuate it over its duration. The attenuation would slow down due to the 4 second pauses, which would update the noise estimate and lessen the attenuation.

---

[32] Noise reduction is the most likely culprit. The VQ system should be able to detect the DTMF and disable all VQ applications immediately. If it fails to do so, the problem application should be disabled until the VQ detection mechanism is repaired.

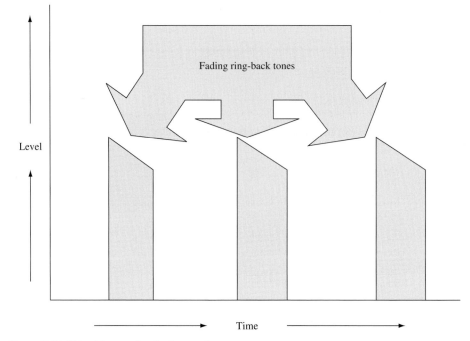

Figure 9.23 NR without a ring-back tone detector.

In the absence of proper detection and disabling, the NR would affect fading of the ring-back tone during each 2 second ring duration. The next ring may be starting at a normal level, only to fade again as illustrated in Figure 9.23. The ability of the VQS to avoid fading ring tones can be tested by using the following steps.

Test sequence

(1) Disable all of the voice-quality applications.
(2) Prepare a digital recording of a relevant ring-back tone.
(3) Play the ring-back tone at the R_{in} port and include white noise over the test circuit.
(4) Record the ring-back tone at the R_{in} and R_{out} ports.
(5) Enable all of the VQ applications.
(6) Replay a ring-back signal including added noise over the test circuit.
(7) Record the ring-back tone at the R_{in} and R_{out} ports.
(8) Check for fading between R_{in} and R_{out} ports and refer to the results in step 4.
(9) A properly designed system does not attenuate the ring-back tone between the R_{in} and the R_{out} ports. Any minor fading that may occur between the ports must not exceed that experienced when the VQ system is disabled.
(10) If the VQ system as a whole failed the test, than isolate the particular problem application by enabling one application at a time while repeating steps 6–7.[33]

---

[33]   Noise reduction is the most likely culprit. The VQ system should be able to detect the DTMF and disable all VQ applications immediately. If it fails to do so, the problem application should be disabled until the VQ detection mechanism is repaired.

Figure 9.22 illustrates the test set-up for the DTMF detection and disabling. Replacing the DTMF signal with a ring-back tone in the figure would satisfy the setup requirements.

Fax and voice-band data

It is important to test the VQS for compatibility with voice-band data. The EC must be disabled upon detecting a phase-reversed tone while the rest of the VQS must be disabled upon both phase-reversed and straight tones.

Test sequence

(1) Enable all VQ applications.
(2) Prepare digital recordings of 2100 Hz tones, one with and the other without phase reversal followed by white noise at -40 dBm0.
(3) Play the recordings through the R_{in} port.
(4) Record the signals at the R_{in} and R_{out} ports.
(5) Emulate an echo signal at 6 dB ERL to be sent into the S_{in} port.
(6) Record the signals at the S_{in} and S_{out} ports.
(7) A properly designed system does not attenuate or amplify the signal between the R_{in} and R_{out} ports. The signal at the R_{in} port must equal the signal at the R_{out} port of the VQ system.
(8) A properly designed system must cancel the echo when a 2100 Hz tone with phase reversal is sent through the R_{in} port. Accordingly, the S_{in} port recording indicates an echo signal that vanishes before exiting at the S_{out} port.
(9) If the system is configured to cancel echo when the 2100 Hz tone is either straight or phase reversed than step 8 must be repeated with a straight tone. If the system is configured to disable upon detection of a straight 2100 Hz tone, then the echo in S_{in} should not fade away at the S_{out} port, but rather stay at the same level.
(10) Repeat the same test without the echo test for the S_{in} and S_{out} ports.

Wireless modems

The ITU-T G.164 and G.165 recommendations prescribe general guidelines for operating, holding, and releasing an echo-canceler "disable state" upon detection of a 2100 Hz tone. One of the key recommendations proposes a maximum of 400 ms "wait-and-see" period after a significant drop in power,[34] before re-enabling the system and continuing with its normal operation of signal processing.

In today's world where wireless modems operate over very long processing delays, in addition to delays caused by data retransmission and dropped packets, there are many incidences where sessions are interrupted and line power drops for periods exceeding 400 ms while the session is still in progress. Not for long. A strict adherence to the G.164

---

[34] The recommendation specifies a drop by a minimum of 3 dB below the maximum holding sensitivity.

recommendation would enable the system after 400 ms and would cause a premature termination due to data corruption.

A non-standard mechanism could be put in place. This algorithm may prevent a re-enabling of signal processing even after there is a period of complete silence, exceeding 400 ms, for an unspecified length of time. We name this procedure voice signal classifier (VSC), and confine its objective to the re-enabling of a signal-processing system after a pure-tone detector has disabled it.

The VSC classifies a signal and assigns it to one of two states – **speech** or **unknown**. Unknown may be either speech or data, while speech is a more specific designation. Since the unknown state may be speech as well, the VSC may not be used for detecting a non-speech signal. It may not be used for disabling a signal-processing system upon data detection. It may only be used for re-enabling a disabled system upon detection of speech.

In the case of the wireless modem that goes silent for a period exceeding 400 ms during an on-going session, the VSC algorithm follows the following logic.

- The signal-processing system is in a disable state following a 2100 Hz tone detection.
- Pure tone is removed (by the modem) while data starts flowing through.
- The line experiences an energy drop for a period exceeding 400 ms while the system stays in a disable state.
- The signal processing system is enabled upon detection of speech.

While in a disable state the NR algorithm continues to update the noise estimate by buffering, windowing, and converting time to frequency domain via an FFT. It does not modify the signal. It refrains from attenuation of any noise suspect frequency before carrying out the IFFT, taking the signal back to time domain.

The key discriminator behind the classification is an inherent property associated with natural-speech signals. When examining the signal envelope over a short interval (conceivably a 16 ms interval), the VSC takes measurements of signal levels and the number of peaks and valleys (local maxima and minima that point to changes in slope) throughout the interval. A high signal level accompanied by a relatively small number of slope changes is an indication that the signal is a voice signal.[35] Anything else may be either a voice signal or a data signal.

Test sequence

(1) Enable all VQ applications.
(2) Prepare digital recordings of 2100 Hz tones, one with and the other without phase reversal followed by white noise at -40 dBm0.
(3) Play the recordings through the R_{in} port.
(4) Record the signals at the R_{in} and R_{out} ports.
(5) Emulate an echo signal at 6 dB ERL to be sent into the S_{in} port.

---

[35]   B. S. Atal and R. Rabiner, A pattern recognition approach to voiced–unvoiced–silence classification with applications to speech recognition, *IEEE Transactions on Acoustics and Speech Signal Processing*, **24**: 3 (1976), 201–212.

(6) Record the signals at the S_{in} and S_{out} ports.

(7) After 10 seconds remove the white noise from the R_{in} port and pause transmission.

(8) After 2 seconds reinsert white noise at the R_{in} port and record the signal on all four ports.

(9) A system properly designed for compatibility with wireless modems may still be disabled. A compatible system would maintain signal equality between R_{in} and R_{out}, and between S_{in} and S_{out}.

(10) Remove the white noise and insert a recording of speech at $-25\,$dBm0 mixed with a noise recording at $-35\,$dBm0.

(11) A system properly designed for compatibility with wireless modems would enable itself.

(12) Evidence of enabling may be unearthed through recordings of signals at R_{in}, R_{out}, S_{in} and S_{out}. After enabling check the following:

echo is to be removed between S_{in} and S_{out},
noise is to be reduced between R_{in} and R_{out},
average speech level is to be amplified between S_{in} and S_{out}.

Voice-band data on the A-interface

Voice-band data sent from a mobile station to the PSTN gets modulated in its analog form after being processed by the interworking function (IWF).[36] Signals traversing between the mobile station and the IWF are transmitted via a digital base-band. Voice-quality systems may not be able to detect $2100\,$Hz tones when placed on the A-interface, simply because tones do not exist in that part of the network. Voice-quality systems placed on the A-interface must be able to detect the commencement of data transmission by resorting to pattern recognition. This methodology is feasible since data present disparate characteristics compared with voice before being transformed into a voice-band signal by the IWF.[37]

Test procedure (lab)

(1) Prepare digital recordings of fax and HSCSD transmissions taken on the A-interface.

(2) Preparing the VQ system, enable all VQ applications.

(3) Transmit the recorded file through the R_{in} port.

(4) Record the output from the R_{out} port.

(5) Compare the input and output files.

(6) Except for the initial segment used for detection purposes, the input and output files should be identical.

(7) Repeat the process for the S_{in} and S_{out} ports.

Tandem-free operations

Network operators, whose networks support TFO, should test any VQS that is to be installed in their network for compliance. Testing a VQS for compatibility with TFO in a

---

36   See Chapter 7.
37   See Chapter 7.

Table 9.3. *All of the defined IS messages*

Message	Header 20 bits	Command 10 bits	Extension 1 20 bits	Extension 2 20 bits
IS_REQ	0x569A9	0x05D	0x53948	None
IS_REQ	0x569A9	0x05D	0x5394B	Possible
IS_ACK	0x569A9	0x0BA	0x53948	None
IS_ACK	0x569A9	0x0BA	0x5394B	Possible
IS_IPE	0x569A9	0x0E7	–	None
IS_NORMAL	0x569A9	0x0E7	0x00000	None
IS_TRANS_1_U	0x569A9	0x0E7	0x044DC	None
IS_TRANS_2_U	0x569A9	0x0E7	0x089B8	None
IS_TRANS_3_U	0x569A9	0x0E7	0x0CD64	None
IS_TRANS_4_U	0x569A9	0x0E7	0x11570	None
IS_TRANS_5_U	0x569A9	0x0E7	0x151AC	None
IS_TRANS_6_U	0x569A9	0x0E7	0x19CC8	None
IS_TRANS_7_U	0x569A9	0x0E7	0x1D814	None
IS_TRANSPARENT	0x569A9	0x0E7	0x22CE0	None
IS_TRANS_1	0x569A9	0x0E7	0x2683C	None
IS_TRANS_2	0x569A9	0x0E7	0x2A558	None
IS_TRANS_3	0x569A9	0x0E7	0x2E184	None
IS_TRANS_4	0x569A9	0x0E7	0x33990	None
IS_TRANS_5	0x569A9	0x0E7	0x37D4C	None
IS_TRANS_6	0x569A9	0x0E7	0x3B028	None
IS_TRANS_7	0x569A9	0x0E7	0x3F4F4	None
FILL	0x569A9	0x129	None	None
DUP	0x569A9	0x174	None	None
SYL	0x569A9	0x193	None	None

laboratory requires a TFO message simulator. A VQS compatible with TFO can recognize the beginning and end of a TFO session by monitoring the messages traversing its ports. When recognizing the beginning of a TFO session the VQS must disable itself and let the message pass through unprocessed.

Tandem-free operations messages (see Table 9.3) conform to the IS_message principles described in Annexes A and B of TS 08.62.[38] All IS messages follow a set of design rules, or a generic structure, designed to assist in identifying them with no need for detailed knowledge of the IS protocol served. Each IS message consists of an IS_header followed by an IS_command_block. Most IS messages have a number of further IS_extension_blocks. Figure 9.24 illustrates two IS_extension_blocks.

The TFO simulator must be able to generate TFO messages. It must include an encoder and a decoder. The TFO encoder program should let the user add TFO type messages to a stream, presented as an 8-bit-word binary file. The actual IS messages may be specified in a command file as a list. The command file may be represented as a text file.

The TFO decoder program extracts TFO messages from a stream, formatted as an 8-bit-word binary file, and writes them to a text file. The decoder performs four tasks: IS header

---

[38]   TS 08.62 *Inband Tandem Free Operation (TFO) of Speech Codecs; Service Description; Stage 3*, release 99 version 8.0.1.

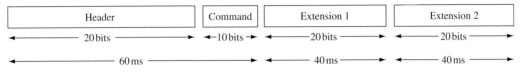

Figure 9.24 Example for IS message with two IS extension blocks.

detection, IS command detection, IS extension-block detection, and IS message identification and writing in the output file. Figure 9.25 presents a state diagram illustrating the operation principle of the TFO decoder.

The test set-up is depicted in Figure 9.26. The two computer icons may represent a single item of physical hardware with an encoder and a decoder program running on the opposite ends of the connection.

The test sequence involves the following steps.

Test procedure (lab)

(1) Prepare digital recordings of TFO text messages.
(2) Preparing the VQ system, enable all VQ applications.
(3) Transmit the recorded file through the R_{in} port.
(4) Record the output from the R_{out} port.
(5) Compare the input and output files.
(6) The input and output files should be identical.
(7) Repeat the process for the S_{in} and S_{out} ports.

Inter-system link protocol (ISLP)

The TIA-728 inter-system link protocol (ISLP)[39] is used for circuit-mode data services, including asynchronous data and group-3 fax.[40] The ISLP provides services that include rate adaptation between the A-interface data rates at speeds as high as 56 kbps. The ISLP synchronizes the output data rate to the physical-layer transmission rate by inserting flags between the protocol-data units (PDU).

A VQS on the A-interface must be able to detect the special structure of rate synchronized PDUs, intended to enable or disable the data service. Upon detection of ISLP rate synchronized PDUs, the VQS must cease processing at the start of a data session or renew processing at the end of it.

Testing for VQS ISLP compliance involves the following steps.

Test procedure (lab)

(1) Prepare digital recordings of ISLP data session, including ISLP rate synchronized PDUs.
(2) Preparing the VQ system, enable all VQ applications.

---

[39]   TIA-728, *Inter System Link Protocol (ISLP)* (January, 2002).
[40]   See IS-99 and IS-135, respectively.

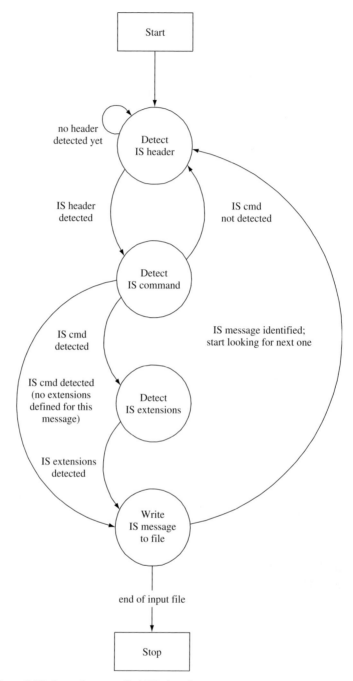

Figure 9.25 State diagram of a TFO decoder.

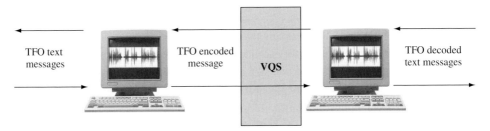

Figure 9.26 TFO test set-up.

(3) Transmit the recorded file through the R_{in} port.
(4) Record the output from the R_{out} port.
(5) Compare the input and output files.
(6) The input and output files should be identical.

Background music from mobile phones

Noise reduction and ALC may interfere with mobile phones that send low-level background music accompanying the phone call. Noise reduction may clip notes, distort, and attenuate phrases, while ALC may amplify the music during speech breaks only, and thus distort the dynamics.

Test procedure (lab)

(1) Prepare digital recordings of speech mixed with background music at the average relative levels used by the service. Add background noise at varying levels.
(2) Preparing the VQ system, enable all VQ applications.
(3) Transmit the recorded file through the R_{in} port, while inserting recorded speech at S_{in} to create double-talk segments.
(4) Record the output from the R_{out} port.
(5) Check for speech distortion, music distortion, and signal clipping resulting from the double-talk condition. Too much of any of these may yield dissatisfaction with the VQS.

9.10   Subjective evaluation through a simulated network

Subjective evaluation of VQ applications may include a laboratory simulation of network conditions. The following is an outline of a test procedure that facilitates an unpretentious feel for what VQ enhancements are capable of. A lab simulation provides a more effective demonstration than a field experience because the simulated environment is under the control of the tester.

Test procedure

(1) Enable noise compensation and noise-reduction applications.
(2) Connect the system to two telephone sets designed for direct connection into test equipment. Designate one side as wireless and the other as wireline.
(3) Ask one participant to speak on the "wireless" phone while another speaks on the "wireline" phone.
(4) Set up a boom box with CD recordings of street noise and let the wireless caller control its volume.
(5) Turn the CD player on and press play to set the traffic noise in motion. Ask the participant on the "wireless" side to vary the level of noise (using the volume control) during conversation to experience the ramping effects associated with the noise compensation feature. The wireline caller would experience the attenuation effect of the noise-reduction application.
(6) Ask the participants to switch sides and phones and repeat the session.
(7) Participants talking over a properly designed system would acknowledge their experience and provide proper feedback on the system performance.

9.11   Live network evaluation

When possible, live network-performance evaluations including voice recordings comprising non-intrusive automated procedures are the best way to determine the effectiveness of a voice-quality system.

9.12   Subjective listening with voice recordings

Subjective evaluations that sharpen the perception of the listener are best performed in a controlled environment where testers route calls through a supervised VQ system, and are able to turn VQ applications on and off throughout a test session. On and off control may be assumed by having the call monitored by an engineer in charge of the VQ system from within the MSC. The engineer monitors and listens in on the test conversation, while responding to verbal enable or disable commands by the testers. The engineer may also be in direct communication with the testers either through third party participation or by bridging onto the line to confirm their command execution. Figure 9.27 depicts the essence of the test set-up.

Recordings made on all four ports of the VQ system may accompany monitoring and controlling of calls. Recordings make for better and more thorough post-call analysis exposing underlying root causes affecting experiences gained by the testers throughout the call.

Field tests of the type described above are slow and tedious. They tie down resources other than the two testers, and they may not be performed in large numbers due to limited resources. An alternative (or supplementary) phase that opens doors to larger scale

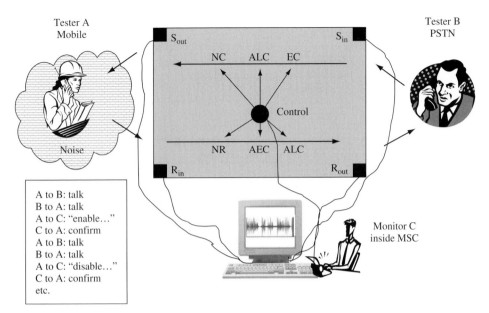

Figure 9.27 Voice-quality system under control of testers.

unsupervised, unmonitored, test calls may be accomplished if the VQ system builds in secret doors letting testers control voice-quality features via their handsets. Secret doors with confidential DTMF keys consisting of two or more DTMF digits may be useful for system testers for reasons other than enabling and disabling VQ applications in real time.[41]

In any event, the main requirement for placing field test calls is an ability to direct these calls into the VQ system.

Test procedure

Prepare a survey form (e.g., as shown in Table 9.4) for testers to fill out after every call. The following is an example of a survey form.

Survey instructions:

1. Place a call, noting on the survey form:
   - number called,
   - mobile-to-land or mobile-to-mobile,
   - time,
   - measured noise level,
   - location.
2. Chat briefly with the VQ application disabled, and rate voice-quality on both ends on a 1–5 scale.

---

[41] See Chapter 13.

Table 9.4. *Survey Form*

Who called whom, No. dialed	Mobile to wireline (MW) or mobile to mobile (MM)	Time of day	Phone manufacturer, location, detailed noise description:	Score with VQ application setting		Comments
				On	Off	

3. Enable the VQ application, and chat briefly. For the record, testers may want to discuss the voice-quality experience while chewing the fat.
4. Rate voice quality on both ends on a 1–5 scale, and record evaluation on the survey form, including user comments. Be sure to note user experiences of both the person placing and the person receiving the call.
5. Deactivate the VQ application.
6. Note additional user experience comments.
7. Toggle VQ application features on and off *at least twice* during the call – it makes for a bigger impact the second time around because both participants have a better idea of what they are listening for.

My personal experience with customer trials throughout the world demonstrated that subjective ratings of voice-quality scores were significantly different when testers were able to compare (in close proximity during the same call) their calls before and after the VQ applications processed the signal. When presented with a noise-reduced signal only, or a level-optimized signal without direct reference to the pre-enhancement, testers could not know what the improvement had been, and they judged the quality by focusing on the speech without understanding the trade-off that had occurred to make it prevail over the noise. Testers tended to rate VQ application effectiveness and the resulting voice quality much higher when they experienced the "before" next to the "after."

I also observed that most users tended to accept poor voice quality on mobile phones because they had not experienced better performance. Once they knew that superior quality was achievable and available, a rating of 3 on the MOS scale plunged to near 2. Once the quality bar moved up, the relative rating of existing (unimproved) voice quality moved down.

The capacity to provide a timely quality reference was a critical way to demonstrate to subscribers that improved wireless voice quality was not only possible, but also achievable. The ability to switch VQ applications on and off was the core tool behind that powerful realization.

10 Service provider's system, management, and delivery requirements

10.1 Introduction

Chapter 10 presents a basic template that may be used by service providers as part of their request for information from vendors. The chapter elaborates on the various elements beyond voice performance that make the VQS easy to manage and easy to integrate within the operation of the network. The information is rather dry, but highly useful as a reference. Readers of the book who are not interested in the system engineering and operational requirements may skip this chapter in their pursuit for understanding of the magic that make voice-quality systems enhance speech communications.

10.2 Management-systems overview

General requirements

Owing to business requirements, a centralized network-management system for VQS is necessary. A VQS management system is an integral part of the operations-support system, which provides the management backbone for the entire service offered by the mobile carrier. As a rule, a management-system architecture operates in a hierarchical structure, as shown in Figure 10.1. The lowest level in the hierarchy is the element-management system (EMS). It allows network operators to monitor the alarm status of individual systems from a centralized (remote) location; it reviews current and historical alarm records; it controls individual systems; it examines equipment-inventory information; it provides equipment parameters; it operates equipment functions; and it analyzes alarm information.

The second level in the hierarchy is the network-management system. This system collects information from all EMS in the network to provide a network view. The NMS, a.k.a. operation and maintenance control (OMC), may still be confined to managing a single element or a single element family.

The highest level in the hierarchy is the operations-support system, a.k.a. centralized OSS. This vehicle collects information from all network elements to offer a comprehensive network view. The system, more often than not, communicates via a standard protocol while the elements and their associated EMS and OMC communicate via a proprietary one. A message translation is frequently provided by way of a mediation device, which takes the proprietary message and converts it to the standard.

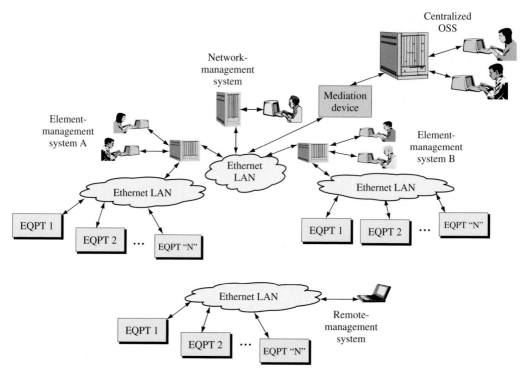

Figure 10.1 Management-system hierarchy.

In the following sections, I provide detailed requirements, which need to be supported by the VQS system, so that it can operate in the complex structure that makes wireless communication work.

Documentation and help tools

Documentation and help tools (e.g., online help) must be available.

Connections

The connection between the network element housing the VQ system and the adjacent external equipment must be implemented via standard interfaces and standard connectors.

Login

Remote and local logins may be identical.

Administration of local user profiles

It must be possible to administer user profiles from the operations and maintenance Control (OMC).

Repair

On-site repair must be done by replacing the faulty hardware unit and not by repairing it on site.

Online hardware changes

Faulty hardware units must be replaceable when the system is in service, i.e. neither power switch off of the affected rack(s), nor restart, nor reset of the affected network elements would be necessary during and after the repair.

Feasibility of hardware changes

A single service engineer should be able to take care of any hardware change.

Impact on adjacent units or functions

All necessary actions on faulty hardware units must not impact other parts of the VQS or any other system.

Inventory update

When replacing a hardware unit the inventory database must be updated automatically by the system.

Test functions

Every hardware unit must be tested by the system automatically before being put into service.

Test results

Test completions must be indicated on the user-interface screen. Failures must be reported to the OMC.

Unblocking

On successful checks and functional tests the system should unblock all blocked units and put them into service.

Procedure documentation

All maintenance procedures must be well documented.

Hardware addition

Automatic test functions must be triggered to verify and ensure that any hardware unit added to the system is compatible with the rest of the system.

Special equipment

Preventive or routine maintenance procedures must not require special equipment.

Software upgrades and replacements

Software upgrades must be able to skip over intermediate releases and be downloaded smoothly.

Software downloads

A VQS must be able to accept new software downloads distributed electronically from the OMC.

Compatibility check

The VQS must prevent incompatible software from being loaded and activated.

Provisioning retention

The VQ system must be capable of retaining all relevant data and parameters during downloads.

Performance during downloads

All uploads and downloads must be able to execute with no performance or functional degradation to the VQ application.

Interruption

If an upload or download is interrupted because of a link outage, then an alarm must be generated. The VQ system must be capable of re-initiating the download operation after connection is established.

Running/fallback version

It is recommended that after successful activation of the new software version the $n - 1$ running version is to be stored in the VQ system and become the fallback version.

Test procedures

Test procedures must be initiated automatically to verify that the activation has been successfully completed at the instant after the software has been activated. The final test results are to be sent to the OMC. Operators should be able to monitor the results as well.

Automatic fallback

If the software activation fails, an automatic fallback to the previous version must take place.

Interruption-free activation

Software upgrade must be accomplished without causing an outage.

Interruption-free fallback

If there is a download failure, fallback to the previous load must be accomplished without affecting any outage.

History log of software upgrades

A history database must contain all available software patches including status and the time stamp of their activation. For special investigations, the system should be capable of tracing and debugging messages. Traced data must be stored on the system disk. The traced data must be exportable. Help functions to decode messages online must be available. Online help must contain useful parameters or examples of messages.

Remote control

All operations including backup and restoration of the VQS databases must be capable of being executed from a remote OMC.

Remote restore and recovery

It must be possible to execute remote restoration and recovery in a single step.

Applicability

Remote backup and restoration functions must be able to address a single VQS or a group of selected systems.

Copying to an external storage system

It must be possible to copy the database from the remote-management system to another system without having to split the operation to more than one system.

Retrieval from an external storage system

It must be possible to retrieve the database from another storage system.

Local restore and recovery

It must be possible to restore a VQS database or to recover a VQS on site in one single working step, e.g., in case of a line outage between the OMC and the VQS.

Version control

The OMC must provide version control of all backups.

Compatibility check

The OMC must prevent incompatible backups from being loaded and activated.

Downtime

Minimum downtime for network monitoring and network operations must be guaranteed when OMC software is activated.

Total time

A complete installation, upgrade, or update measure may take no longer than an hour.

Reboot protection

Operation and maintenance control reboots must maintain integrity of all software patches.

Software listing

A tabular list of all installed software on the OMC must be easily accessible. This must include:

- operating system, including installed patches and the relevant patch levels;
- application software;
- third-party software (such as Netscape, SNMP agent, etc.) provided with the application software;
- software and hardware release levels of any firmware.

Software status

The status of all installed software components must be accessible, including status information and fallback version.

Procedures: upgrade, installation, migration, and fallback

The VQS must include a procedure for each individual upgrade or update including fallback. The procedures includes:

- general upgrade and update description;
- all detailed hardware changes, if applicable;
- all necessary preparations;
- each single working step.

Diagnosis procedure

The VQ system must be capable of performing automatically triggered as well as user-controlled diagnosis procedures. Diagnosis procedures take account of every hardware unit, databases,[1] and the OMC.

Impact-free operation

Running diagnostic processes must not interrupt the normal operation of any resource.

Automatic diagnosis

All diagnostic processes must run routinely in the background during normal operation.

Logging

Self-diagnostic processes are to be logged and to be available for post-processing at a later time.

Abnormal system behavior

An event message to the OMC must be initiated in case of any abnormal system behavior.

Self healing

Some administrations may insist that any detected fault initiates a fault bypass so as not to interrupt service. The OMC must be alerted for further action. Other administrations may

---

[1] The database diagnosis must comprise, but not be limited to, an integrity check of the data.

not opt for a bypass, but would rather let the line go down, and have the switch route calls to healthy circuits.

Alarms

An alarm must be generated if:

- an automatic recovery is not possible;
- an automatic recovery measure has been failed;
- a faulty unit has to be replaced after automatic switch over to a redundant one.

Active alarm list

The VQS must store a list of all active alarms in a non-volatile memory.

Automatic synchronization of potential data inconsistency

After link recovery, any potential data inconsistencies between the VQS and the operations and maintenance control (OMC) must be synchronized automatically.

Transaction logging

The OMC must provide data-transaction logging of user-activity logs and equipment activity.

Log file content

Log-file entries must clearly identify the source and the reason of the activity.

Manual deletion

The OMC must provide the ability to manually delete log files.

Log-file retention under restart

The VQS and the OMC system restarts may not have any impact on existing log files.

Retrieval function

The OMC and VQS must support flexible retrieval of data log for each of the required data-log types.

Data export

The OMC must provide an export facility with a common data format (e.g., ASCII) for all data-log files.

Search functions

The OMC must support search tools for all available data-log file types.

Sort function

The OMC must support a sort function on all data-log file types.

Sort active alarms

The OMC must support sort routines for all active alarms, using alarm attributes as sort criteria. The user's profile should include the ability to contain the sort criteria configuration.

Availability and resilience

Faulty line cards must not affect other line cards.

Restarts

A restart of a VQS must not take longer than 10 minutes.

Supervision of external equipment

External equipment (such as router, modems) need to be fully capable of being integrated in the fault-management architecture of the VQS (e.g., by providing a standard SNMP interface).

Internet-protocol fault management

Alarm and event notifications must be provided on the OMC in case of an IP fault within the VQS or within its IP connections.

Internet-protocol configuration application

The OMC must provide the ability to display all IP configurations of the VQS and its related transmission lines.

Heart beat

Transmission lines between an OMC and its connected VQS must include a heart beat mechanism with a configurable period.

Loss of heart beat

The detection of a heart-beat failure and the generation of the alarm notification must be available at the OMC within a maximum of 5 seconds.

Recovery

In the event of recovery from line or power loss, the VQS must automatically re-establish normal operation and connection. The recovery procedure must be completed within 5 seconds for short or unsteady-time loss and within 5 minutes for long-time loss. After the automatic re-establishment of a link or power, the related alarm must be cleared autonomously.

Access via logical OMC

The VQ system must allow the operator to perform all management functions connected to different physical OMCs from the one logical OMC.

Redundancy concept

All essential components must be redundantly implemented so that all tasks of a failed component can be taken over by another one automatically and instantly.

Configuration during operation

All configuration work, e.g. replacement of CPUs or extension of memories, may be performed while the OMC is in operation.

Environmental conditions

The upper limit for the permissible operating temperature of any hardware component must be at least 40 degrees Celsius. The lower limit for the relative humidity for any hardware component must be a maximum of 20%.

Operation and maintenance control capacity and performance requirements

The OMC must be capable of serving up to ten clients simultaneously.

- The OMC must support a minimum of ten user sessions simultaneously, without distinguishing between application and physical OMC. The VQS performance requirements must not weaken as a result of the type of sessions used, the number of active sessions, or the number of user terminals connected to the OMCs.
- The OMC capacity must allow connection all network elements to one logical OMC while maintaining the network architecture and performance requirements.

Bulk-data downloads and activates

Bulk-data downloads and activates must be terminated in less than 15 minutes.

Background operation for mass data

Mass-data handling jobs must be performed in the background, without impacting the operator on any other action.

Database swapping

An OMC database swap must be performed in less than 30 minutes.

Network-parameter configuration

The OMC must allow for modification and allocation of all network parameters of an OMC (e.g., IP address, node name, sub net mask, DNS, etc.) without any major operational restrictions.

Domain-name service (DNS) function

Applications residing on the OMC must operate under the domain-name service (DNS) function.

Peripheral administration

The OMC application must support the administration (e.g., addition, modification or deletion) of peripheral equipment (HMI servers, printers, clients, etc.) such that the equipment is fully integrated into the OMC application.

Administration of protocol interfaces

The OMC must let its operator administer the configuration files of used protocols (e.g., FTAM, CMISE, ATM, etc.) without any restrictions on the OMC.

Descriptions

Descriptions of all of the above-mentioned administration functions, in a task-oriented format, must be provided, including the layout of configuration files and their description.

Operation and maintenance control trace tools

The OMC must provide trace mechanisms for tracking faults within the interfaces (internal and external) and process communication (standard UNIX and vendor specific processes).

Operation and maintenance control external interface communication

The OMC must include an external interfaces trace and decoding mechanism for each layer of the used protocol.

Transcoder and internet-protocol tracing and decoding

The OMC must include TCP and IP trace tools and a decoding of the data-stream capability.

Tracing and decoding of application layer

The OMC must include a trace function, a decoder, and a configurable filter function for the application layer independently of the OMC protocol.

Indication collection

A script designed for collection of all relevant information in case of faults on the OMC must be included.

Operation and maintenance control user-interface functions

Response times for common OMC user-interface functions, e.g., the pop-up time of images or window masks, must be less than one second (<1 s).

Operation and maintenance control application functions

The OMC response times for application functions, including changes in the related VQS, must follow:

- user command or action must be concluded in less than 5 s,
- It must take less than 10–15 s to conclude a user command or action that sets off several changes within the OMC and the VQS databases,
- running command or action must have no impact on any other command or action performed in parallel.

Graphical user interface (GUI) for all applications

The VQS user interface should be easy to operate and provide effortless access to maintenance features. A commercially available PC platform (e.g., laptop) running the MS Windows operating system that is used to administer application specific VQ software is a good example of a user-friendly VQS user interface. The user interface should have a secure remote-access capability via the internet.

Access rights

The OMC application must be capable of listing all network elements, masks, windows, etc.

Task orientation

The most frequently used applications must be implemented in a task-oriented manner. Task orientation in this context means that the operator is automatically guided through the related windows, masks and menus without requiring expert knowledge.

Syntax, semantic and range check

For each space requiring a user entry, the system must perform syntax, semantic, and range checks, and pop up a message to the operator in case of non-compliance.

Multitasking

The OMC must support multi-tasking operations.

User environment customization

Users must be able to customize their own personal desktop (window arrangements and size, color, type size) and save the relevant settings.

Tree structure

The system must support the use of tree structures (e.g., analog MS explorer) to show dependencies.

Window title

The OMC operator must be able to administer a window's title.

Command-line interface

The OMC must be provided with a command-mode oriented (non-GUI) interface. This interface must support all GUI commands.

Operation and maintenance control jobs

The OMC must support the offline creation, modification, and deletion of jobs on the system for either the GUI or the command-line interface.

Job modification

The OMC jobs and batch jobs must be supplied with names or other identifiers. They must be capable of being retrieved, changed and then saved under a new name or other identifier.

Access rights

Jobs may only be created, modified and deleted by OMC users with adequate access rights.

Interrogation

The OMC may be able to interrogate and interrupt status of a running job manually and at any time.

Alarms

Interruption of a running job on the OMC must generate a related alarm.

Scheduling of jobs

The system must support job scheduling, i.e., the automatic activation at a predefined time.

Syntax, semantic and range checks

The OMC must perform a syntax, semantic and parameter range check for any job before activation.

Online help

The OMC must provide context sensitive help facilities on the user interfaces for all applications and functions.

Help information

The help information must include but not be limited to:

- descriptions of functions;
- descriptions of forms, fields, and related entries;
- task procedures for fault management, performance management, and configuration management;
- interactive messages, e.g., warnings.

Installation

The installation of updates or corrections must have no impact on running OMC processes.

Access to electronic documentation

The entire electronic system documentation must be capable of being accessed directly from the OMC user interfaces. This is independent of whether the documentation is located on the OMC itself or on an external documentation server.

Internet links

It must be possible to access the following information related to the VQS directly via the internet through the user interface:

- software releases;
- known faults;
- hardware restrictions;
- deviations from or restrictions to normal handling;
- frequently asked questions (FAQs).

Operation and maintenance control design requirement

The OMC must be designed so that the applications and the GUI do not require any hardware or software to be installed on the autonomous workstation.

Network printers

The OMC needs to control network printers connected via LAN.

Screen copies

The OMC applications must support screen copies performed by the user at the autonomous workstations.

Full and incremental backup

The OMC must support full and incremental backup procedures. The full backup must be a complete backup of all OMC servers, e.g., the backup of the operating system, the database, the file system, and the OMC application files.

Interworking of client with OMC

The OMC must include an interface to the client's networking software to ensure consistency of the backup of the OMC data. The OMC documentation must include scripts and procedures to support backup and restore.

Online backup

The backup procedure must include an online capability such that performance of other applications is not degraded during the operation.

Backup of clusters

The client module may be installed on all hosts or single hardware entities, which form a cluster. This way, all backup and restore functions can be applicable on a cluster configuration as well.

Provision of OMC northbound interface specification for configuration management (CM)

The OMC must support the integration of configuration management data by providing an OMC northbound open interface for CM so that external third party OSS may access its data.

Filtering

The OMC must support filtering on configuration data.

Availability of all configuration parameters

The OMC must support uploading and downloading of all managed configuration parameters.

Accuracy

All data retrieved by an upload operation must accurately reflect the current configuration and all previously performed changes of the VQS.

Fallback

The OMC must be able to fall back to the previous active configuration setting. The fallback must affect only those VQS that were subject to the previous operation. All other VQS must not be affected by the fallback.

File transfer

The file transfer must be compatible with standard protocols (e.g., FTP).

Documentation

The VQS and OMC offer must include a process description, indicating all activities (on site, and remote on the OMC) performed, including provisioning of mandatory parameters, provisioning of operator-defined resource identifiers, and online interactions during installation of new VQS.

Auto-discovery

The OMC must discover automatically the appearance of new VQS as soon as it becomes connected to the network.

Report details may be configured by the user.

Data retrieval

The OMC must include the facility to collect the configuration data from all VQS concerning time schedules, operator initiations, and controlled events.

Historical storage of retrieved configuration data

The OMC must retain a historical log of retrieved configuration for as long as required by the service provider.

Displaying configuration data

Data tables displaying the configuration data of the VQS must be provided.

Displaying network resources (NR) topology

A graphical overview of the different NR topologies (physical and logical) must be provided.

Data consistency

The application must perform syntax, semantic and range verification for all data changes. The application must verify dependencies between different parameters.

Generating delta files

The application must support the generation of delta files between the current data and the planned or modified data.

Plausibility verifications

The application must perform a plausibility or consistency verification before the activation of modified data. Warnings and errors that occur during a verification must be displayed and recorded in a log.

Generating command files

The application must generate editable command files directly based on delta files while command files must be stored in separate folders.

Plausibility check for command files

The application must perform a plausibility or consistency check before the activation of command files.

Downloading and activation of command files

It must be possible to download and activate command files in parallel into all or selected VQS connected to an OMC in an operator-defined sequence.

Import and export function

An import and export function of any file or directory from or to external devices (e.g., server, database, floppy disk) must be implemented, e.g., import of planning data, export of actual data. This functionality must be triggered via a time schedule or it must be operator initiated.

Support of software releases

The CM application must be compatible with the configuration data of the current and the previous release.

Performance and storage requirements

All configuration activities must have no impact on the traffic handling, on any non-affected VQS, on other management applications, and on operational activities.

Parallel processing

All operations on configuration data from one VQS to several connected ones must be performed in parallel.

Inventory management

All hardware, software, and firmware equipment must be uniquely identified.

Inventory information

The inventory information must consist of

- serial number,
- vendor's part id,
- mnemonic,
- physical location,
- time stamp of changes,
- voice-quality system generation.

Labeling of units

Hardware units must be identifiable by labels, which must be clearly visible and accessible by an operator while the unit or system is in use or operation (e.g., using a bar-code scanner and reading textual information).

The textual information must represent the exact content of the barcode, i.e., including spaces, special characters, etc.

The mnemonic on the label must be presented in plain text.

The label or labels on the unit must contain information about a serial number, a vendor's part ID and a separately labeled mnemonic. These labels must be easily distinguished from each other.

The bar code format must be a supplier independent and one-dimensional standard code, which must be mutually agreed between the seller and the buyer.

Common identification

The identification of all hardware, software, and firmware items must be commonly used in all written and electronic documentation, and in all manually or electronically produced inventory data (e.g., data files, compatibility matrix).

Compatibility matrix

A compatibility matrix must be supplied in machine-readable format (e.g., ASCII, XML), whereas updates must be supplied when changes occur to the matrix (e.g., with new releases, modified hardware, software, or firmware).

System handover

The following document and information should be provided.

- The entire system backup, including all necessary configuration data sufficient for total system recovery.
- System-design document.
- System-integration test cases and results.
- Operation and maintenance manuals.
- Procedures for checking system health.
- Materials (hardware and software) with appropriate licenses and serial numbers.
- Equipment inventory with appropriate licenses, serial numbers, and quantities.
- Contact information for technical support and maintenance.
- Any additional software must be available free of charge and without license restrictions for the customer, including all software revisions and upgraded versions.

Storage at the OMC

The OMC must be capable of storing the inventory data of all its connected VQS.

Automatic update of inventory data

Inventory information must be compiled automatically during installation of equipment. In the event of any repair or modification, the inventory must be automatically updated.

Acquisition of inventory information

It must be possible to distinguish between active and passive units in the inventory database and all related documentation.

Automatic update

In case of an VQS modification (e.g., repair) all inventory-data changes must be updated automatically (in the VQS, OMC) and forwarded to a supplier-independent OSS/NMC.

Time stamps

Every update of inventory information must be time stamped.

Data synchronization

Synchronization between all VQS connected to an OMC and the OMC itself must take no longer than 1 hour.

Performance

Inventory functions must not impact the performance of the OMC or the traffic handling capacity of the NE.

Remote access to inventory

The export inventory interface must provide a mechanism for uploading inventory data from the OMC/VQS in a standard file format (e.g., XML), whereas the file format must be independent of the data-transfer protocol.

The fault management (FM) northbound interface of the enterprise OSS and the VQS/OMC functions

Data recovery

In the event of communication loss between the OMC and OSS, the OMC must store events in a log file to upload those event notifications in chronological order as soon as the communication is recovered. This log must hold data for 48 hours.

Active alarm list

At any time, the OMC must provide an "active alarm list" (a list of the current alarm conditions on the VQS). If requested by the external OSS, the alarm list must be made available to the OSS. Such events must be identified as coming from the alarm list.

Recovery notification

To support the detection of the link outage the OMC must supervise the link to the OSS and generate an event notification in case a link outage occurs. After link recovery, the OMC must notify the OSS about the recovery, providing time stamps for the beginning and end of the link outage in order to retrieve the alarm log files.

Heart beats

The OMC must provide a heart-beat event to the enterprise OSS. The period must be configurable with a nominal setting of 15 minutes.

Latency

All event notifications from the VQS or OMC must be received by the OSS within 10 s of being generated on the VQS. In cases of alarm correlation within the OMC, the event notifications must be received by the OSS within 10 s of being generated on the OMC.

Event forwarding and filtering

The definition of event forwarding discriminators from the OSS must be supported by the OMC.

The OMC must allow the operator to filter events forwarded to the OSS selectively.
If specific events are not forwarded over the external OSS FM interface, the OMC must support the creation of system alarms (e.g., based on the occurrence of threshold crossing in relation to certain events), which should be forwarded over the external OSS FM interface.
The event filtering from the OMC towards the OSS must be independent of any alarm or event filtering on the OMC.

Event logging

The OMC must support the creation and storage of log files triggered from the OSS. The creation and configuration of one or more logs must be supported.

Alarm clearing

For every alarm on the northbound OSS FM interface, an alarm clear has to be provided on that interface in the event that the alarm has been cleared on the VQS/OMC.

- Alarm clears generated at the OMC must be conveyed to the OSS and vice versa, to keep the clear states synchronized between OMC and OSS.
- The alarm clear must contain the notification identifiers of the associated alarms in the attribute "correlated notifications."

State management

The states of resources kept in the OMC and forwarded to the OSS must allow in case of redundancy a clear distinction between the active, online, busy, and standby states of the resource.

Alarm reservation

The application must allow the OMC operator to mark single or multiple alarms as acknowledged.

- The alarms must be identified as being owned by a specific OMC operator.
- The allocation must be achieved by a single action.
- The system must allow authorized staff to cancel the alarm allocation.

Alarm-severity modification

The application must allow the modification of the alarm severity (indeterminate, warning, minor, major, or critical) for individual or multiple alarms by an authorized OMC operator.

- The alarm severity must be modified by a single command.

Configurable command-acknowledge button

The application must support the definition of thresholds (e.g., parameter value, alarm severity) so that if those thresholds are crossed a request for acknowledgement is presented to the operator at the OMC.

Note: it must be possible for the OMC operator to configure the thresholds.

User comments on alarms

The system must allow the OMC operator to create, modify, or delete specific alarm comments, where the comments must be associated with the alarm, e.g. each time the same fault occurs the comment may be linked.

Latency

The OMC must reflect all fault detection and state changes on the VQS within two seconds of the event occurring.

Event filtering

The application must support the filtering of events by type, severity, and state, VQS or NR, where combinations and specific rules must be supported (e.g., time between subsequent faults, number of subsequent faults, etc.).

Event suppression

The application must support event suppression.

- All event suppressions must be indicated for the related VQS, e.g., special symbol, color, etc.
- Suppressed events must be retrievable.

Event correlation

Alarms derived from subsequent events should be suppressed. Only root-cause events must be displayed.

Event thresholds

The application must support the definition of thresholds for each event category. An alarm must be generated if such a threshold is exceeded. Thresholds must be configurable by the OMC operator.

Active alarm counter presentation

The application must present summary counters to the user for all alarms that exceed users defined alarm severity.

Unambiguous alarm description

The alarm text must clearly identify the problem and the cause.

Unambiguous fault localization

The fault source must be clearly identified as part of the alarm description.

Alarm notification procedures

The VQS should categorize alarms as either equipment alarms or facility alarms. Alarm conditions should be displayed locally on the VQ equipment in a logical manner (e.g., alarm indicators physically mounted on corresponding circuit packs) while also providing an interface to drive remote or centralized alarm displays. Alarm information should be conveyed in a meaningful manner that adheres to standard colors, terminology, and conventions. The VQ equipment should also allow network operators to provision alarm severity levels. In addition to visual alarms, the VQ equipment should include an audible alarm capability, an alarm cut-off (ACO) feature for silencing audible alarms, and an alarm test feature for exercising the alarm notification functions.

Task-oriented fault-clearance procedure

The task-oriented fault-clearance procedure must provide sufficient guidance for the OMC operator.

- For each alarm it must be possible to directly access predefined task-oriented fault-clearance procedures.
- The specified procedure must guide the OMC operator in successfully clearing the fault.

Link from alarm to help text

For each alarm, a direct access to a related help text must be provided. The help text should provide support and advice to the operator concerning the steps required to clear the alarm.

Direct access from alarms to maintenance functions

There should be access to related maintenance functions for each alarm.

Automatic fault-clearing (self-healing) statistics

A summary count of automatically initiated fault-clearing actions must be provided.

Local access

The system must incorporate a security function whereby unauthorized access is prohibited. Security must be provided by means of user account and password. It is desirable to have different authority levels.

Remote access

Maintenance personnel must be able to access the VQS remotely. User identities must be verified before allowing access to the system. It is desirable to have different authority levels.

Diagnostic test

The VQS must be able to run automatic diagnostic tests periodically, without service interruption on each individual traffic channel in the whole system, regardless of whether it is in the busy or idle states. In case the test fails, an associated alarm must be raised. Other diagnostic features, such as a power-up diagnostic test and a loop-back test are also desirable. Maintenance personnel may initiate the diagnostic tests manually for particular parts of the system. A detailed result must be provided at the end of the diagnostic test.

Alarm reporting

When the VQS generates some alarm messages, they should appear on the system consoles. Different classes of visual or audible alarms must be available and extendible to the network management center for alarm supervision. The network management center collects all alarms from different systems in real time. It allows the establishment of remote access to different systems via itself.

All system alarm messages are to be stored and retrieved for fault tracking. Housekeeping is necessary and should be performed regularly, to minimize their size, providing sufficient hard-disk space for system operation.

Interfacing and protocol

The VQS may provide both ethernet and serial interfaces for communication with the network-management system. The communication protocols between them should be transcoder or internet protocol and cannot be proprietary.

Network-analysis system

The VQ system should provide a reporting feature measuring the voice-quality performance on the input and output interfaces of the equipment. The measurements should include echo-tail statistics, ERL statistics, speech levels, and noise levels (before and after processing). In particular, the network-analysis system should:

- allow telecom operators to monitor and analyze the performance of their network;
- collect and store data based on parameters retrieved from VQS products;
- generate reports that summarize network operational characteristics;
- provide "clues" for identifying or resolving errors;
- allow equipment provisioning and configuration errors;
- indicate unusual network usage patterns;
- point to degraded performance or quality of service (QoS) issues.

The network analysis system management capabilities should be capable of:

- collecting VQS parameters;
- provisioning the monitoring of particular T1/E1 facility signals for a desired time interval;

- collecting and storing relevant parameter values extracted from VQS products;
- communicating with VQS products over the TCP/IP;
- exporting data files for long-term storage;
- generating tabular and graphical reports for performance analysis and troubleshooting.

Upgrade and enhancements

Voice-quality technology continues to evolve at a rapid pace. Given this condition, an important feature required by network operators is a flexible VQ platform that can easily be upgraded after field installation. The field upgrade should not be restricted to changes in parametric values. It should also include easy deployment of new features and enhanced algorithms for existing applications. The presence of robust digital signal processors (DSP) in the VQ system's architecture makes this feasible.

The VQS should support field upgrades in either software downloading or hardware upgrading, to allow easy deployment of new features and product enhancements. It should be capable of solving any known faults while providing feature enhancements. It should cause no service interruption due to hardware and software upgrading.

Power and environment requirements

The VQS must meet the following requirements.

- A single power supply failure should not take down more than four T1/E1 lines, if any.
- A stand-alone system must accept dual-feed power sources.
- A stand-alone system must include fuse protection for each power source.
- Average power consumption cannot be greater than 130 mW per channel including the fan tray.

Physical-damage test results

The VQS vendor must supply test results for the following physical tests to verify that the VQS is compliant with standard procedures. Tests include:

- **Office vibration** – equipment operation during office vibration.
- **Sine survey** – equipment operation before and after vibration.
- **Seismic vibration** – equipment operation during seismic vibration.
- **Transportation vibration** – equipment operation before and after vibration.
- **Package drop** – equipment operation before and after being dropped during shipment.
- **Installation drop** – equipment operation before and after being dropped during installation.

Table 10.1 provides a list of standard documents that specify the test procedures and the criteria used.

Table 10.1 *List of standard test documents*

Environmental compliance	
ETS 300 019-1-2, CLASS 3.1	*Environmental Conditions and Environmental Tests for Telecommunications Equipment*
GR-63-CORE	*Network Equipment Building System (NEBS) Requirements: Physical Protection (Seismic Testing and Temperature and Humidity Tests)*

Table 10.2 *List of EMC tests*

EMC compliance	
ETS 300 386-2 V1.1.3	Electromagnetic compatibility and radio-spectrum matters (ERM); telecommunication-network equipment; electromagnetic-compatibility (EMC) requirements. part 2: product-family standard
ETS 300 386-1	Equipment engineering (E); public telecommunications-network equipment electromagnetic compatibility (EMC) requirements. part 1: product-family overview, compliance criteria, and test levels
EN 55022	Limits and methods of measurement of radio-interference characteristics of information-technology equipment
EN-61000-4-2	Electrostatic compatibility (EMC) part 4: testing and measurement techniques. section 2: electrostatic-discharge immunity test
EN-61000-4-3	Electromagnetic compatibility (EMC); part 4: testing and measurement techniques. section 3: radiated radio-frequency, electromagnetic-field immunity test
EN-61000-4-4	Electromagnetic compatibility (EMC) part 4: testing and measurement techniques; section 4: electrical fast transient/burst immunity test
EN-61000-4-5	Electromagnetic compatibility part 4: testing and measurement techniques; section 5: surge immunity test
EN-61000-4-6	Electromagnetic compatibility part 4: testing and measurement techniques; section 6: conducted disturbances induced by radio-frequency fields

Electromagnetic compatibility

Voice-quality system vendors must provide EMC test results. The test results should confirm that the VQS equipment has passed all environmental tests, and reflect the extent to which the VQS has met each requirement (i.e., indicate margin with respect to the minimum threshold). The specific tests that should be performed are described in Table 10.2. The results of environmental testing should be a prime consideration for network operators with networks deployed in areas prone to earthquakes or similar environmental conditions.

The EMC test report should specify the requirements and the VQS performance for both emission and immunity tests. Complete test configuration descriptions should be provided, and the test results should reflect the extent to which the VQS has met each requirement (i.e., indicate margin with respect to the minimum threshold).

Table 10.3 *Standard procedures for temperature and humidity tests*

Environmental compliance	
ETS 300 019-1-2, CLASS 3.1	*Environmental Conditions and Environmental Tests for Telecommunications Equipment*
GR-63-CORE	*Network Equipment Building System (NEBS) Requirements: Physical Protection (Seismic Testing and Temperature and Humidity Tests)*

Electrical safety test results

Electrical safety test results should follow the standard procedures outlined in EN 60950 (*Safety of Information Technology Equipment, Including Electrical Business Equipment*).

Temperature and humidity

Temperature and humidity test results must follow standard procedures outlined in Table 10.3.

10.3   Customer support

The period must include 24 hours per day; 7 days per week, and response time should be within 4 hours. The VQS vendor must describe in detail the support agreement and fault handling procedures.

The VQS vendor must recommend a spare quantity for a certain amount of hardware purchased and pricing quotation for the recommended spares.

The VQS vendor must provide replacements over the local support center for the defective units within the warranty period. The replacements must carry the same warranties as the defective items.

The VQS vendor must provide sufficient information about supporting, e.g., customer reference, number of supporting staff and company financial status, etc.

10.4   Training

The following items must be included in the training:

- VQS basics – hardware and software,
- system operation and troubleshooting,
- system backup and restore skill,
- training documents,
- additional information request by trainee.

11 Making economically sound investment decisions concerning voice-quality systems

11.1 Introduction

Chapter 11 discusses key differences between stand-alone and integrated systems. It points to the pros and cons of investing in a full VQS suite versus a minimal set containing a hybrid-echo canceler only, and it closes with a simple model providing guidelines for assessing return on investment.

In recent years, stand-alone voice-quality systems have been losing share to voice-quality and echo-canceler modules that have been integrated within mobile-switching centers (MSCs) even though some of these integrated systems have demonstrated poorer performance, and most have not included voice-quality applications beyond echo cancelation. Still, there have been pockets of trend-defying markets that continue to pursue stand-alones, and some companies have even installed them in place of existing integrated systems.[1]

This chapter brings to light business considerations associated with investment decisions by service providers who look for return on their investment in VQS. An important step in the determination process is whether the VQS is to include functionality in addition to hybrid-echo cancelation, and whether that functionality is to be delivered via a stand-alone or an integrated system.

Accordingly, the first part of the chapter contrasts the merits and drawbacks of using stand-alone versus integrated systems. It describes the various trade-offs that must be considered during the process of determining which configuration (stand-alone versus integrated) is best suited for a specific application. The analysis involves considering state of the art performance and feature richness, cost, network growth, operations, and codec implementations. The second part of the chapter analyzes the potential business benefits brought about by a full voice-quality suite versus a minimal functionality consisting of echo cancelation only.

11.2 State-of-the-art performance and feature richness

Although it has not been a rule, it has been generally true that stand-alone voice-quality systems performance used to be superior to their integrated cousins. This has been because

---

[1] T-Mobile has done so throughout Europe.

vendors of stand-alone systems have been viewing their product as the heart of their venture, while those who incorporate them into their switch view them as a secondary-feature set. The effort and testing that go into the algorithm design and implementation by the stand-alone vendors is usually more substantial than the effort allocated to the same function by the switch vendor. There are, however, exceptions. When a stand-alone system vendor buys core technology and algorithms from another vendor,[2] the other vendor may make the technology available to anyone who would pay their price. That includes switch vendors who integrate voice-quality applications into their system.

Yet, there is one more crucial factor behind the fact that stand-alone systems have been one or two steps ahead of switch-integrated voice-quality applications. Switch-feature upgrades are generally performed as part of a new switch generic release. New technology and new features take a relatively extended pilgrimage through design, implementation, testing, and integration prior to being incorporated into a new switch generic. Furthermore, every new feature must be evaluated and approved before it is declared available for commercial use. This strict adherence to process is necessary because of the highly complex nature and sheer magnitude of switching systems. Since a switch-integrated echo canceler and voice-quality enhancement is considered an inherent switch feature, it must endure the same ratification process before it is deployed in field applications. In comparison, stand-alone VQS are not encumbered by the overhead associated with large switching complexes and they tend to champion newer technology, innovative designs, and unique features that are delivered to customers in a more efficient and timely fashion.

It is also a matter of record. As of this writing, the majority (with some notable exceptions)[3] of the existing switch-integrated systems contain limited functionality confined to hybrid-echo cancelation, whereas all major brands of stand-alone systems embrace an entire suite of voice-quality applications, including hybrid-echo cancelation, acoustic-echo control, noise reduction, and level optimization.

11.3   Cost considerations

Stand-alone EC configurations

Stand-alone VQS can be connected to a switch in either a dedicated or pooled (shared) resource configuration. A pooled configuration may have cost advantages in cases where the number of calls (through the switch) that require voice-quality processing is, statistically, a good deal below 100%. That is, the characteristics of the traffic handled by the switch will never require echo cancelation or other voice processing for every call simultaneously. Accordingly, a dedicated echo-cancelation function will be significantly underutilized. However, as the number of calls requiring echo cancelation increases,

---

[2] Ditech Communications acquire their echo-canceler algorithm technology from Texas Instruments.
[3] Lucent has introduced a broader set of voice-quality applications in their 5ESS wireless switch. Nortel has incorporated (into their wireless switch) technology from Tellabs that include many voice-quality applications in addition to echo cancelation. Nokia added voice-quality applications as an option on their base station. Mobile-switching centers from Siemens, Alacatel, Ericsson, Samsung, and older releases of Lucent, Nortel and Nokia still limit their integrated voice-quality offering to echo cancelation.

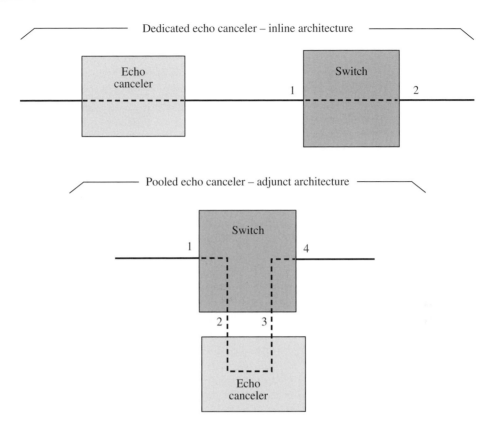

Figure 11.1 Dedicated versus pooled; stand-alone EC architectures.

a corresponding higher level of shared-resource utilization occurs and the pooled configuration loses its advantage. This is primarily because the pooling arrangement requires additional (expensive) switch resources. As shown in Figure 11.1, every pooled echo-canceler function requires two additional switch ports. In summary, the cost savings gained by deploying fewer echo cancelers in a pooled arrangement may be overshadowed by the cost of providing more switch resources (i.e., expensive switch ports) needed to support this configuration.

Switch-integrated EC or VQS configurations

While stand-alone VQS offer flexibility in the way they are connected to a switch, a switch-integrated function is subject to relatively static configurations. Two common architectures are used: either pooled or dedicated.[4] It should be noted that the dedicated (fixed) configuration could become costly for some applications.

---

[4] For example, most Ericsson switches employ a shared echo-cancelation function. Lucent 5ESS wireless switches have a dedicated echo-cancelation function.

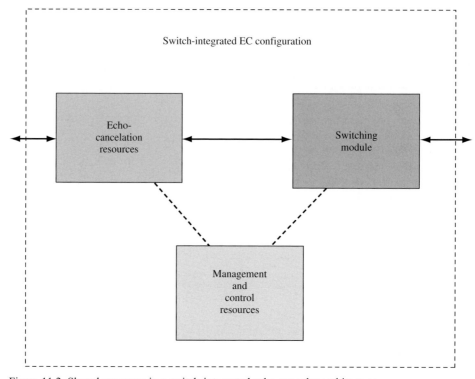

Figure 11.2 Shared resources in a switch-integrated echo-canceler architecture.

In some cases, switch-integrated echo cancelation bears the disadvantage of displacing switch or signal-processing capacity (i.e., echo-canceler and VQS hardware may replace switching or signal-processing functionality), which has a negative impact on overall switching-system cost, efficiency, and performance. That is, decreasing switching or signal-processing capacity, while maintaining the overall system expense as a constant, results in an increased per-channel cost factor.

Even when switch-integrated echo cancelers or VQS utilize a dedicated architecture, they may share other pooled functions (e.g., management and control), as shown in Figure 11.2. In some particular cases, the sharing of pooled functions can become saturated due to growth and can set off an internal contest for resources. Consequently, the opportunity cost of these particular switch-integrated echo canceler or VQS configurations is essentially equivalent to the cost of expanding switch capacity.[5] Furthermore, when fully analyzed the switch-expansion cost may prove to be higher than the cost of a stand-alone echo-canceler arrangement.

In general, some voice over packet (VOP) terminals are designed with universal slots that can accommodate different circuit-pack types on a mutually exclusive basis, within the

---

[5]   Ericsson MSCs provide one particular example. An older version of Lucent 5ESS is another.

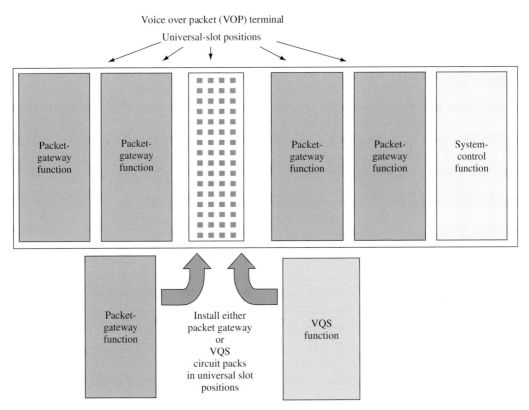

Figure 11.3 Typical VOP terminal with universal-slot architecture.

same scarce physical space. Innovative technology has made this efficient and flexible system architecture possible. Accordingly, each VOP terminal can be configured with a customized mixture of circuit packs that is tailored to deliver the specific functions needed for a prospective customer's application.

It is possible that one of the potential functions residing within the universal slot(s) of a VOP terminal is echo cancelation or a fully featured VQS. For example, a VOP terminal configured with echo cancelation could be deployed during initial installation when network growth is often limited (i.e., in the start-up phase). As the network expands and VOP-terminal capacity begins to approach exhaustion, the universal slots containing the VQS functionality could be freed up and replaced with more packet-gateway functions to support continued growth (see Figure 11.3). In this scenario, an external stand-alone VQS could handle the displaced VQS functions. This practice may bring about an economic benefit in the form of a lower average cost per channel,[6] while also deferring the high cost of installing a second VOP terminal to handle network growth (i.e., the additional traffic is carried by the original VOP terminal).

---

[6] In this example, the cost of the VOP terminal is distributed over a greater number of channels, which is a result of replacing the VQS function with a packet-gateway function to increase system capacity.

11.4   Network-growth considerations

Network growth may be accommodated with less effort and may be less expensive when a stand-alone system configuration is deployed. Network planners are not compelled to acquire excess switch capacity before it is needed. Building a new network or expanding an existing infrastructure using stand-alone systems may be better phased-in over time, whereas switch-integrated features are typically purchased and installed at the outset even though the start-up traffic load may be relatively light. Consequently, capital spending might be more efficient in some cases of stand-alone VQS. Of course, as always, the conclusion above is based on particular topologies. There are exceptions to the rule, and a different case may dictate an unlike judgment.

11.5   Operations and space

Operations happen to be the category where integrated VQS shine over the stand-alone systems. Switch-integrated VQS are managed by a common management system and a common user interface. They are fully integrated within the operations system (OS), and do not require a separate workstation. The switch vendor is responsible for supporting them as part of the switch support, and the possibility of finger pointing in case of trouble is minimized. What's more, stand-alone VQS devote far less than 25% of the space they use to core VQS functionality. The rest is used for system operations, transmission, interfaces, general management, power supplies, and housing. When integrating VQS inside a switching system, most of the transmission overhead is shared with the switching system. Accordingly, the total space used for switching and VQS is reduced considerably when the VQS is integrated inside the switch.

11.6   Conclusions

As technology becomes better and more compact, and performance attains a reasonably high-quality steady state, incorporating voice-quality applications into an MSC or wireless codec is becoming less costly and more effective. Although the performance of integrated VQ applications used to lag behind that of stand-alone cousins, it has been approaching parity with them, while achieving great advantages when it comes to ease of operations, cost, and space savings. As DSP technology becomes highly dense, enabling many more millions of instructions per second (MIPS) on a single chip, they have become capable of processing more traffic and more features at the same time. Many implementations of integrated VQ applications have been sharing DSP resources with codec as well as other applications, and many MSC designs have not found it necessary to dedicate complete switch modules to voice quality. They tend to mix the application with other functions on the same hardware.

At the same time, the latest codec implementations, as well as extra features inside many mobile handsets, offer noise reduction, adaptive level control, and acoustic-echo control.

Once again great advances in DSP technology make it possible to package these VQ applications in a teeny corner of a small appliance.

Integration makes it simple for the operator and, naturally, it becomes increasingly popular and less costly when considering total network cost.

11.7   Fully featured VQS versus echo-cancelation only

In survey after survey, it has been shown that when asked about voice quality people tend to place it at the top of their list when ranking wireless-service providers. At the same time, service providers obtain feedback on their customers' perceived voice quality by analyzing customer complaints.

When customers take notice of their own echo, they tend to blame their service provider for not taking care of the problem. When customers are subjected to noise accompanying speech from the other end, they blame the environment. By and large, they do not issue a complaint. And when speech levels are too low, customers tend to blame the mobile phone rather than the service provider.

Most wireless-service providers do not see evidence of either noise reduction or level control being an important component of their customer-satisfaction matrix, simply because customers do not register many complaints addressing these problems. At the same time, they do see evidence of echo control being a major contributor to customer satisfaction because customers do issue immediate feedback when echo is not controlled.

When service providers introduce voice-quality applications in addition to echo control, they had better market it at full volume if they look for appreciation. Return on investment may only be achieved if the introduction of the VQS other than echo control is accompanied by a considerable promotion.

Increasingly, voice-quality applications other than hybrid-echo cancelation have been placed at the endpoint, either inside a handset or as a standard feature part of the codec (also inside the handset). Noise reduction has been a feature of CDMA since the introduction of EVRC. It has also been introduced to many of the GSM mobile phones by the phone vendors as a non-standard feature of the EFR and the AMR codecs. Acoustic-echo control has been incorporated into many mobile phones. Some of these implementations have not been well designed, and when surrounded by high noise some mobile-phone built-in AEC have caused more harm than good, due to noise clipping (see Chapter 12).

Hybrid-echo cancelation is best accomplished inside the network since it is caused by a network element (hybrid) that is beyond the reach of the endpoint. This echo-cancelation application – in contrast to NR, AEC, ALC, and NC – may not move into the mobile phone as time goes by.

One of the major factors impacting the quality of the transmitted voice is the amount of noise surrounding the end-points of the conversation. Noise has been shown to have a far-reaching impact on HR versus other codecs. The perceived impact is amplified considering that noise surrounding an HR–HR (mobile-to-PSTN) enabled session reduces DMOS to the range of "annoying" to "very annoying" (≤ 2; without and with errors, respectively) and from highs of "not annoying" to "slightly annoying" (3+). It should be noted that noise interferes with other codecs, but to a lesser degree, as the perceived DMOS may deteriorate to a rather higher range bordering on "slightly annoying" (3+ on the DMOS) in the absence of transmission errors, and to "very annoying" in the presence of errors and noise.

Even a slight enhancement in DMOS may be perceived as exceedingly worthy provided that voice quality is already at a low level due to noise or transmission errors. Evidently, this conclusion applies to situations involving mobile-to-mobile communications in general, particularly where one end employs HR.

European service providers have been, by and large, resisting temptations for deploying HR even though great savings in spectrum cost could have been realized. The reduced voice quality in the presence of noise was a price too high to pay for these service providers. Some Asian countries, on the other hand, have experimented with HR while complementing its deteriorating performance under noisy conditions by incorporating NR algorithms that tend to elevate the voice quality to a more acceptable range. The process may continue until the AMR codec assumes a more ubiquitous hold over the subscriber's landscape. The AMR codec (as detailed in Chapter 1) includes several half-rate versions, most of which perform better than HR under a variety of conditions as well as those that include acoustical noise.

In a survey commissioned by NMS Communications,[7] and run by Isurus Market Research, a research firm that specializes in helping technology providers, it was shown that network sound quality might have a notable impact on customers' perception concerning the importance of voice quality. In a phone survey of 440 active mobile subscribers throughout the USA, respondents cited voice and sound quality as differentiators that could affect their mobile-phone usage and choice of operator. NMS also reported on their web site that 72% of respondents indicated that they would take and place more calls in noisy environments such as their cars, in restaurants, or on city streets if the quality of those calls was improved.

NMS reported that other research findings included the following.

Thirty-eight percent of the study's participants feel their mobile-phone quality is not as good as the quality of their landline phone.

When bothered by poor voice quality, most people choose not to complain, as it isn't "worth the hassle," or presume there are technology limitations that can't be addressed.

One quarter of respondents indicated that they would consider paying more for improved voice quality.

It may be possible to derive a return-on-investment (ROI) model that generates an estimate of present value, given certain assumptions concerning customers' behavior at the conclusion of upgrading network voice quality with noise reduction and level-optimization applications.

The ROI model needs to take into account an estimate of the additional revenue due to an increase in average rate per user (ARPU), estimates of the potential reduction in churn, and any increased utilization in spectrum, thanks to improved voice quality.

---

[7] See www.nmscommunications.com.

An increase in ARPU may be converted to an increase in revenue per T1/E1 transmission facility. Since the cost of enhancing voice quality is best measured against the cost of implementing new technology per T1/E1, it is possible to derive direct net margin off the upgrade by multiplying the margin figure by the number of upgraded T1/E1. Indirect gains derived from churn reduction and spectrum savings may then be added to obtain an overall rate of return.

Part V
Managing the network

12 Trouble shooting and case studies

12.1 Introduction

This chapter presents an extensive account of conditions that must be accommodated for healthy voice communications to result. The presentation is carried out as if the reader is in charge of running a mobile-switching center where all equipment, including voice-quality assurance gear, has been operating satisfactorily up to that moment in time when the specific trouble at issue has been reported. The specific troubles are analyzed for root cause and remedial actions.

The chapter includes a list of 40 cases. Some of these cases are supported by example recordings (www.cambridge.org/9780521855952), which illustrate certain aspects of the case in more detail.

12.2 List of troubles

Just installed a new upgraded EC and there is echo

Probably one of the most common complaints: "I have just upgraded my network and installed stand-alone echo cancelers and voice-quality enhancement applications. Immediately after, customers from all over started complaining about echo. My older EC did not have all the bells and whistles, but was canceling the echo most of the time."

A common root cause: the newly installed echo cancelers were mounted backwards. The receive and send port connections to the network were reversed. Echo returning from the PSTN enters the receive side instead of the send side (see Figure 12.1).

Just moved the stand-alone EC from the PSTN interface to the A-interface and there is echo

A common mistake: "I replaced the echo canceler to PSTN connection with a connection to the TRAU. I followed the same procedure and steps I had gone through when installing it on the PSTN side."

On paper, the high-level view makes the connections seem the same (see Figure 12.2). However, relative to the MSC, the wires must be reversed. When installing an external echo canceler on the PSTN interface, the MSC connects to the receive side of the external

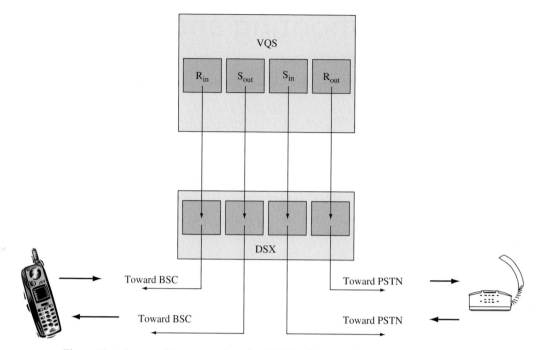

Figure 12.1 Proper wiring connections for PSTN and A-interfaces.

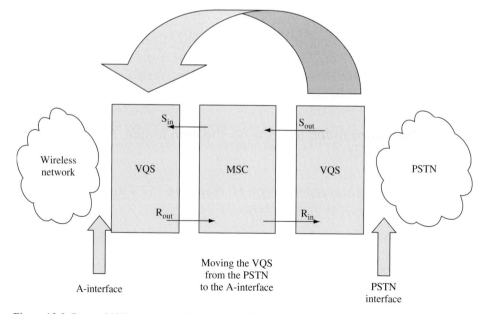

Figure 12.2 Proper VQS port connection to the MSC on either side of the switch.

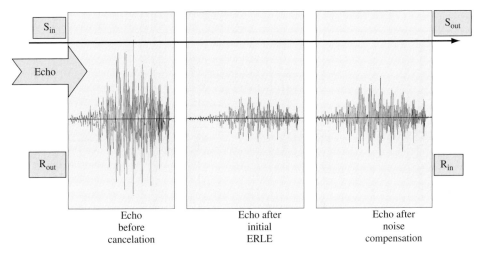

Figure 12.3 EC convergence process with noise compensation inside the VQS.

EC. When installing it on the A-interface, the MSC must be connected to the send side of the EC. In both cases, the EC send side must be closer to the PSTN while the receive side must be closer to the mobile phone.

When surrounded by noise I can hear my own voice coming back, especially at the beginning of a call

It is time for fine tuning some parameters. The mobile subscriber is experiencing echo. He is hearing his own voice coming back at some delay.

When the noise-compensation application is enabled and it is configured to be aggressive, it may amplify any remaining residual echo as well. At the beginning of a call when the echo canceler tries to converge, an aggressive NC may slow down the effect of the ERLE, the NLP may not hit its target threshold soon enough, and echo may last longer before it fades away (see Figure 12.3).

When surrounded by noise on the mobile side, PSTN speech gets saturated

It is time for fine tuning the NC-algorithm performance. Noise compensation amplifies the signal without regard to a saturation point. Instead of slowing it down to a halt as the gain boost approaches average levels of -5 dBm0 and above, the NC application continues to apply gain disregarding the risk of saturation (see Figure 12.4).

Is it acoustic or electrical echo?

When hearing your own voice bouncing back at an irritating pace how do you tell whether it is acoustic echo or hybrid echo?

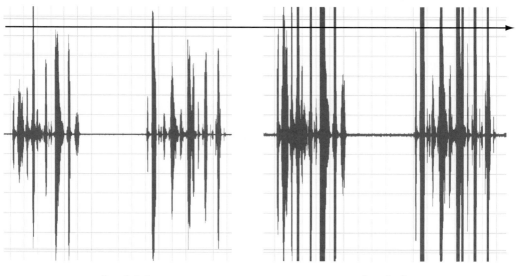

Speech before
noise compensation

Speech after
noise compensation

Figure 12.4 PSTN speech before and after noise compensation.

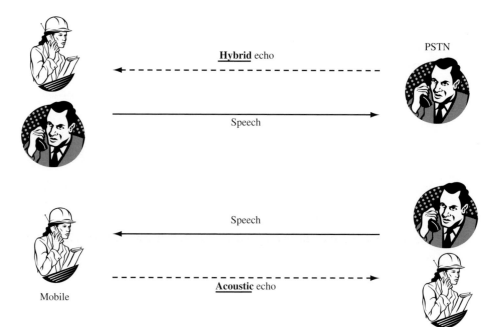

Figure 12.5 What type of echo.

If you talk to a PSTN person and hear your own voice bouncing back at a consistent or a diminishing level, you experience hybrid echo. If you talk to a mobile person and hear your voice bouncing back (mostly at an inconsistent power level) then it is acoustic echo (see Figure 12.5).

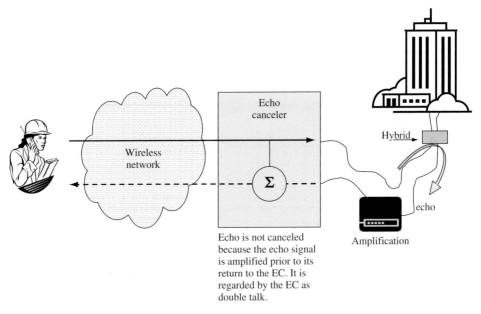

Figure 12.6 Amplification of echo prior to its arrival at S_{in}.

When calling certain numbers only there is echo

"When I call the water company, there is loud echo on the line. When I call everyone else, there is no echo. In fact, everyone I talked to has similar experiences."

Apparently, the echo returned from the water company is amplified to the extent that the ERL is a great deal lower than 6 dB. Either the hybrid in the vicinity of the water company is out of spec, or some equipment along the path to the central office introduces additional gain. The echo canceler mistakes the loud echo signal for speech and fails to cancel it. Potential remedies include finding the amplification source and correcting it. If this is not a near-term option, then lowering the EC threshold from 6 dB ERL to 3 dB or even 0 dB may remedy the water company situation. The trade-off, of course, is a higher likelihood of error type II – erroneously classifying some double-talk speech as echo and trying to develop an echo estimate based on the wrong input for calls that share the same line interface to the central office (see Figure 12.6).

Noise gets louder when the far end stops speaking

Apparently, the ALC algorithm amplifies noise. Since the ALC algorithm tries to set the signal level to a given target, it may not need to amplify it when the far end speaks, because the speaker's signal level is sufficiently high. However, when the far end stops speaking the only signal left on the line is noise, the average signal level of which is lower than the preceding speech, and the ALC feels obligated to move it closer to the target level.

The remedy is to set the noise floor to a higher value and prohibit the ALC from boosting signals exhibiting levels lower than the noise floor. See Figure 12.7.

Without ALC

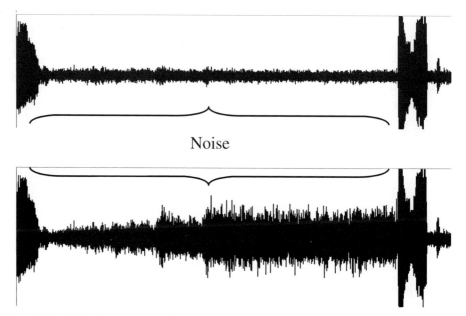

Figure 12.7 ALC with a low noise floor.

Loud echo from a wireless phone when enabling the EC

This phenomenon is not very common, but when it happens it is not always well understood. The echo canceler generates echo on its own where no echo is present. A probable cause is high and very colorful noise being transmitted from the far-end into the S_{in} port of the EC. Erroneously, the echo canceler treats the noise as if it were echo. The echo estimate constructed by the EC is a linear function of the actual speech, but in this case, the correlation between the speech and the noise is absent, and the resulting error after the h-register summation operation is extremely high. Excessive errors prevent the NLP from operating and the error signal gets through the echo canceler and into the listener's ear.

The error signal is the outcome of the actual difference between the noise and the echo estimate, which is based on the actual-speech signal. The result is a signal resembling the echo estimate or a signal resembling the actual speech.

When noise is sufficiently high, it may emulate double talk. At this point, the double-talk detector stops the echo-canceler adaptation and the erroneous frozen value inhabiting the h register is then subtracted from the noise signal entering the S_{in} port. The h register value may contain an echo estimate that is a linear function of the actual speech at R_{in}. It has no correlation to the noise, and again, the resulting error after the h register summation operation is extremely high.

Obviously, the EC is not performing fittingly, and it requires tuning. The EC should not confuse noise with an echo or a double-talk signal. It should either elevate the noise floor or

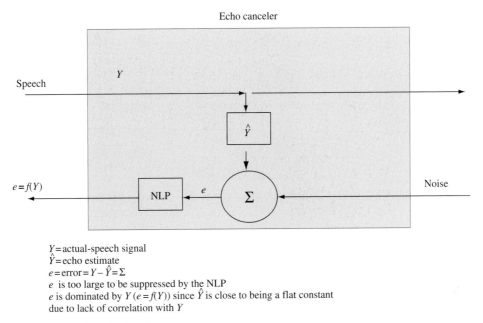

Y = actual-speech signal
$\hat{Y}$ = echo estimate
e = error = $Y - \hat{Y}$ = Σ
e is too large to be suppressed by the NLP
e is dominated by Y ($e = f(Y)$) since $\hat{Y}$ is close to being a flat constant
due to lack of correlation with Y

Figure 12.8 Noise mistaken for echo.

include indicators that trigger preventive actions when the high error magnitude refuses to trim down as the convergence process persists. See Figure 12.8.

Strange disturbing noises at the beginning of the call

"It takes a minute to two before the noise goes away and before I can comfortably listen to the far-end speaker. The noise transmitted from the far end at the beginning of the call is horrendous, and its high pitch makes my ears fall off."

Most echo cancelers do not reset their h-register at the commencement of a new call. Echo cancelers may not be aware that a new call passes through its circuitry. The previous call supposedly contained a fitting h-register value suitable for the previous call, but no more. Most echo cancelers are equipped with leak rates that weight recently obtained data much heavier than old data, while older data is discarded entirely. When the leak rate is set to a slow pace, the previous state of the h-register dominates the beginning of the call, and it takes a long time for a new, correct value to take hold. When leak rates are too low, convergence time is slow and high-level noise reflecting the existing value of the h-register is sent through the S_{out} port to the poor listener on the other end. See Figure 12.9.

Clipped music

"Whenever I am on hold and the music is playing back while I am waiting for the attendant, any little noise in my vicinity interrupts the music. The constantly clipped sound is highly annoying."

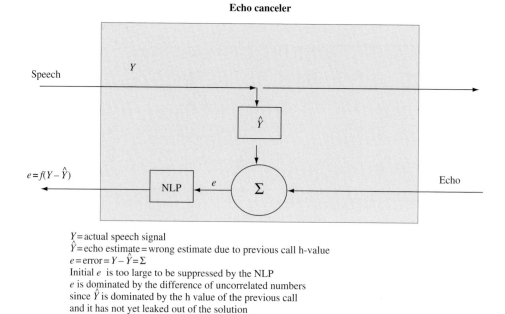

Echo canceler

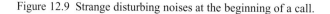

Y = actual speech signal
$\hat{Y}$ = echo estimate = wrong estimate due to previous call h-value
e = error = $Y - \hat{Y} = \Sigma$
Initial e is too large to be suppressed by the NLP
e is dominated by the difference of uncorrelated numbers
since $\hat{Y}$ is dominated by the h value of the previous call
and it has not yet leaked out of the solution

Figure 12.9 Strange disturbing noises at the beginning of a call.

Audio transparency is an echo-canceler feature that prevents the NLP from blocking the path of the music even upon getting triggered. Noise, speech, or other sounds generated on the waiting end may trigger an NLP action that prevents echo from returning to the waiting end.

An audio-transparency feature employs a modified NLP mechanism based on a center clipper and a floor threshold that lets music sneak through even when the NLP is actively blocking louder echo signals. See Figure 12.10.

Hot signal with ALC

Figure 12.11 depicts a situation where the ALC function attenuates a hot signal. It does so gradually to maintain natural dynamics. However, since the signal was hot before applying the ALC, its peaks were clipped by the codec. The cooling of the signal by the ALC could not restore the clipped peaks. Nevertheless, it did restore proper listening level.

It takes longer to converge due to long, persistent echo at the beginning of the call

"I can hear echo for a minute or two at the beginning of the call before it fades away."

When the ALC at the R_{out} port is aggressive, it may amplify the signal that would be reflected back by the hybrid. As a rule, it takes longer to eliminate loud echo as compared

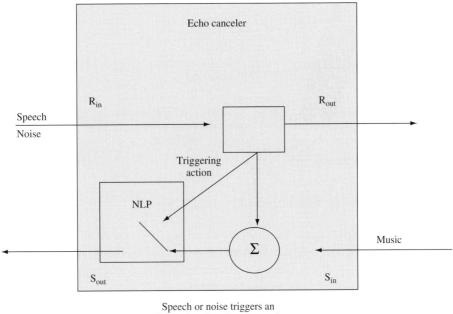

Speech or noise triggers an
NLP action that clips the music

Figure 12.10 Music on hold gets interrupted.

Speech at R_{in}

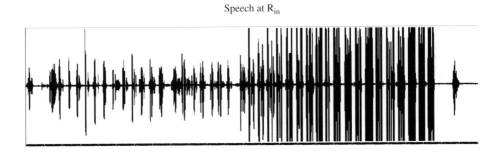

Speech at R_{out} after ALC

Figure 12.11 ALC function attenuates hot signals.

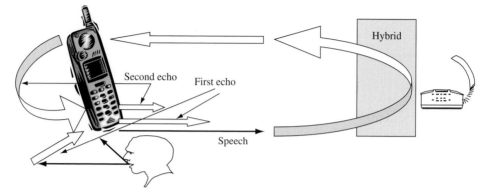

Figure 12.12 Listener's echo.

with a subdued signal. The main reason for the rule is the NLP. Most EC algorithms tend to trigger an NLP action only when the echo signal has been reduced to a level threshold of −15 to −20 dBm0. Since a loud speech signal spawns a relatively loud echo, reaching the EC NLP threshold may take a bit longer.

In some cases the echo-signal level is boosted by a transmission plan or an ALC device placed between the hybrid and the EC (see Figure 12.6). The ALC device may contribute to a difference smaller than 6 dB between the speech and its corresponding echo. Accordingly, the EC may treat some parts of the echo signal as if it were double talk. It would freeze its convergence process and would take longer to reach a steady state.

Evidently, a combination of aggressive ALC, an ALC preceding the EC, and a less aggressive NLP would lengthen the time it takes to suppress an echo signal.

Listener's echo

Network echo cancelers are designed to eliminate talker's echo. Although the definition of talker's echo may seem straightforward – the speaker hears her own voice returning at some perceived delay exceeding 100 ms – many confuse it with listener's echo.

There are two types of listener's echo. The first type is brought about by room acoustics, where a speaker at the far-end uses a speakerphone and her voice bounces off the walls before entering the microphone behind the actual speech. The second type is created by a loud talker's echo that makes one extra round as illustrated in Figure 12.12.

Echo cancelers are not designed to eliminate listener's echo of the first type. The best remedy in this case is to avoid using the speakerphone if possible, or stay closer to the microphone and use a softer voice. Echo cancelers should be able to eliminate the second type of listener's echo. The second type of echo in question is a problem for both sides of the connection, and it should be controlled better at both ends.

Distorted speech especially with noise

"When speaking on my mobile phone from my car, my spouse, at home, complains that my voice is distorted, sounds tinny, and at times, she does not even recognize it as mine."

Obviously, the chief suspect is the noise-reduction application. It may be a bit aggressive. The spouse does not appreciate the benefits it brings since she is not able to sense the quality of the call before applying the NR and reducing the noise. An overly aggressive treatment of noise by a noise-reduction algorithm may yield excessive distortion, and since most of the noise concentrates in the lower frequencies of the voice spectrum, attenuating those may result in leaving a tinny voice behind.

An aggressive NR setting may attenuate the lower-band speech frequencies where noise may be abundant. The potential side effect of noise reduction is the attenuation of speech-frequency bands that share the same range occupied by noise. The direct result of aggressive attenuation may yield a tinny speech sound where bass frequencies are toned down significantly enough to fashion a noticeable distortion.

Making the NR less aggressive is a solution that reduces the severity of or eliminates the issue.

Fading ring tone

"When calling my boyfriend on his mobile phone I experience a fading ring tone. The ring volume starts at a normal level but fades towards the end of each ring."

This phenomenon indicates that the noise-reduction application views the ring-back tone as noise. The ring tone has similar characteristics to noise due to its exhibition of high stationarity, as discussed in Chapter 9. Consequently, the NR application tries to attenuate it. However, since rings are separated by silent intervals the NR algorithm may reset before the start of the following ring, which would once again exhibit an identical diminishing tone level while starting a normal-point level.

The NR algorithm should be fitted with a ring-tone detector and disable itself upon detection. See Figure 9.23.

Ring tone fades while noise between rings ramps up

When placing the VQS on the PSTN side, the VQS is exposed to ring tones carrying characteristics typical of a noise signal. Unlike speech, the ring is stationary, periodic, and consists of a thin spectrum. A VQS not trained to detect the ring may treat it as noise. Consequently, the noise estimate would incorporate it in its level estimate and would increasingly overestimate the noise-signal level. It would also include the spectral content of the ring in its estimate of the noise spectrum. As a result, the NR function may continue to attenuate the ring-tone signal, while, at the same time, the spectral noise matching of the AEC would include it in its noise estimate and continue to amplify the noise between the rings.

Figure 12.13 presents a view of a ring-tone signal. As seen in Figure 12.13, attenuation (set off by an untrained VQS) via NR would diminish the power of the ring tone as it moves forward, while the AEC noise matching would keep on amplifying the noise between rings.

When the VQS includes ring-tone detection, it would freeze the noise estimate upon detecting the ring. The ring tone would not be included in the noise estimate and there will be no ring attenuation and no added noise between rings.

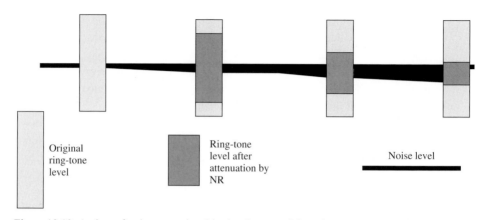

Figure 12.13 A view of a ring-tone signal in the absence of detection.

Note: when placing the VQS on the A-interface the ring tone on the receive side is non-existent, and the special ring detection and disabling may not be required.

Clipped first syllable

"Every time I talk to my friend he needs to repeat words because the first syllable gets clipped."

The comfort-noise matching applied by the DTX feature in GSM smoothes out transitions between speech and speech pauses (see Chapter 5). As transmission disconnects upon speech pauses (and reconnects upon speech) the comfort-noise level replacing the speech during a pause is set at a level high enough to mask imperfections during the transition. If the NR algorithm attenuates the comfort noise, the transition becomes noticeable, and the sharp changeover is perceived as speech clipping.

The best remedy is to curb the NR maximum attenuation with a hard floor by limiting it to the DTX comfort-noise matching level. As long as the NR algorithm does not drive the noise signal below the comfort-noise level, the transition between speech and speech pauses may not be noticeable and may not cause any speech-clipping perception.

DTMF not going through

Trying to avoid talk-off, detection and disabling algorithms must always consider the trade-off of being too aggressive by leaving wider margins around the specific tone frequency, duration, and level. Tone transmitters may not always be perfect and some receivers may be overly sensitive to small deviations from the standard. At the same time, detection mechanisms within VQ systems may vary in quality as well. Very robust algorithms may require more scarce resources than somewhat less robust ones, and when these resources are not allocated, NR may impose some damage to a perfectly shaped DTMF digit, enough to prompt a strict DTMF receiver to misinterpret it.

Dual-tone multi-frequency detection must be considered a higher priority than NR. If both are incapable of co-existence then NR must be disabled. If the DTMF detection can be made fully proof, then, and only then, may NR be enabled.

Wireless-fax failure

"Frustration over the inability of an ambulance to send out a fax of an ECG reading to the hospital."

Some modems do not follow G.164 and G.168 ITU-T standard recommendation. Fortunately, I propose a way to correct it with a feature I name voice-signal classifier (VSC). See Chapter 9 for further details.

High-speed circuit-switched data failure

When placing the VQ on the A-interface it may block or corrupt data transmission if it is not equipped to detect high-speed circuit-switched data. See Chapter 7 for further details.

Packet-data failure

When placing the VQ either on the A-interface or on the PSTN interface, it may block or corrupt data if it is not equipped to detect and disable upon detection of ISLP-type transmission.[1] See Chapter 9 for further details.

Fax failure

When placing the VQ on the A-interface it may block or corrupt fax transmission if it is not equipped to detect and disable upon detection of V.110-type transmission.[2] See Chapter 7 for further details.

Persistent echo throughout calls

More and more, it keeps happening. A salesperson using the service for a long time starts complaining, all of a sudden, about frequent occurrences of echo when calling back to her office from a particular customer's location. "This is new. It started a month ago," she explains.

"It started happening after we installed ATM cross-sections," recalls the service provider CTO.

A telecommunications manager in a long-distance telephone company started receiving echo complains from one of his largest customers. "Every time we call outside our private network we have echo on the line. When reconfiguring the lines with another service provider the problem is gone. We are thinking of switching to the other service provider."

Quite threatening, but why now? What was working before that is not working in the present?

---

[1] TIA-728, *Inter System Link Protocol (ISLP)* (January, 2002).
[2] ITU-T Recommendation V.110 *Support of Data Terminal Equipment's (DTEs) with V-Series Interfaces by an Integrated Services Digital Network*.

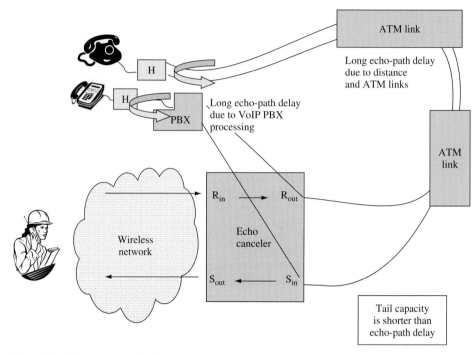

Figure 12.14 Long echo-path delay.

Both cases mark a change in infrastructure architecture. Voice over ATM and voice over IP sections in the echo path contribute to significant delays that may exceed the present capabilities of the existing echo canceler.

The delay associated with a wireless application is primarily a result of voice-operated codec equipment that is used to process voice signals for transmission over the air interface. Similarly, a typical voice-over-internet protocol (VoIP) connection incurs up to about 200 ms of round-trip delay. The delay associated with a VoIP application is distributed among coding and decoding, packet assembly and disassembly (including variations associated with packet sizing), buffering (and buffer sizing), and routing (including network-congestion issues) functions. In comparison, a typical voice-over-asynchronous transfer mode (VoATM) connection experiences about 50 ms of round-trip delay. The delay associated with a VoATM application is distributed among coding and decoding, buffering, and cell assembly and disassembly functions.

When service providers install multiple ATM links in the echo tail, they inadvertently increase the requirement. If the present tail capacity is 64 ms[3] and the tail delay exceeds that measure, then the echo canceler will not cancel the resulting echo (see Figure 12.14).

When companies replace existing analog or digital PBX with a VoIP PBX on their premises,[4] unless the VoIP PBX contains a built-in echo canceler, they involuntarily increase the requirement for tail capacity in the echo canceler of their service provider.

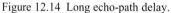

[3] This provision used to be a standard requirement until about 2003.
[4] This architecture has become increasingly popular since 1999.

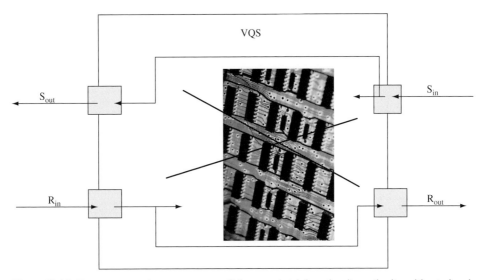

Figure 12.15 Bypass processing upon system failure: maintaining circuit continuity without signal processing.

Although the service provider may not be aware of the reason, the company that installed the VoIP PBX may experience echo when its staff call a PSTN number. At the same time, people who call in either from a mobile phone or a PSTN phone may experience echo as well. In some cases, only one of the call ends may experience echo while the other end may not.

Once again, an insufficient tail capacity of an echo canceler may be the reason for not canceling the echo. The other carrier's echo canceler (in the complaint above) may have a shorter or longer tail capacity.

Persistent echo may result for other reasons. It is, therefore, necessary to verify the tail-capacity assertion by placing calls and measuring the tail delay before finalizing the conclusion. If the tail-delay assumptions prove true, then there is a need to replace the existing echo canceler with one that assumes a longer tail delay.

In today's market, the unofficial standard has been set to 128 ms. There are products on the market with tail capacities of 192 ms and more.

A sudden burst of echo complaints

Occasionally, a circuit pack that carries several hundred circuits fails. In some cases when the failure is limited to software, it may not even generate an alarm. Many echo cancelers are equipped with a bypass feature, which maintains circuit-pack connectivity in the face of hardware or software failures. When connectivity is maintained in spite of a failure, existing calls are not terminated and future calls may be able to set up through the failed equipment. Nevertheless, echo may not be canceled and complaints would start flowing in. See Figure 12.15.

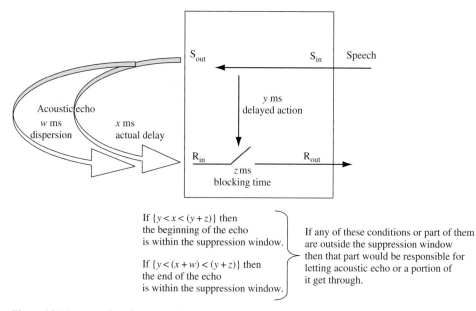

If $\{y < x < (y+z)\}$ then
the beginning of the echo
is within the suppression window.

If $\{y < (x+w) < (y+z)\}$ then
the end of the echo
is within the suppression window.

If any of these conditions or part of them
are outside the suppression window
then that part would be responsible for
letting acoustic echo or a portion of
it get through.

Figure 12.16 Acoustic-echo control system.

Partial acoustic echo

Acoustic echo in digital-wireless communications experiences tail delays ranging from about 140 ms to 250 ms. Tail delays are mainly a function of codec type and performance, and may vary from one mobile phone to another.[5] Acoustic-echo control algorithms employ a delay window before triggering the NLP signal blocker that prevents the acoustic echo from returning to the speaker. It is difficult, if not impossible, to adapt to the acoustic echo delay in real time because, unlike electrical echo, acoustic echo is inconsistent. It may or may not trigger in response to a particular far-end word or sentence. What's more, new wireless phones include a hands-free mode. Under this mode of operation (if the phone supports full duplex mode), acoustic echo is more likely to occur and the delay can be higher. Its intensity may vary with each sentence, and its delay may fluctuate with each phrase as well. The common practice is to guesstimate the delay window during set-up and provisioning, and let it ride. If the setting is incorrect for a particular circumstance, then the NLP may miss either part of the echo or the entire signal.

Another possible option is to adjust the NLP state so that, once triggered, it stays open for an interval long enough to cover the range of the potential echo-path delay. The risk associated with a long NLP open interval is blocking of potential speech that may barge in from the near end. See Figure 12.16.

---

5   The codec performance inside a mobile phone may vary among different mobile-phone vendors due to dissimilar implementations.

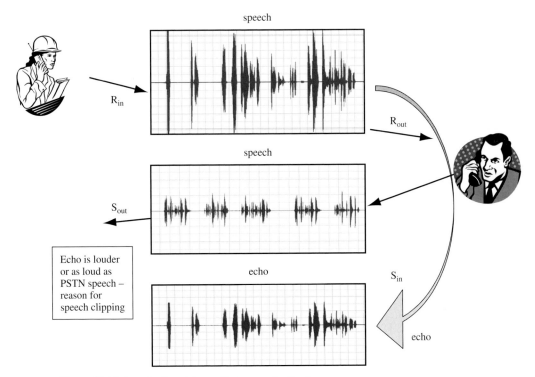

Figure 12.17 Level mismatch results in speech clipping.

Speech clipping

"Every time I talk to my mom, many of her words get chopped off, and I find it hard to understand her. I checked my mobile phone and the signal strength is at its maximum. I don't understand the reason for the terrible reception."

Large signal-level differences between mobile and PSTN that far exceed 6 dB may cause echo cancelers to misinterpret some parts of the low-level speech and regard it as echo of the louder end. While attempting to cancel the speech signal, the echo canceler may suppress some of its valleys and bring about speech clipping. See Figure 12.17.

Although some are, echo-canceler algorithms should not be confused by level differences between the two ends. Added intelligence may help echo cancelers adapt to speech levels on each end and interpret it correctly when making a distinction between speech and echo.

Echo bursts in the middle of a conversation

Echo bursts may result from having an echo signal amplified before entering the S_{in} port of an EC.

When an ALC function is placed upstream of S_{in}, it would change the ERL value perceived by the EC. If it is a fixed and consistent gain, then the ERL threshold may be set to a lower value to compensate for the louder echo. If the gain were time-variant, resembling a standard ALC function, then the EC would have to re-converge at every gain adaptation.

Since echo is usually less loud than the source triggering it, an ALC would amplify it most of the time when the echo signal is hot enough and is above the internal VAD threshold. In some cases, the amplification may result in an echo signal louder than the source. The EC may regard it as double talk, and may not try to cancel it, adapt, or converge on it.

When the ERL threshold setting is adaptive, it resets itself at the beginning of a call. This works well, so long as the ERL does not change throughout a particular call. Having an adaptive ALC before the S_{in} port would make the ERL setting erroneous since the actual ERL would be smaller than the one used for the EC setting. Under these conditions, the echo signal would be treated as double talk. The EC may freeze its convergence state and cease adaptation.

One way to remedy the problem is to eliminate, if possible, the ALC function upstream of the S_{in} port.

Another potential reason for having echo bursts in the middle of a call is the existence of very hot signals. When signals are hot, they may spawn hot echo as well. Hot signals may have their peaks saturated and clipped by the codec, and the difference between the clipped peaks and their corresponding echo may be below the ERL threshold setting of the EC. Under these circumstances, the echo signal would be treated as double talk. The EC may not operate the NLP, freeze its convergence state, and cease adaptation. The direct result would be bursts of echo.

Hot signals must be treated differently than the rest. When an EC receives a hot signal, it should anticipate the circumstances. Accordingly, it ought to adapt its internal ERL threshold and lower it for the duration of the hot-signal experience. By so doing, the EC is less likely to confuse echo with double talk, and echo bursts are less likely to take place.

Echo bursts

Continuous complaints from customers concerning bursts of echo in the middle of a phone conversation could very well be a direct result of a disabling tone talk-off. Echo cancelers are generally set to detect certain narrow-band signals and disable their processing upon detection. Margins around the particular frequency are set to insure that even when the particular signal deviates a tad from the precise frequency, the detection and disabling may still work properly. When these margins are set exceedingly wide, the chances of a person inadvertently emulating a disabling tone increase. When an echo canceler erroneously detects a disabling tone it would cease its echo cancelation, and would let echo return to the talker and interrupt the quality of the session.

One common remedy is to re-evaluate the detection margins for a possible tightening.

Noise bursts

Continuous complaints from customers concerning noise bursts in the middle of a phone conversation could very well be a direct result of a disabling tone or DTMF talk-off.

Voice-quality systems are generally set to detect certain narrow-band signals and disable their processing upon detection. Margins around the particular frequency are set to insure that even when the particular signal deviates slightly from the precise frequency, the detection and disabling may still work properly. When these margins are exceedingly thick, the chances of a person inadvertently emulating a disabling tone increase. When a VQ system erroneously detects a signal emulating a disabling tone, it would cease its VQ processing, and would let noise through with the speech. It would enable the VQ processing at the first speech pause and disable it again when detecting a disabling signal. The effect of noise switching between on and off would be exceedingly disturbing to the listener.

One common remedy is to re-evaluate the detection margins for a possible tightening.

Far end EC works some times, but not others

In cases where the echo tail is long and variable, echo cancelers may not possess sufficient tail capacity to cover some of the variant architectures. Examples of cases where tail delays may be subject to variations, some of which are longer than the EC tail capacity, are depicted in Figure 12.14. These are:

- far-end echo cancelation with satellite or ring back-ups,
- VoIP cross sections behind the tail,
- ATM cross sections behind the tail.

Calls don't get set up

"Once we installed the new VQ system, we were unable to set up calls. We removed the VQS and were able to set up calls again. Clearly the VQS is in the way of call set-up."

Signaling tones may be blocked when trying to pass through a VQ system if they are not properly detected. Specifically, echo cancelers may block transmission of a response signal that bears a resemblance to the initiating signal since the response may appear to emulate a corresponding echo. A C5 signaling tone (2400 Hz and 2600 Hz) and a voice path assurance (2010 Hz tone) are transmitted by both end switches as part of a more complete protocol. If not disabled, echo cancelers may block transmission of a returning signal and get in the way of calls set-up. See Figure 12.18.

Noise reduction does not reduce noise

When low-level speech originates at noisy surroundings, a noise-reduction algorithm may remove some of the noise only to have it amplified afterward by a subsequent ALC. In fact, the VQ system may work properly by improving the SNR through amplification of the speech without a corresponding action concerning the noise. The complaint may not have taken into account the SNR effect, and therefore it missed the fact that voice quality has been improved due to NR. See Figure 12.19.

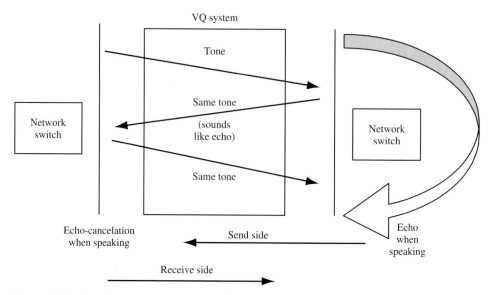

Figure 12.18 Signaling and VQS systems.

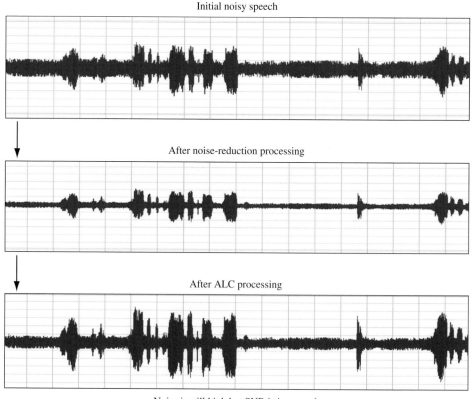

Figure 12.19 Notable improvement in SNR with small improvement in noise level.

Send-side speech

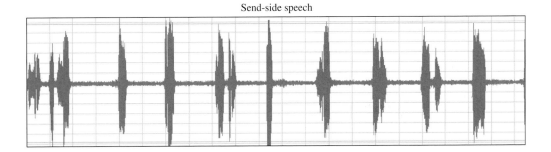

Receive-side noise reacting to speech

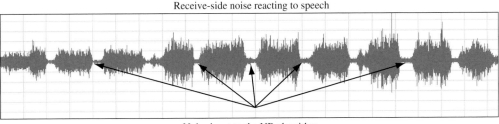

Noise input to the NR algorithm

Figure 12.20 Noise gaps caused by AEC on the receive side used as input for the NR algorithm.

There is a need to investigate this hypothesis by having "before" and "after" recordings of calls made by the complaining party. There may be other reasons for having under-performing NR, as we see in the next case.

When placing an acoustic-echo control[6] function in front of the NR application, the noise input to the NR algorithm may be severely misrepresented. When the acoustic-echo control algorithm suppresses the signal at the receive side in response to sensing far-end speech emerging through the send side, it may not fill in the follow-on gap with matching noise.[7] The NR algorithm may take in the consequential silent period as the input to its noise estimate and conclude that no noise needs to be treated. In fact, since noise is interrupted by silent gaps caused by an AEC, it does not look to the NR algorithm like noise, since it loses its stationary characteristic. It looks more like speech and the NR may not consider the noisy segment as input to the algorithm at all. See Figure 12.20.

A remedy that does not include the immediate noise input following a speech segment that originates on the send side would derive a proper noise estimate and would provide the NR algorithm with an opportunity to perform properly. The procedure may have to trade off a potentially more accurate noise estimate for situations where the mobile phone is either not equipped with an AEC, or its AEC function generates an accurate comfort noise when suppressing the acoustic-echo signal.

---

[6] Acoustic-echo control in a mobile handset that does not employ proper comfort-noise injection may distort the input going into the NR algorithm.

[7] Many mobile phones perform that way.

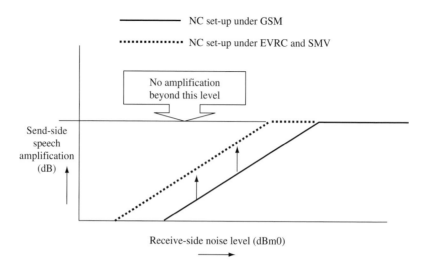

Figure 12.21 NC with EVRC and SMV.

Artifacts like waterfalls in the background

Aggressive treatment of noise by a NR algorithm may leave a residue of low-sounding artifacts in its trail – sometimes referred to as musical noise. A common artifact may sound like running water in the background. This musical noise is an artifact of the spectral subtraction used by an aggressive NR algorithm.

A potential treatment for remedying the negative sensation is to have NR provisioned for a less-aggressive treatment.

Noise compensation does not work

"Although I provisioned the noise-compensation function to an aggressive level, the outcome, even with high-noise background, is barely noticeable."

The network engineer who has tested the NC feature has been disappointed with the results. The NC showed little reaction to noise. The engineer did not realize that CDMA networks that deploy the EVRC codec (and later the SMV codec) dilute the effect of the NR and NC in the VQ system. The CDMA codecs include a noise-suppression application. Consequently, the noise information reaching the VQ system is significantly reduced. The noise information reaching the VQ system misrepresents the true ambient surroundings of the listener. Since the NC application amplifies the signal in response to noise – more noise, more amplification – and since the EVRC attenuates the noise before it reaches the VQS, the NC is not reacting to the true noise.

A potential remedy when operating in CDMA networks is to shift the amplification starting point to a lower noise level (see Figure 12.21). When designed to operate at lower noise levels, the NC may amplify the signal even when noise is low. As long as the NC does not amplify hot signals, the trade-off would be worthwhile.

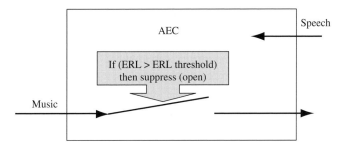

When the level difference of (speech–music) is substituted for the ERL, then the level difference must be greater than the set ERL threshold for the music to move through uninterrupted.

Figure 12.22 AEC and mobile-phone music.

Why is there acoustic echo?

Although the VQS is installed in the network, customers complain about echo when in a mobile-to-mobile conversation. Why is the VQS not taking care of the echo?

Echo in a mobile-to-mobile conversation is likely to be of the acoustic type, unless the connection is routed through a long-distance PSTN section that is steering its way through two to four-wire conversion.[8] Assuming a pure mobile-to-mobile connection, the acoustic echo may not be treated if the architecture calls for TFO. Under TFO, the VQS is either disabled or bypassed entirely.

Music from mobile handset gets clipped

Some mobile services support an option that lets users send low-level background music from their mobile phone throughout a communications session. Notes and phrases get clipped in reaction to far-end speech. It reaches a point where the far-end speaker becomes irritated, and does not stop complaining about the unpleasant experience.

Obviously, the AEC algorithm tries to protect the far-end speaker from acoustic echo by suppressing the signal following the far-end speech. When the ERL threshold is set to a relatively small value, the AEC suppression mechanism operates more frequently, and it may have a higher likelihood of erring by suppressing signals other than acoustic echo. The advantage of a low ERL setting is a higher likelihood of preventing all acoustic echoes from returning to the speaker.[9]

Setting a higher ERL threshold may prevent the AEC from suppressing signals louder than a potential acoustic echo. See Figure 12.22.

---

[8] Not a likely architecture.
[9] When acoustic echo is very loud, there is a chance that it would not be suppressed because the AEC may regard it as double-talk speech.

Music from mobile handset gets distorted

Some mobile services support an option that lets users send low-level background music from their mobile phone throughout a communications session. Notes and phrases may get distorted as the music plays on. It reaches a point where the far-end speaker becomes irritated at the distortion, and does not stop complaining about the unpleasant experience. Distortion on the receive path is most likely due to an aggressive NR algorithm, which regards parts of the music as noise. Parts of the music may carry stationary characteristics that set noise apart from speech. When attempting to attenuate noisy frequency bands notes and phrases may get distorted.

The problem may be remedied by changing the existing NR algorithm setting to less aggressive.

The echo canceler failed to converge in a timely manner when tested

Some test engineers start testing the echo canceler with a simulation of double talk. They watch as the echo canceler fails to converge in a timely manner.

Echo cancelers must be able to converge before they perform their function properly. Convergence is dependent on the ability of the EC to construct mathematical relationships between speech and its corresponding echo. The EC then applies the echo estimate by subtracting it from the returning actual echo signal. The learning process takes less than a net second of single speech and echo.

When starting a new call with double talk, the EC has no immediate input for constructing its echo estimate, and it may take a relatively long time before sufficient data is collected to construct one. When testing an EC, engineers should allow the echo canceler to converge on a single-talk signal before subjecting it to stressful tasks.

Although the EC specs clearly state that its tail-delay capacity is 128 ms, it does not cancel any echo with delay greater than 124 ms. Are the specs wrong?

Many network echo-canceler products are designed with an echo-path (tail)-delay capacity of 64 ms or 128 ms. A "64 ms tail capacity" provides adequate echo cancelation for up to approximately 4000 miles of cable distance ($64 \times 125/2 = 4000$) between an echo canceler and the echo source (i.e., the reflection point or electrical hybrid). If transmission cables were routed on a point-to-point basis employing the shortest-path principle, then a 64 ms echo canceler deployed anywhere within the continental USA would be able to control echo originating at any point on the same continent properly.

If propagation delays were the only factor effecting echo characteristics, then echo cancelers would not be required for transmission routing distances shorter than 1600 miles. However, most modern-day networks (especially wireless and voice over packet [VoP] applications) introduce excessive delays during signal-processing activities, hence physical distance in these cases becomes almost a secondary consideration.

The term tail delay or echo-path delay refers to the time it takes a speech signal to travel from the R_{out} port of an echo canceler to the point of reflection (e.g., the 2/4 wire hybrid) and back (as an echo signal) to the S_{in} port of the echo canceler as shown in Figure 3.11.

Tail-delay components

The delay in question is the sum of three components:

- propagation delay,
- processing delay, and
- dispersion delay.

The propagation delay is a direct function of distance. It could be approximated as 500 air miles per 8 ms (or 4000 miles ($\times$ 2) per 64 ms).

The processing delay is independent of distance. Processing is particularly notable in wireless and voice-over-packet communications. A typical wireless-codec processing is about 30 ms (20 + some looking ahead), with a total network round-trip delay of 180 ms (round trip through codec).

Dispersion delay is a function of the PSTN hybrid. Dissimilar hybrids reflect echo signals in various ways, so that parts of the reflection are delayed a few milliseconds more than other parts. This phenomenon is referred to as dispersion.

The ITU G.168 recommendation defines dispersion of an echo path as an integral part of the echo path delay. The following is a direct quote from the standard.

The delay from the R_{out} port to the S_{in} port due to the delays inherent in the echo path transmission facilities including dispersion time due to the network elements. In case of multiple echo paths, all delays and dispersions of any individual echo path are included. The dispersion time, which varies with different networks, is required to accommodate the band limiting, and hybrid transit effects.

Dispersion time may vary from null to over 20 ms. Average dispersion time is about 4 ms.

Tail-delay settings on test equipment must take into account potential hybrid dispersion in addition to the propagation and processing delays. Dispersion delay is somewhat tenuous because the onset of the echo signal (most of the reflected portion) is returned immediately, while the remaining segments are reflected later. Since dispersion is an integral part of the tail capacity delay, propagation and processing delays must not exceed the specific tail-capacity limits after affecting an allowance for dispersion delay. For example: an EC with memory-capacity limits of 64 ms should be able to satisfy tail delays of 60 ms (average) of propagation and processing delays, while allowing for an average of 4 ms additional dispersion delay.

Speech sounds unnatural and overly robotic

"Did I just wake you up? You sound like a computer chip."

When the ALC algorithm attack rate and step size are too aggressive and fast reacting; when the algorithm tries to reach its target level too fast; it loses the natural speech dynamics that typifies live, emotional, expressive speech. The ALC algorithm must restrain its advance toward its set target and move the average speech level slowly and moderately, as outlined in Chapter 6.

Part VI

Afterthoughts and some fresh ideas

13 Tracer probe

13.1 Introduction

For the entire telecommunications industry, one of the most difficult maintenance and troubleshooting tasks is pinpointing a single channel or trunk degradation in voice quality. Troubleshooting T1 or E1 problems are somewhat easier because only 24 or 32 channels are involved, degradation of the T1 or E1 facility will almost always result in an alarm indication and, using a T1 or E1 test-set, the trouble can be detected by error per second and bit error-rate testing.

However, finding a single noisy or high-loss 64 kbps trunk or voice channel can become a very long-winded task. With many carriers going to higher primary facility rates, STS-1, STM-1, OC-3 and even packet or voice-over-IP (VOIP), the problem of uncovering a single 64 kbps voice-channel degradation is an overwhelming task in most cases. The technique of switch-tracing a call involves entering the calling and called phone numbers to a number of switches, sometimes of different telecom administrations, in order to determine which voice channel from within a particular STM-1, OC-3 or an STM-1 is degraded.

The concept presented here, "tracer probe,"[1] combines common-product implementations with a unique testing sequence to produce an easy way of determining the trunk or voice channel experiencing the problem, either in real time during a degraded test call, or after a degraded test call has ended, using a logging feature. Echo cancelers happen to be excellent candidates for implementing this application where, in addition to isolating degraded voice channels, this concept is used to find echo-canceler channels that are implemented in the wrong direction (see Chapter 12, trouble number 1).

13.2 Case study

The case study presented here illustrates how difficult the task of isolating a particular 64 kbps voice channel in the present environment is. Figure 13.1 illustrates a typical application of an STS-1 echo canceler in a cellular application.[2]

In this example, the cellular provider encounters an echo problem perceived by wireless customers on a small percentage of calls. The complaint log points to an echo canceler that

---

[1] The tracer-probe concept was conceived by Ron Tegethoff in 2003.
[2] The problem becomes three times more tedious when the STS-1 is replaced with an OC-3 or a STM-1.

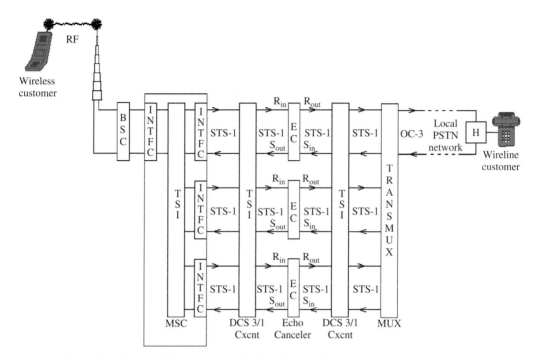

Figure 13.1 Typical STS-1 echo canceler in cellular application.

is either incorrectly provisioned, malfunctioning, or erroneously wired. Only a small percentage of customers complain. Accordingly, this problem may involve less than a full STS-1 of echo cancelation. It could also be a single 64 kbps voice channel of echo cancelation that is either broken, provisioned incorrectly, or a virtual tributary (VT) member, 1.544 MB T1, orientated in the wrong way. The latter may happen due to provisioning errors of the digital cross-connect system (DCS3/1) that connects the two sides of the STS-1 echo canceler to the MSC and to the MUX. Regardless of the source of the echo problem, correcting the problem may become a considerable effort, starting with isolating the root cause.

The trouble-shooting procedure would have the cellular provider's technician place calls to a wireline customer in anticipation of encountering echo. In many cases this wireline customer is a second employee. Once an echoing circuit is found, that technician must ask the wireline customer to stay on the line and begin the process of tracing the call. The process begins by determining on which specific 64 kbps voice channel or trunk of which STS-1 interface (INTFC) the mobile-switching center (MSC) is routing this partic- ular call. Next, the trunk of the STS-1 is routed to a digital cross-connect system (DSC3/1). Here the trunk that is part of T1 and STS-1 is cross-connected virtually at the T1 level and the T1 becomes member a new STS-1 that will be routed to the R_{in}/S_{out} side of the echo canceler (EC). This process is equivalent to a second trace at the DCS3/1 and often requires some time and effort. Now the technician has identified the trunk and T1 within the STS-1. At this point, the technician must determine the particular channel and T1 of the STS-1 this call is going through. In many cases, this means translations between the MSC, the DCS3/1, and the STS-1 EC to determine exactly which channel of the echo canceler is processing this call. From the above description, one can appreciate that this is a complicated and time-consuming process with many translations of voice channels, T1s and STS-1s, which make it extremely prone to errors. This is an example of where a tracer probe can significantly improve the troubleshooting procedure outlined above.

13.3 Description of tracer probe

The basic concept of the tracer probe is to equip a product, in this case the STS-1 EC in Figure 13.1, with a unique DTMF or touch-tone detector positioned at the S_{in} and R_{in} ports. The block diagram shown in Figure 13.2 shows the application of the tracer probe to the STS-1 echo canceler described in Figure 13.1.

    The echo canceler is the same as shown in Figure 13.1 but now a DTMF or touch-tone detector continually monitors the S_{in} and R_{in} ports of the canceler. The operate time of both

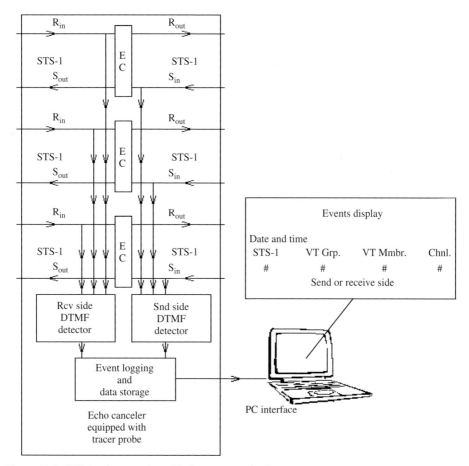

Figure 13.2 STS-1 echo canceler with the tracer-probe feature.

detectors is extended from the typical milliseconds to as much as six seconds so that any DTMF digital that is present at the S_{in} or R_{in} ports of the echo canceler for less than six seconds may not be detected. However, if the DTMF digit is depressed and held for ten seconds, the tracer probe logs the date and time of the event and the STS-1, VT group, VT member, and channel.

When using a tracer-probe feature, the trouble-shooting procedure consists of the cellular provider's technician placing calls to a wireline customer until the technician perceives the echo. At that point, the technician merely asks the person on the wireline end of the call to press and hold a DTMF digit, for example the digit "5," for more than six seconds. At this point the relevant data (see Figure 13.3) would be logged and stored at the echo canceler.

The tracer probe has avoided a hold and trace on the MSC, and channel translations for the MSC, the DCS/3/1, and the EC. Instead, it requires technicians doing "drive tests" to depress and hold a DTMF digit for ten seconds when perceiving echo on a test call. The logged data unearths the particular echo-canceler channel that has been used for the bad call. In addition to the particular channel that the echo canceler employed, a tracer probe indicates which side to the echo canceler detected the special DTMF digit 5. Often the source of the echo complaint results from erroneous wiring of the ports of the echo canceler. If in this example, the data indicated that the DTMF digit was detected on the receive side, this would imply that the wiring of the echo canceler is backwards, explaining the reason for the inability to cancel echo.

Figure 13.3 Tracer-probe capture and recording of "bad" channel.

13.4 Requirements and options

The tracer-probe implementation shown in Figure 13.2 is merely an example. Although an echo canceler is a natural and excellent home for the feature, a tracer probe could be utilized as part of almost any digitally multiplexed piece of equipment. These could include digital cross-connects, switches, VoIP gateways, digital-conference bridges, and even portable test equipment. The DTMF detector can be a standard detector with two levels of digit recognition. In level-one mode it detects DTMF digits in the typical milliseconds time period and provides the normal response. In addition, when a single digit is present for >6 seconds the DTMF detector could recognize that digit as a level two and log the source all the way down to the 64 kbps voice-channel level, while determining the path through the product where it was detected, and time stamp the event. In products where the complexity of a full DTMF detector is not already employed and cannot be cost justified, a fixed single-digit detector could be implemented in DSP code and the associated logging function could be utilized.

13.5 Conclusion

The concept presented in this chapter could provide any piece of telephony equipment with the ability to isolate a problem down to a channel level, where the only other equipment required is a common telephone, with two people on a call. In a network where multiple pieces of equipment all have their own tracer-probe capability, isolating troubled circuits could be accomplished without the technology, manpower, and extra time, that a traditional hold-and-trace across multiple switches would require.

14 My sound

14.1 Introduction

My sound is a concept that provides a preferred-user segment (business-class segment) with control over customized sound atmosphere, flavor, clarity, and quality.

Speech and sound quality are an essential part of wireless communications. Unadvertised step-ups and upgrades in network voice-quality capabilities go, for the most part, unnoticed by subscribers, particularly when these capabilities address insufficiencies rather than impairments. For example, echo is a clear impairment while inadequate speech level in the presence of listener noise would be viewed as insufficiency. Intrusive background noise and poor signal-to-noise ratio (SNR), more often than not, would be considered as an insufficiency, while punctured radio coverage and even echo would seize top awareness on the impairment list.

Voice-quality insufficiencies can cross the threshold of a subscriber's awareness (or lack of it) if control over their intensity is given to users throughout or prior to a communication session. The exercise of control may attract users' attention and have them notice and appreciate the enhancement, while simultaneously provide a specific subscriber segment with the pleasure of being in command of their sound output.[1]

One way of letting a subscription-based class of preferred subscribers assume control over certain parameters affecting speech levels and noise-reduction aggressiveness may be implemented via a DTMF sequence, which these users may apply throughout a phone conversation. When striking a DTMF sequence during a call, an authorized user may assume control over the VQS channel he or she is seizing. The VQS may be programmed to interpret the DTMF special sequence,[2] and act to satisfy its intent.

Control over speech and sound quality is not the only trait marking the potential value of voice-quality solutions. Coloring of a communication session by way of atmosphere formation, vocal aids, on-demand announcements and jingles, could very well capture a notable portion of the air spectrum in an attempt to add an entertainment factor to the information transfer.

This concept specifies fresh applications, which may be implemented as an integral part of a VQS system. It includes a proposed network architecture intended to support a

---

[1] Just like drivers who prefer manual transmission over automatic when price is not the issue.
[2] The DTMF special sequence may not resemble any standard sequence used to control generally available features such as *67, *69 or voice mail.

universal service capability brought to bear by a creative mobile-communications service provider.

14.2 Controlling sound flavor

Voice effects

Do you wish you could borrow Barry White's voice, if only for the next phone call? Of course, you would, if you are a young male trying to impress a girlfriend.

Using special voice effects could transform a flat-sounding timbre and a boring content into a conversation piece thick with entertainment. For example, you may dress up your voice with flavors to sound (like):

- **Helium** – sounds like you sucked the air out of a balloon.
- **Spooky** – converts your voice to a ghostly sound – perfect for Halloween.
- **Ducky** – makes you sound like a duck.
- **Chipmunks** – makes you sound like – Hey, watch out, Alvin.
- **Masculine** – gets a deep bass sound.
- **Feminine** – get high pitch. It's easier to sing soprano.
- **Deep** – use reverb to emulate a conversation at the Grand Canyon.

If you are a preferred-class subscriber, you could use touch-tone digits throughout the call, fashioning a particular voice effect you wish to impress your listener with.

Virtual places effect

Talking to your friend about your next vacation place, you may illustrate what it feels like over there in dreamland by way of activating a background sound effect with proper flavor and feel as in the following examples:

- **Beach** – seagulls calling; waves crashing.
- **Park** – birds chirping; a gentle breeze resonating.
- **Ball game** – a bat cracking; organ playing.
- **Circus** – clown music; elephant sounds; tigers roar.
- **Forest** – wolves howling; crickets chirping; wind blowing through the trees.

If you are a preferred-class subscriber, you could use touch-tone digits throughout the call, fashioning a particular sound effect you wish to impress your listener with. What's more, you may edit your user profile ahead of the call and have the effect take place with or without any real-time control.

My jingle and my singing telegram

We all appreciate your inexcusable singing voice, but you are keen on singing "Happy Birthday" to your mom on the eve of her 50th. You want to keep her from being nauseated, so you borrow a voice and an arrangement rather than use your own. As she picks up the ringing phone receiver she hears *Happy Birthday To You*, fully arranged with proper harmony, followed by "Hi, mom" in your own voice. It will make her day, and she would even forget about her nagging arthritis.

A singing telegram is an option a privileged subscriber may have in his or her bag of tricks when signing up for a preferred status. Another choice is a 30 second jingle commercial; one is triggered at the commencement of a call, providing a specified number of free minutes of long-distance calling.

If a subscriber is a preferred-class client, he or she could use touch-tone digits through-out the call, fashioning a particular jingle with which to impress his or her listener. What's more, the privileged subscriber may edit his or her user profile ahead of the call and have the jingle go off at the start of the call with or without any real-time control.

A conference-call mute button

You must have experienced an annoying and intrusive background noise on a conference bridge set off by a single leg, troubling every other participant. You can't help it. "Hey, Joe, could you turn your phone on mute?" But he is unable to. His mobile phone does not have a mute button. But, wait. If Joe is a preferred class, or if all legs on the conference bridge are of the preferred type, then a simple DTMF code sent to the VQS would do the trick, and every conference-bridge participant would feel liberated. Relief would be the right word.

14.3   Architecture

Implementation of a preferred-subscriber class in a mobile network could take place if the voice-quality system (VQS) can take cues from

- users, during a call,
- network elements (MSC, a signal-processing terminal (SPT), an MSC adjunct), before the start of a call, and possibly during the call.

The communications apparatus exercised by callers and put forward in this scheme is an exclusive DTMF signal detection and decipherment capability built into a ubiquitously deployed VQS.

The communications mechanism exercised by intelligent network elements could involve any in-band signal, including a DTMF type.

At the start of a call the HLR/VLR or any other intelligent network (IN) element would access the user's profile and advise the MSC as to whether the user is (or is not) a preferred-class subscriber. The MSC (or the IN element) would arrange to precede the standard ring with a specific indicator, signaling to the VQS the start of the call and subscriber type (preferred or standard). It could, just as easily, employ a different signal to provision the circuit according to the subscriber's profile (or lack of it).

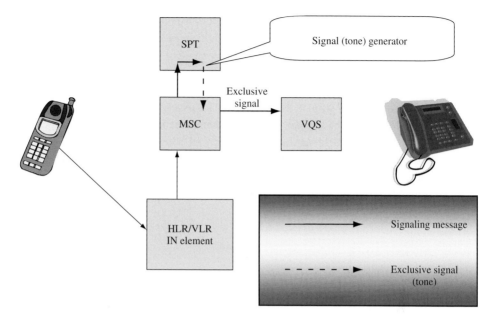

Figure 14.1 Call set-up preceding ringing.

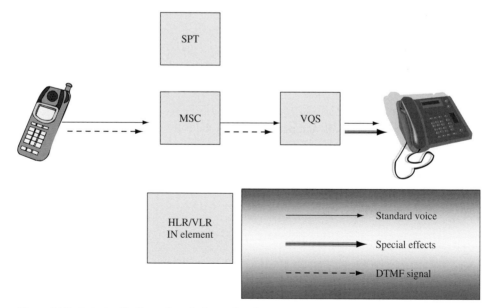

Figure 14.2 Actual call of a preferred-class subscriber.

Once the call starts, the preferred-class subscriber may be able to override and fine tune the initial provisioning via direct DTMF sequences from his or her cellphone.

The sound effects would be generated by some of the stored effects (the ones chosen by the caller) within the VQS positioned in the path of the call at all times.

Figures 14.1 and 14.2, respectively, illustrate the particular implementations.

The VQS would store in its memory sounds associated with the virtual places, including a limited number of specific jingles and announcements, to be potentially triggered and pulled by preferred subscribers during the call. (In the case where the SPT is activated throughout the call, there would be a larger library of jingles and announcements, potentially triggered at the start of the call). Jingles and announcements may be downloaded periodically into the VQS or the SPT in an attempt to provide extended variety while refreshing the special sound library.

15 A procedure for evaluating and contrasting new wireless codecs' performance with and without VQS

15.1 Introduction

In Chapters 2 and 9, I discussed studies that illustrate a considerable weakness of the GSM half-rate codec in the face of noise. I also brought to light study results showing that an effective NR algorithm may be able to remedy some of the HR weakness and lift the voice-quality performance back to par for signals with a relatively poor SNR (see Figure 9.13).

As new wireless codecs are introduced at the market place, many service providers wonder whether VQ systems are effective in enhancing quality, as higher compression ratios are used in an attempt to reduce spectrum requirements and augment air capacity.

In this chapter, I introduce a procedure for testing the hypothesis that a new codec with a higher compression ratio offers an inferior (or equivalent) voice-quality performance in comparison to an existing codec, specifically under noisy conditions.

If the hypothesis proves to be true, then there is a need to test the hypothesis that VQS can lift the performance level of the new codec to a close proximity (or beyond) of the existing codec (under noisy conditions and without VQS).

15.2 Procedure

The procedure outlined here involves two phases.

Phase 1 (see Figure 15.1)

(1) Send pre-recorded speech signals from point B without noise. Capture and record speech signals (pre-recorded) at point A.
(2) Send pre-recorded speech signals from point E with noise. Capture and record speech signals (pre-recorded) at point D.
(3) Send pre-recorded speech signals from point B without noise. Capture and record speech signals (pre-recorded) at point A while sending speech signals from point A (double talk).

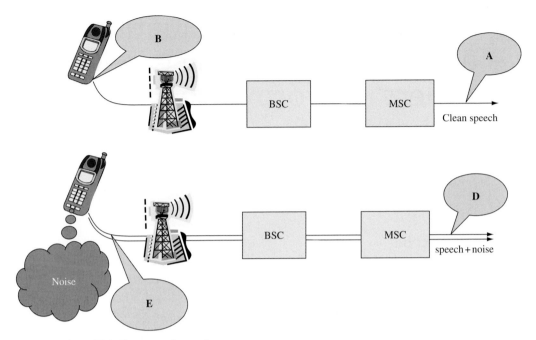

Figure 15.1 Capture and record set-up.

(4) Send pre-recorded speech signals from point E with noise. Capture and record speech signals (pre-recorded) at point D while sending speech signals from point D (double talk).
(5) Capture and record speech signals of natural conversations at point A without noise.
(6) Capture and record speech signals of natural conversations under a variety of noisy conditions (street noise, car noise) where noise is originating at point E.
(7) Note: steps 1–6 must be performed for the existing codecs and repeated for the new codecs.

Phase 2 – lab analysis

(1) Employ PESQ (see Chapter 2) to get preliminary indications of performance directions.
(2) Employ audiovisual subjective evaluation with Cool Edit[1] or other signal-processing software.
(3) Insert a VQS with all voice-quality applications turned on in front of point D and repeat steps 1 and 2.
(4) Generate analysis-and-conclusions document.

---

[1] Popular signal-processing software by Adobe (formerly Sintrillium Software Corporation) www.adobe.com/special/products/audition/syntrillium.html.

16 The theory of sleep

Most people view sleep as a biological necessity designed to rejuvenate and invigorate body and mind so that human beings can function effectively when they are awake. While awake we tend to eat, dispose of waste, spawn the new generation, and take care of business, so that when the day is over we can go back to sleep.

When asked: "What is life's purpose?" we immediately think of life as the time we spend on earth while being awake.

Now let's reverse the paradigm. Let's pretend for a brief moment that life's purpose is a good sleep, and that all of the energetic activities above are designed to sustain our ability to go back and "live" (while asleep) the next day (see Figure 16.1).

I know. This is a weird thought, but let's see if we can apply the concept to voice-quality engineering in wireless networks, so that it makes sense.

Voice-quality systems are designed to operate on voice and to disable themselves on detection of data transmission. In other words, when turning on the VQS, voice-enhancement applications are enabled automatically while the system goes on guard, watching for any data that might require a disabling action.

Now let's reverse the paradigm. Let's disable all or a particular VQ application upon system turn-on, and continue to monitor the channel until voice is detected. When this occurs, the system enables the VQ application, and then puts it back to sleep when the monitored signal no longer looks like voice.

Is this a feasible approach? before I answer, I owe the reader an explanation. I need to give reasons for suggesting this unthinkable way of thinking. I need to substantiate why the paradigm reversal that enables voice processing upon its detection rather than the one that disables voice processing upon data detection is even under consideration.

As the reader went through Chapter 12, it became clear that voice-quality systems must require an exceptionally careful tuning and a comprehensive catalog of any form of non-voice signal and its corresponding detection algorithm, lest subscribers become dissatisfied. Anything less than perfect coverage may set off incompatibility that would break a connection or an attempt to establish it. At the same time, a noise-reduction application that is turned off, even in the presence of acoustic noise, may not cause an equivalent level of discontent. Subscribers may not even comprehend the impairment in the current environment.

Consequently, if voice signals could be detected reliably, then logic reversal that leaves noise reduction disabled until the system detects the emergence of voice may be applied.

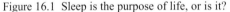

Figure 16.1 Sleep is the purpose of life, or is it?

Unfortunately, VQSs do not currently possess a simple algorithm that can classify voice signals as definite non-data. The best it can do is, as described in Chapter 9, to assign a signal to one of two states – speech or unknown. Unknown may be either speech or data, while speech is a more specific designation. Since the unknown state may be speech as well, the classification may not be used for detecting a non-speech signal.

Employing the signal classification for detecting voice may seem fine so long as detection is done only once after a data session. An attempt to repeat detection and use the outcome for turning VQ applications on and off continuously may yield a very annoying sequence of on and off noise bursts when the system disables the application on detection of an unknown signal, which may just as well be voice.

Although the idea of logic reversal may be an interesting one, it can only be applied once signal classification that detects speech can do so without leaving any doubt that the other possible classification is non-speech rather than unknown. Presently, it is much easier to enable voice applications and disable them upon detection of a non-voice signal, rather than having them disabled as a default, and enable them only when voice is detected as the active signal traversing the channel.

Part VII

Recordings

17 Audio illustrations

The accompanying website (www.cambridge.org/9780521855952) contains audio illustrations of various aspects discussed in the book. These examples are presented in the wav format; they may be listened to on a personal computer with Microsoft Windows Media Player,[1] RealPlayer,[2] or other audio programs that play the wav format. Specialized programs such as Cooledit[3] are best suited for the demonstrations. They offer an audio-visual experience and the ability to analyze and manipulate the signal.

The audio illustrations are divided into the following groups:

- Codec illustrations, where the same speech segment is played through a variety of particular codecs.
- Hybrid echo, where the same speech segment produces echo as a function of delay and ERL combinations.
- Various recordings of noise types, including different crowd noises, wind, seashore, busy street, and white noise.
- Acoustic-echo control under noisy conditions with different comfort-noise matching.
- Recordings that demonstrate the "before" and "after" VQS processing of a conversation taking place under noisy conditions.
- Examples related to cases discussed in Chapter 12.
- DTMF and ring-back tones.

The intent of the audio illustrations is to highlight certain elements that can be better depicted by having them listened to. They can also serve as raw material for those who wish to use the demos in their own presentations and test platforms.

17.1 Codec illustrations

The codec illustrations on the website are all produced without background noise or transmission errors. Consequently, the listener may not always find significant differences in performance between some of the better codecs as compared with the ones with higher compression ratios, or those incorporating noise reduction as an integral part of the codec.

---

[1] www.microsoft.com.
[2] www.real.com.
[3] www.adobe.com/special/products/audition/syntrillium.html.

Still, in spite of this less than complete account, a sensitive ear can detect differences in voice quality.

Although I recognize the insufficiency in this section, I realize that, due to the infinite number of adverse-condition combinations, it is still worthwhile to present the codecs' performance under a common denominator excluding noise and errors. Even though some of the performance differences may not be as sharp, they still serve as a fair illustration of the particular circumstance.

The following is a list of the files containing the recordings in this group[4]:

- CodecTest.wav – unencoded speech segment,
- CodecTest_Alaw.wav – A law encoded and decoded speech segment,
- CodecTest_ulaw.wav – μ law encoded and decoded speech segment,
- CodecTest_AMR475_VAD1.wav – speech segment encoded and decoded with AMR using VAD1 using 4.75 kbps rate,
- CodecTest_AMR475_VAD2.wav – speech segment encoded and decoded with AMR using VAD2 using 4.75 kbps rate,
- CodecTest_AMR515_VAD1.wav – speech segment encoded and decoded with AMR using VAD1 using 5.15 kbps rate,
- CodecTest_AMR515_VAD1.wav – speech segment encoded and decoded with AMR using VAD2 using 5.15 kbps rate,
- CodecTest_AMR59_VAD2.wav – speech segment encoded and decoded with AMR using VAD1 using 5.9 kbps rate,
- CodecTest_AMR59_VAD2.wav – speech segment encoded and decoded with AMR using VAD2 using 5.9 kbps rate,
- CodecTest_AMR67_VAD1.wav – speech segment encoded and decoded with AMR using VAD1 using 6.7 kbps rate,
- CodecTest_AMR67_VAD2.wav – speech segment encoded and decoded with AMR using VAD2 using 6.7 kbps rate,
- CodecTest_AMR74_VAD1.wav – speech segment encoded and decoded with AMR using VAD1 using 7.4 kbps rate,
- CodecTest_AMR74_VAD2.wav – speech segment encoded and decoded with AMR using VAD2 using 7.4 kbps rate,
- CodecTest_AMR795_VAD1.wav – speech segment encoded and decoded with AMR using VAD1 using 7.95 kbps rate,
- CodecTest_AMR795_VAD1.wav – speech segment encoded and decoded with AMR using VAD1 using 7.95 kbps rate,
- CodecTest_AMR102_VAD1.wav – speech segment encoded and decoded with AMR using VAD1 using 10.2 kbps rate,
- CodecTest_AMR102_VAD2.wav – speech segment encoded and decoded with AMR using VAD2 using 10.2 kbps rate,
- CodecTest_AMR122_VAD1.wav – speech segment encoded and decoded with AMR using VAD1 using 12.2 kbps rate,

---

[4]   See Chapter 1 for an overview of codec implementations.

- CodecTest_AMR122_VAD2.wav – speech segment encoded and decoded with AMR using VAD2 using 12.2 kbps rate,
- CodecTest_EFR.wav – speech segment encoded and decoded with EFR,
- CodecTest_EVRC.wav – speech segment encoded and decoded with EVRC,
- CodecTest_FR.wav – speech segment encoded and decoded with FR,
- CodecTest_HR.wav – speech segment encoded and decoded with HR,
- CodecTest_SMV.wav – speech segment encoded and decoded with SMV.

17.2   Echo as a function of delay and ERL

This section of audio illustrations contains combinations of delay (in milliseconds) and ERL. The longer the delay, the more noticeable the echo. The lower the ERL, the louder the echo. Figure 17.1 maps out the recording combinations.

The following is a list of files containing the recordings in this group:

- no_echo.wav – original speech with no echo,
- echo_d50_ERL3.wav – speech with echo at 50 ms delay and ERL of 3 dB,
- echo_d50_ERL6.wav – speech with echo at 50 ms delay and ERL of 6 dB,
- echo_d50_ERL12.wav – speech with echo at 50 ms delay and ERL of 12 dB,

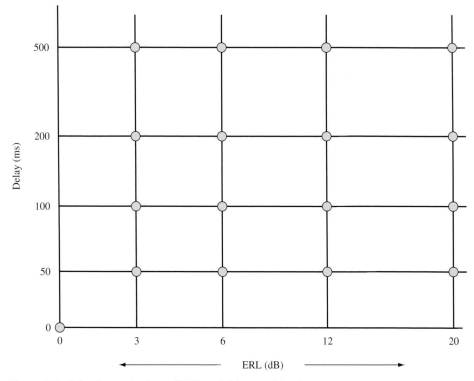

Figure 17.1   Echo demonstrations of ERL and delay combinations.

- echo_d50_ERL20.wav – speech with echo at 50 ms delay and ERL of 20 dB,
- echo_d100_ERL3.wav – speech with echo at 100 ms delay and ERL of 3 dB,
- echo_d100_ERL6.wav – speech with echo at 100 ms delay and ERL of 6 dB,
- echo_d100_ERL12.wav – speech with echo at 100 ms delay and ERL of 12 dB,
- echo_d100_ERL20.wav – speech with echo at 100 ms delay and ERL of 20 dB,
- echo_d200_ERL3.wav – speech with echo at 200 ms delay and ERL of 3 dB,
- echo_d200_ERL6.wav – speech with echo at 200 ms delay and ERL of 6 dB,
- echo_d200_ERL12.wav – speech with echo at 200 ms delay and ERL of 12 dB,
- echo_d200_ERL20.wav – speech with echo at 200 ms delay and ERL of 20 dB,
- echo_d500_ERL3.wav – speech with echo at 500 ms delay and ERL of 3 dB,
- echo_d500_ERL6.wav – speech with echo at 500 ms delay and ERL of 6 dB,
- echo_d500_ERL12.wav – speech with echo at 500 ms delay and ERL of 12 dB,
- echo_d500_ERL20.wav – speech with echo at 500 ms delay and ERL of 20 dB.

17.3   Noise types

The different noise types may be useful for those who may wish to incorporate these into their own sample recordings for demonstration purposes. The included types are recorded in the following files:

- W_Noise.wav – white noise,
- airport.wav – airport noise,
- café.wav – café noise,
- crowd-noise.wav – crowd noise,
- seashore.wav – seashore noise,
- train.wav – train noise,
- windy.wav – wind noise,
- noisy street.wav – noisy street.

17.4   Acoustic-echo control and comfort-noise matching

To demonstrate how important proper comfort-noise matching is, I produced five recordings. The first one exemplifies an untreated acoustic echo; the others depict suppression techniques accompanied by four different types of noise fill ranging from silence, through white and colored, to a spectral match.

Figure 17.2 exhibits the various different scenarios illustrated in the recordings. The following is a list of the files containing the recordings in this group.[5]

- echo and noise.wav – untreated acoustic echo with background noise,
- AEC and no NM.wav – acoustic-echo control with no comfort-noise fill,

---

[5]   For further details on the characteristics of comfort-noise matching see Chapter 4.

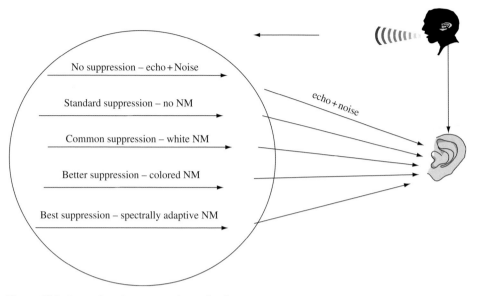

Figure 17.2 Acoustic-echo suppression technology.

- AEC and W NM.wav – acoustic echo control with white-noise fill,
- AEC and colored nm.wav – acoustic echo control with colored comfort-noise fill,
- AEC and spectral nm.wav – acoustic echo control with spectral comfort-noise fill.

17.5 Before and after VQS

This group of recordings demonstrates a broader scope comprising two voice-quality applications – noise reduction and noise compensation. Figure 17.3 depicts a scenario where a woman is surrounded by high acoustic noise. The displayed signals on both sides of the VQS equipment illustrate what the other party is listening to. The woman's side transmits speech and noise, but after it is processed by the VQS the noise is reduced, and the man is listening to noise-reduced speech. On the other side, without VQS, the woman would have listened to the man's speech, as it would have been masked by the noise surrounding her. After the VQS NC processing, the man's speech is amplified without affecting the noise level to make it stand out more clearly.

The following is a list of the files containing the recordings in this scenario.

- Female before NR.wav – represents the top-right signal in Figure 17.3,
- Female after NR.wav – represents the top-left signal in Figure 17.3,
- Male before NC.wav – represents the bottom-left signal in Figure 17.3,
- Male after NC.wav – represents the bottom-right signal in Figure 17.3.

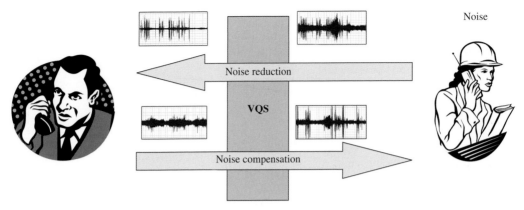

Figure 17.3 VQS before and after.

17.6   Trouble shooting

Chapter 12 comprises examples of common troubles associated with voice quality and malfunctioning voice-quality systems. This section contains examples supporting cases discussed in the chapter.

The following is a list of the files containing audio demonstrations.

- example_12.1.wav – This file contains a simple demonstration of consistent hybrid echo. It may be used as a reference for the next examples.
- example_12.3a.wav – This file illustrates echo at the beginning of the call before convergence takes place.
- example_12.3b.wav – This file illustrates a louder echo (in comparison with 12.3a) at the beginning of the call due to NC.
- example_12.4.wav – This file illustrates how NC may cause saturation of speech when not controlled properly.
- example_12.5.1.wav – This file illustrates hybrid echo. It ought to be contrasted with example_12.5.2.wav.
- example_12.5.2.wav – This file illustrates acoustic echo. It ought to be contrasted with the previous example to sense the difference between the two echo types.[6]
- example_12.6.wav – When echo is returned with no loss (0 dB ERL), it is as loud as the speech source, and echo cancelers find it extremely difficult to cancel. The file illustrates echo at 0 dB ERL.
- example_12.7.1.wav – This file ought to be used as a reference for the next file.
- example_12.7.2.wav – This file illustrates how an improperly tuned ALC may amplify noise, and may even cause periodic speech saturation.
- example_12.11.wav – This file illustrates how clipped peaks sound, even though level has been restored to a comfortable listening.

---

[6]   The main differences are consistency, relationships to the speech source, and level.

- example_12.14.1.wav – This file ought to be used as a reference for the next two files. It is the signal entering the R_{in} port of the VQS.
- example_12.14.2.wav – This file contains the signal in example_12.14.1.wav after it has passed through a noise-reduction application.
- example_12.14.3.wav – This file contains the signal in example_12.14.1.wav after it has passed through an overly aggressive noise-reduction application.

17.7   Ring-back and DTMF signals: example test tools

This directory includes recordings that may be used to verify the correct operation of VQ systems during ring-back and DTMF signaling. For a more comprehensive test, a robust suite of tests ought to include varied power levels, frequency deviations, and different durations to test the complete specifications of the standards.

Ring back

In the United States, a large majority of carriers use a de-facto ring-back standard during call initialization. This ring-back tone is the sum of a 440 Hz and a 480 Hz signal for two seconds duration. Each tone pulse has a four second silence period between pulses. Research has shown that, in most cases, a common ring-back tone exists at levels from −20 dBm0 to −15 dBm0.

Test and verification

Three files are used to test ring-back through VQS systems:

- network_ring.wav – series of ring-back pulses that were recorded from an actual US commercial carrier network,
- ring.wav – series of ring-back pulses that were generated using commercial signal-processing software,
- ring_low.wav – series of ring-back pulses that were generated using commercial signal-processing software at a lower power level.

Each file can be played independently through the S_{in} or R_{in} ports of a VQS. The test is conducted by evaluating the output recording subjectively. There should be no indication of distortion, changes in level, or changes in frequency content. This is verified both audibly and visually using commercial software.

Test and verification of DTMF

Four files are used to test DTMF through VQS systems. Each file consists of the following string of digits "1234567890*#".

- dtmf_50_50.wav – series of DTMF digits with a 50 ms duration and 50 ms silence between digits,
- dtmf_50_50_low.wav – series of DTMF digits with a 50 ms duration and 50 ms silence between digits, sent at a lower power level,
- dtmf_500_500.wav – series of DTMF digits with a 500 ms duration and 500 ms silence between digits,
- dtmf_500_500_low.wav – series of DTMF digits with a 500 ms duration and 500 ms silence between digits, sent at a lower power level.

Glossary of common voice-quality systems terminology

1X multi channel (1XMC)   An advanced wireless technology that is part of the cdma2000 standard.

2G   2nd generation wireless network, technology, or standard.

2.5G   2nd $+ \frac{1}{2}$ generation wireless network, technology, or standard.

3G   3rd generation wireless network, technology or standard.

3GPP   3rd generation partnership project. 3G wireless standards are being developed jointly by several product vendors to ensure compatibility.

3XMC   3X multi channel. This is an advanced wireless technology that is part of the cdma2000 standard.

8-PSK   8-state phase-shift keying. A PSK modulation technique that uses eight unique states.

A-interface (or A-link)   The link between the mobile-switching center (MSC) and the TRAU leading to the base station in a GSM network. Echo cancelers are sometimes placed here to remove acoustic echo and for voice enhancement in mobile-to-mobile calls.

Acoustic echo (AE)   Echo generated by sound reflection in a telephone subscriber's environment, such as walls or the interior of a car, which is mechanically coupled to the transducer of a telephone handset.

Acoustic-echo control (AEC)   Acoustic echo control refers to the capability of a voice-quality system to reduce voice-signal degradation caused by the environment surrounding a subscriber (e.g., ambient noise in an airport, factory, restaurant, etc.). Acoustic echo may be present in both wireline and wireless applications. Most wireline applications that encounter delay are equipped with echo cancelers. Therefore, standard echo-canceler algorithms and non-linear processing (NLP) techniques used to eliminate electrical echo also control acoustic echo in wireline applications. Originally, it was assumed that echo cancelers were not needed for wireless applications. Wireless subscribers were not expected to experience either electrical or acoustic echo because (1) hybrids (the primary source of electrical echo) are not present in digital-wireless networks, and (2) standards for wireless-equipment manufacturers require mobile-handset designs to have sufficient loss to eliminate acoustic echo. Unfortunately, the equipment manufactured by many wireless-phone vendors does not fully comply with the official standards that specify adequate isolation between the receiver (earpiece) and the microphone (mouthpiece) of digital-mobile phones. In addition, many mobile subscribers operate their equipment in moving vehicles (configured with hands-free features and speaker-phone arrangements), noisy public areas, and in a variety of environmental conditions. The uncontrolled acoustical characteristics and changing noise levels surrounding mobile subscribers returns a significant amount of acoustic echo (and noise) via the voice path. As a result, acoustic echo is a primary concern for wireless-service providers. Furthermore, the acoustic-echo problem is more pronounced in the case of digital-wireless applications because of long processing delay times (>200 ms round-trip delay) introduced by speech compression techniques and the non-linearity of speech-compression algorithms. Two techniques, noise reduction (NR) and noise compensation (NC), used to reduce or eliminate acoustic echo, are highly desirable for wireless applications.

Adaptive intelligent near-end speech detection   Detection of "double-talking" (i.e., the condition when both subscribers on a call speak simultaneously) is enhanced so that near-end and far-end speech signals are accurately distinguished. This feature prevents "clipping" and distortion of voice-band signals during double talking.

Adaptive multi rate (AMR)   A GSM and W-CDMA codec comprising eight source codecs ranging from 4.75 kbps to 12.2 kbps. The AMR employs extensive error control, which greatly improves its performance under severe to moderate error conditions.

Air-interface user rate (AIUR)   The user rate is the rate seen by the mobile terminal (e.g., modem). This data stream can be broken into sub-streams before transmission over one or more radio channels. This is analogous to the fixed-network user rate (FNUR) on the PSTN side.

Algebraic CELP (ACELP)   a voice-coding procedure, which was adopted for the GSM enhanced full-rate (EFR) codec. The ACELP reduces the complexity of the codebook search and removes the need for fixed codebook storage at both the encoder and decoder. It trades it off by resorting to a more exhaustive series of nested loop procedures, which determine critical components of the excitation pattern algebraically.

Alternate mark inversion (AMI)   A line code that employs a ternary signal to convey binary digits, where successive binary ones are represented by signal elements, normally of alternating positive and negative polarity of equal amplitude, and with binary zeros represented by signal elements that have zero amplitude.

Amplitude   The strength or volume of a signal, usually measured in decibels.

Analog signal   A continuously varying physical quantity (such as voltage), which reflects the variations in a signal source with respect to time. For example, changes in the amplitude and pitch of the human voice during speech.

Analysis by synthesis (AbS)   A voice-coding methodology, which merges a linear prediction scheme that models properties of the vocal tract with an adaptive excitation signal chosen via an algorithm minimizing the error (using the least-squares algorithm) between the input speech and the reconstructed version.

American National Standards Institute (ANSI)   Coordinates the United States voluntary consensus standards system, and approves American national standards. In addition, ANSI also influences the development and content of international standards to meet current technological needs and to facilitate competitiveness in the marketplace.

Asynchronous transfer mode (ATM)   A facility signal-formatting standard.

Audio transparency   A technique that allows actual low-level background sound (music) to pass through a connection even when the NLP function of an echo canceler "opens" the circuit to block residual echo. This method of eliminating residual echo while permitting unique audio signals to "sneak through" improves continuity and enhances the quality of highly compressed speech and voice signals that have long delay characteristics.

Automatic level control (ALC)   An application that adjusts the listening level to a comfortable region. It does so by comparing the power level of an incoming voice signal with a "pre-provisioned reference level." Based on the difference between the incoming signal level and the reference level, either gain or attenuation is applied to the incoming signal to produce a "corrected" voice signal that has an average speech power level that approximately matches the "standard or reference" level.

Bandwidth   The frequency range used by a telecommunications channel. For narrow-band (telephone) applications, this is defined as 300–3000 Hz. For wide-band applications, (such as video conferencing) this is defined as 50–7000 Hz.

Base-station controller (BSC)   A land-based control and management point for a wireless communications system. The installation houses the equipment needed to set up and complete calls on cellular phones, i.e. FM radio transmitter and receiver, antenna, and control elements. The BSC is positioned between the subscriber's handset and the mobile-switching center.

Base Transmitter Station (BTS)   The transmitter and receiver point between the mobile terminal and the BSC.

Bipolar 3-zero substitution (B3ZS)   A line-coding method that replaces a string of three zeros with a sequence of symbols having special characteristics, to ensure an adequate density of ones for signal recovery.

Bipolar 8-zero substitution (B8ZS)   A line-coding method that replaces a string of eight zeros with a sequence of symbols having special characteristics, to ensure an adequate density of ones for signal recovery.

Bit-error rate (BER)   A measure of transmission accuracy expressed as the ratio of bit errors received to bits sent. A common voice and data BER benchmark is 10^{-9} or one error in a billion.

Broadband   A high-rate facility signal with sufficient bandwidth to provide all the DS1/E1 and 64 kbps channel-synchronization references required for proper operation in a telecommunications network. Typically STS-1, DS-3, OC-3, STM-1, and higher-rate signals.

Built-in self-test (BIST)   The ability for a system to verify automatically the proper operation of its functions, typically performed during the initial testing on system power-up.

Carrier signal   An electromagnetic wave that can be modulated, as in frequency, amplitude, or phase, to transmit speech, music, images, or other signals.

Carrier-to-interference ratio (C/I)   An index indicating the severity of error conditions in the transmission channel. The lower the ratio the poorer the channel quality.

Cdma2000   A 3G CDMA wireless-system standard succeeding cdmaOne.

CdmaOne   A 2G CDMA wireless-system standard.

CdmaTwo   A 2G or 2.5G CDMA wireless-system standard succeeding cdmaOne.

Cellular networking   A mobile-communications system that contains radio-transmission and telephone-switching components. Cellular networking permits telephone communication to and from mobile subscribers within a defined geographical area. Each large area is divided into cells, each of which has its own mobile base station (BS) and mobile-switching center (MSC) connected to the public switched-telephone network (PSTN).

Central office (CO)   Usually used to refer to a local telephone switching system, which connects "lines-to-lines" and "lines-to-trunks." This term is also generally applied to network-switching systems, and sometimes loosely refers to a telephone-company building in which switching or other types of equipment are located.

Channel   A single path in a transmission medium. A channel forms a connection between two nodes in a network. The term channel typically refers to a 64 kbps signal (DS0 or equivalent), which carries a single subscriber's service connection.

Circuit   The complete path between two end points that carries information.

Clear channel   A channel in which the entire bandwidth is available for carrying user-payload information. A clear channel is not subjected to any type of processing (such as echo cancelation). Often referred to as "64 (kbps) clear channel" or "transparent channel" operation.

Code-division multiple access (CDMA)   A digital multiplexing method used for wireless telephony signals. The methodology is popular in North America with major carriers such as Verizon Wireless and Sprint PCS, and in Korea.

Code-excited linear prediction (CELP)   A voice-coding platform, which uses a code-book (look-up table) for constructing the finest match, and which contributes to a reduction in algorithm complexity and the required bandwidth.

Convergence rate   The convergence rate is the speed, as measured in dB per second, applied by an echo canceler when deriving a best estimate of the echo signal. An echo canceler with a fast convergence rate will remove echo quickly. To meet the initial convergence time of G.168, an echo canceler must have a convergence rate of at least 20 dB/s. In some implementations, a fast convergence leads to a fast divergence, which may introduce instability or additional noise. An echo canceler with dual convergence rates benefits from both characteristics. The initial convergence rate has "fast onset" characteristics at the beginning of a call that quickly remove the echo, then a slower continuous-echo-adaptation convergence rate is used for the remainder of the call. This approach prevents noise bursts and maintains echo cancelation stability throughout the call.

Degradation mean opinion score (DMOS)   This measurement tool is equivalent to the MOS metric but it is more specific, as it provides for characterizations introduced by variations in noise level. Like the MOS scale, the DMOS metric is based on a five-point degradation category scale.

Delay   The amount of time it takes a signal to travel between two points in a network.

Demultiplexing   The process applied to a multiplexed signal that is used to recover (extract) the signals embedded within it. The term is also used to describe recovery of individual channels (e.g., DS0s) embedded in a signal.

Digital – advanced mobile-phone service (D-AMPS)   A wireless technology used in the North American TDMA.

Digital (data) signal   A signal composed of discrete quantities that have well-defined states at any point in time. Common examples of digital signals are those used to transfer information between computers and data terminals.

Digital-signal level 0 (DS0)   A 64 kbps channel, typically used in North American telecommunications networks.

Digital-signal level 1 (DS1)   A 1.544 Mbps channel. It is equivalent to 24 DS0 channels, typically used in North American networks. See T1.

Digital-signal processor (DSP)   An integrated circuit device, which can be programmed to perform a variety of signal processing functions.

Digital subscriber line (loop or link) (DSL)   This is a general term describing the electrical (e.g., cable or wire) connection between a subscriber and a telephone network.

Disable   The technique used to prevent the processing of signals by applying a control signal to inhibit the operation of a device.

Discontinuous transmission (DTX)   A method (used mostly in GSM) of powering down, or muting, a mobile or portable telephone set when there is no voice input to the set. This procedure optimizes the overall efficiency of a wireless voice-communications system.

Dispersion   This term describes characteristics that determine the width of a signal impulse response. In general, electrical (hybrid) echo signals have relatively short reverberation times and small dispersion characteristics. Conversely, acoustic echo signals tend to have long reverberation times and large dispersion characteristics.

Dissipation   The act or process of consuming power or energy.

Double talk   The state of the echo canceler when there is speech on both ends of the connection. This is the most difficult state to detect accurately, and most problems with audio quality occur during double talk.

Downstream/upstream   The transmitted-signal-path direction relative to the received-signal-path direction. For example, from the send in port to the send out port is considered the downstream direction. In contrast, from the receive in port to the receive out port is considered the upstream direction.

E1   In global applications (other than North America and some Central American countries), a signal format with a rate of 2.048 Mbps that contains the equivalent of 32 64 kbps channels. Usually, 30 channels are used for subscriber information (typically voice) and 2 channels are used for signaling purposes. A 31-channel configuration (31 channels of voice and 1 channel for signaling) is also used for specific applications.

EC-1   Electrical-carrier signal level 1. It is a basic logical building-block facility signal with an electrical interface that has a rate of 51.844 Mbps. It is equivalent to an STS-1 signal that is implemented with an electrical interface.

Echo   An audible signal in a telephony connection that resembles speech and has sufficient magnitude and delay to be perceived as interference. Electrical echo is caused by impedance mismatches at the four-wire to two-wire hybrid in telecommunications networks. Acoustic echo is caused by sounds (other than the subscriber's voice) being reflected into the transducer (mouthpiece) of a telephone handset.

Echo cancelation (EC)   The process in which the echo signal in a telecommunications network is detected and eliminated by subtracting a replica of the echo, thereby "canceling" it. An EC is a.k.a. **electrical-echo cancelation (ECC)**.

Echo canceler (EC)   A product or system in a telecommunications network that performs echo cancelation. Typically applied to electrical (hybrid) echo cancelation, but may also contain features for handling acoustic echo.

Echo-path change detector   The EPC detector monitors the echo path and automatically adjusts the convergence rate accordingly. Echo-path characteristics can change when a new call is initiated (also called the call boundary), or during a call that is in progress when a subscriber is transferred to a different connection or bridged onto a conference.

Echo-path delay   Commonly known as tail delay, tail length, or echo-tail delay. The delay incurred by a signal traversing the path between the R_{out} and S_{in} terminals of an electrical (hybrid)-echo canceler as viewed from the echo canceler located at the far end (e.g., network or transport) side of a network. The echo canceler physically located at the far-end point in a network cancels the echo signal that would have been returned to the near-end subscriber (i.e., located at the hybrid or switch side of the network). Electrical (hybrid) echo cancelers are traditionally designed to handle an echo-path delay of approximately 64 ms.

Echo-return loss (ERL)   ERL is the hybrid loss characteristic that affects the level of the echo signal returned towards the subscriber that is speaking. A low ERL characteristic results in a greater amount of objectionable echo being returned to the subscriber who is speaking. Conversely, a high ERL reduces the amount of objectionable echo that is returned to the subscriber who is speaking.

Echo-return loss enhancement (ERLE)   ERLE is the level of signal attenuation that the linear convolution processor of the echo canceler inserts in the echo return path to reduce or eliminate the echo signal heard by the subscriber who is speaking. Echo cancelers that have "**stable ERLE**" characteristics typically provide signal attenuation that is greater than 34 dB when the non-linear processing (NLP) feature is disabled.

Echo suppression   The process in which echo signals in a telecommunications network are eliminated (removed) by opening the circuit path in one direction. An echo suppressor detects speech signals in either direction on a four-wire circuit and introduces loss (e.g., blocks or opens the connection) in the opposite direction of transmission to suppress the echo. In general, echo suppression technology has been replaced by echo cancelation.

EDGE   Enhanced data rates for GSM evolution. Standards that apply when evolving GSM to support 3G wireless applications. See global system for mobile communications.

EEC   Electrical-echo cancelation. See echo cancelation.

Electromagnetic capability   In communications, electromagnetic capability is the relative immunity of a device or devices to the effects of externally generated electromagnetic fields.

Electromagnetic compatibility (EMC)   This is the ability of an electronic system to operate without disruption in the presence of external electromagnetic fields, without radiating electromagnetic fields large enough to disturb other equipment.

Electromagnetic interference (EMI)   Disturbances in the operation in an electronic system caused by electromagnetic fields originating from outside sources. An electromagnetic field is a combination of alternating electric and magnetic fields.

Electrostatic discharge (ESD)   The discharge of static electrical energy that is potentially harmful to electronic equipment or components.

EMC   See electromagnetic compatibility.

Enable   The activation of a function by applying or removing a controlling signal.

Enhanced full rate (EFR)   A GSM codec, which improves voice performance relative to the full rate (FR), while employing the ACELP procedure.

Enhanced variable-rate codec (EVRC)   A CDMA codec that was the first to introduce RCELP as a methodology for voice coding. The EVRC operates at rates of 8 kbps, 4 kbps, 1 kbps, and 0 kbps – rate 1, rate $\frac{1}{2}$, rate $\frac{1}{8}$, and blank, respectively.

EPC (detector)   See echo-path change.

Errored seconds (ES)   A performance monitoring parameter used in telecommunications systems to indicate the stability of a transmission path.

ESD   See electrostatic discharge.

Extended CELP (eX-CELP)   A coding algorithm that is based on a combined closed-loop–open-loop analysis (COLA). In eX-CELP the frames are first classified as: silence or background noise, non-stationary unvoiced, stationary unvoiced, onset, non-stationary voiced, stationary voiced. The algorithm includes voice-activity detection (VAD) followed by an elaborate frame-classification scheme.

Extended super frame (ESF)   A signal format used for DS1 facilities. See superframe.

Facility bypass   In some echo cancelers, when an error condition is detected, service is maintained (i.e., the call remains active) by "bypassing" (routing around) the echo canceler. This technique prevents service interruption of calls that are in progress, but can also temporarily result in degraded voice quality while the trouble condition persists (i.e., echo cancelation is not performed during facility bypass).

Far end (far side)   The connection point farthest away from the local switch that is terminated on the R_{in} and S_{out} ports of the echo canceler. The far end is also known as the network side, line side, or transport side. The preferred term for identifying the R_{in} and S_{out} port orientation of an echo canceler is called the "receive side" (equivalent to far end) for both wireline and wireless applications.

Four-wire interface   The basic transmission interface where four wires carry message signals. Two wires (a pair) are used for each direction of transmission.

Frame   (1) The box-like structure that contains telecommunications equipment. (2) A collection of bits arranged in a recurring pattern in a facility signal. See frame structure.

Frame-alignment signal (FAS)   The first bit in time-slot 0 (TS0), intentionally encoded as a binary one. This intentional violation of the line-code format is used to indicate the beginning point of the frame structure.

Frame structure   For E1 facilities, the organization of the 256 bits in each frame, which contains 32 eight-bit time slots (64 kbps channels). For T1 facilities, the organization of the 193 bits in each frame, which contains 24 eight-bit time slots (64 kbps DS0 channels) plus one framing bit.

Frequency   The number of oscillations (vibrations) in one second. Frequency is measured in hertz (Hz), which is the same as "oscillations per second" or "cycles per
second." For example, the alternating current in a wall outlet in the USA and Canada is
60 Hz. Electromagnetic radiation is measured in kilohertz (kHz), megahertz (MHz)
and gigahertz (GHz). See wavelength, frequency response, and carrier signal.

Frequency response   In an audio system, the accuracy of sound reproduction. A totally
flat response means that there is no increase or decrease in volume level across the
frequency range. Measured in decibels (dB), this would be plus or minus 0 dB from
20 Hz to 20 000 Hz. A high-end audio system can deviate by $+/- 0.5$ dB, and a
CD-ROM drive should not be off by more than $+/- 3$ dB.

Full rate (FR)   An early GSM codec that employs the RPE-LTP approach.

Gateway   The point where two or more networks are connected. A gateway is the
boundary between the internal protocols of different networks that are connected
together.

Gateway application   The use of an echo canceler in a transmission system that includes
one or more gateways.

Gateway MSC (GMSC)   See mobile-switching center (MSC).

General packet-radio service (GPRS)   A GSM (2.5G) high-speed packet data transmission standard.

Gateway SGSN (GSGSN)   A centralized SGSN element in a 3G wireless network that
usually has broadband-facility interfaces. See service GPRS support node.

Global system for mobile communications (also group special mobile) (GSM)   A form
of TDMA protocol multiplexing for wireless-telephony signals that uses minimum shift-
keying (MSK) modulation technology.

Graphical user interface (GUI)   An interface used by technicians and maintenance
personnel to operate equipment. A GUI usually has graphic displays and menu-driven
commands.

Half duplex   Behavior of most speaker phones, which prevents howling and acoustic
echo by allowing only one party to talk at a time.

Half rate (HR)   A GSM codec, which compresses voice signals into an 8 kbps digital
signal while employing the VSELP procedure.

High-level compensation (HLC)   The high-level compensation (HLC) feature is
intended to ensure that hot signals are cooled off before being processed by the
other voice-quality enhancement applications.

High-speed circuit-switched data (HSCSD)   The transmission of data signals over
circuit-switched facilities, which has a theoretical speed of 57.6 kbps.

Home-location register (HLR)   The home-based registration point for a mobile subscriber in a wireless network that contains user-identification information.

Howling   Pitched, squealing tones that occur when hands-free systems at both ends of a
connection have open speakers and microphones. This is similar to microphone feedback in an auditorium. Howling can cause damage to audio components if it is not
attenuated.

H-register   The register in an echo canceler that is used to store impulse response sample
values detected in the echo path.

Hybrid   A piece of equipment (typically consisting of passive components) used as a transition between two-wire and four-wire circuits in a telecommunications network.

Impedance   The opposition to current flow in an electrical circuit, or a device that performs this function.

Initial convergence time   The time it takes an echo canceler to converge to a specified amount of cancelation at the beginning of a call. In G.168, the canceler must have an initial convergence time of 1 second, and must achieve 20 dB of cancelation within this time. Manufacturers may alternatively refer to the convergence rate of the echo canceler on their data sheet.

Integrated Services Digital Network (ISDN)   An advanced telecommunications network that provides "end-to-end" digital connectivity, customer-controlled features, upward compatibility of services, and standardized user-network interfaces.

Inter-exchange carrier (IXC)   (1) The category of service providers that specialize in competitive long-haul network interconnection. (2) The network facilities that are used to interconnect exchange switches to form large backbone networks.

International gateway   A telecommunications system that typically performs bi-directional conversion of signal formats and protocols to allow communication between different countries (e.g., E1/T1 conversion with A-law/μ-law encoding).

International Mobile Telecommunications-2000 (IMT-2000)   This is the global standard for third-generation (3G) wireless communications, defined by a set of interdependent ITU recommendations. IMT-2000 provides a framework for worldwide wireless access by linking the diverse systems of terrestrial- and satellite-based networks. It will exploit the potential synergy between digital-mobile telecommunications technologies and systems for fixed and mobile wireless access systems.

International Telecommunication Union (ITU)   A United Nations agency that sets international standards for telecommunications (ITU has replaced the CCITT name).

International Telephone and Telegraph Consultative Committee (CCITT)   This is an international advisory committee under the United Nation's sponsorship. It issues standard recommendations for international telecommunications. Its name has changed to the International Telecommunications Union Telecommunications Standards Sector (ITU-TSS). See International Telecommunications Union.

Internet protocol (IP)   A packet-transmission technology used in telecommunications and computer-data networks.

Jack   A device into which a plug is inserted to complete an electrical connection. For example, the craft interface terminal jack is the physical connection point between a PC (or terminal) and an echo canceler.

Key   (1) A manual switch or signaling device. (2) A physical interlock, such as a "keyed" circuit pack connector.

Least mean squares   A method for obtaining a best fit for correlated variables by minimizing the sum of mean (average) squared errors resulting from linear mathematical relationships.

Least squares   A method for obtaining a best fit for correlated variables by minimizing the sum of squared errors resulting from linear mathematical relationships.

Line   A device for transferring electrical energy (signals) from one point to another in a network. In telecommunications, a line refers to a "transmission conductor" (i.e., wire, cables, or facilities) that carries two-way signals between subscribers.

Line format   A variety of encoding schemes used for transmitting data bits over telecommunications transmission systems.

Line-format violation (LFV)   The term used to describe a variety of errors in the line-code format.

Linear predictive coding (LPC)   A short-term predictive scheme that generates the next speech sample by resorting to a linear combination of the preceding 16 samples.

Listener   The subscriber in a two-way telephone conversation who is not speaking at a particular interval in time.

Local-area network (LAN)   A network that typically interconnects data terminals using bus or ring topologies and routers, and often relies on the availability of leased transmission facilities.

Local-exchange carrier (LEC) (in the USA)   Providers of telecommunications interconnection and switching services to local subscribers.

Long-term prediction (LTP)   A coding methodology, which employs samples from between 20 and 120 apart from the predicted one. The predictive coefficients are chosen by the least-squares method, providing the greatest correlation to the actual speech sample.

Loss   The difference (measured in decibels) between the power applied directly to a load and the power applied to a load through a network element (apparatus). Also known as insertion loss.

Maintenance-tone detection and disable   When performing maintenance routines on telecommunications equipment, it is often necessary to disable echo-canceler equipment and other VQ applications that are in the transmission path. When the maintenance tone is detected, the VQS disables the voice-quality applications (e.g., typically providing a 56 kbps clear channel connection). The typical default maintenance tone is a continuous 2225 Hz signal. In addition, some VQS allow customers to provision the maintenance-tone detection-and-disable feature as "on" or "off," while selecting a variety of frequency tones for different testing applications.

Major alarm (MJ)   A fault condition that requires immediate maintenance action. A major alarm typically indicates a service interruption or the failure of a significant function or feature of a telecommunications system.

Mean opinion score (MOS)   MOS is a standardized method used to evaluate the sound quality of telephone signals subjectively using a scale of 1 to 5 (1 indicating poor quality and 5 indicating excellent quality).

Mean time between failures (MTBF)   For a particular time interval, the total functioning life of a population of items or components divided by the total number of failures within the population, during the measured interval (usually calculated using hours or years).

Minor alarm (MN)   A relatively low-priority fault condition that does not require immediate maintenance action. A minor alarm typically indicates a failure of non-essential functions or features in a telecommunications system that does not interrupt or seriously degrade service (i.e., the ability to carry calls or traffic is not affected).

Mobile-switching center (MSC)   Usually refers to a wireless-switching system used to route calls between base stations and providing an interface to the PSTN for handling wireless long-distance calls.

Modem (modulator–demodulator)   A device that allows a computer or terminal to transmit data over a standard telephone line. It converts digital pulses from the computer to audio tones that an analog telephone line is set up to handle and vice versa. The term usually refers to 56 kbps modems, of which V.92 is the latest.

Modem tone-detection and disable   When a facility is used to carry voice-band data (e.g., modem or signals), echo cancelers that are in the transmission path must be disabled to insure bit and byte integrity of the transmission. Echo cancelers are required to disable echo cancelation upon detection of a continuous 2100 Hz tone with periodic phase reversal (as per ITU recommendation G.165 and G.168), and a 2100 Hz tone without periodic phase reversal (as per ITU recommendation G.164).

Modulation is the process of varying a *carrier signal*, typically a sinusoidal signal, in order to use that signal to convey information. One of the three key characteristics of a signal are usually modulated: its phase, frequency or amplitude. A device that performs modulation is known as a *modulator* and a device that performs demodulation is known as a *demodulator*. A device that can do both operations is called a modem, a contraction of the two words.

Multi-frame alignment   A signal format scheme that is used to indicate the first frame position of a multi-frame structure.

Multiplexer (MUX)   A device capable of multiplexing signals into a single stream of binary data.

Multiplexing   The process of combining several distinct digital signals into a single composite digital signal with a higher rate of transmission.

Near end (or near side)   The connection point closest to the local switch that is terminated on the S_{in} and R_{out} ports of the echo canceler. The near end is also known as the switch side, hybrid side, or echo-source side. The preferred term for identifying the S_{in} and R_{out} port orientation of an echo canceler is called the "send side" (equivalent to near end) for both wireline and wireless applications.

Network equipment-building system (NEBS)   A generic standard defining environmental and structural requirements for equipment that is typically installed in central office applications.

Node   In SONET applications, a node is a line-terminating element.

Noise compensation (NC), a.k.a. adaptive level control   If the surrounding noise level entering a subscriber's ear in a noisy environment (e.g., airport, restaurant, factory, or moving vehicle) is exceptionally high, he or she may be required to physically block the exposed ear canal (cover the exposed ear with a hand) to hear the conversation. This is necessary because the level of the voice signal being produced by the telephone's transducer (the handset's earpiece) is not sufficient to be easily heard. The technique called noise compensation (NC) measures the noise level surrounding the subscriber, and automatically adjusts (either increases or decreases) the volume of the voice signal accordingly. This method of *compensating* for ambient noise, by adjusting the volume of the voice signal delivered to the earpiece, allows a subscriber to hear the conversation clearly under varying conditions. As a result, the subscriber perceives an improvement in the voice quality. This feature is especially important for wireless applications because of the wide range of conditions encountered by mobile subscribers.

Noise matching (NM)   Traditional echo cancelers perform noise matching (NM) by substituting Gaussian white noise with the same amplitude as the existing background noise. However, the spectrum of Gaussian noise is not correlated with the original noise spectrum. Although this type of noise discontinuity is not objectionable to the human ear, it can severely impact the function of wireless vocoder equipment that relies on linear-signal characteristics. The noise discontinuity can produce erroneous linear predictions in the vocoder that could generate noise bursts. *Smooth noise matching* improves the sound quality of voice-band signals by measuring both the power and frequency spectrum of background noise, and then generating accurate noise-matching signals that are used to remove (i.e., cancel) the noise. In addition, true background noise is allowed to pass through the echo canceler during silent periods (i.e., intervals when only one subscriber is speaking), thereby preventing noise bursts and offering superior performance, especially in digital-wireless applications.

Noise reduction (NR)   When a subscriber is making a call from a noisy environment (e.g., airport, restaurant, factory, moving vehicle, etc.) the surrounding noise level entering the telephone's microphone (i.e., the handset's mouthpiece) can degrade the speaker's voice quality. A high level of background noise can make it difficult for the subscriber on the other end of a call to understand the person located in a noisy environment. The technique called noise reduction (NR) isolates the speaker's voice from the surrounding noise, and automatically attenuates the background noise signals being sent as part of the call. This method of *reducing* only the background noise (by selective signal attenuation) allows the subscriber at the other end of the call to hear the conversation clearly, and as a result, he or she perceives an improvement in the voice quality of the person located in a noisy environment.

Non-linear processing (NLP)   Residual echo is caused by imperfect μ-law or A-law quantization of voice signals, which introduces noise in the voice path. Traditional non-linear processing (NLP) eliminates residual echo by "cutting off" the near-end signal when speech is not present. This approach eliminates residual echo, but the "cut-off" NLP-switch action may generate audible clicking sounds. In addition, silent periods embedded in human speech during double talking can cause the NLP function to toggle between on and off states, which causes speech clipping. *Soft non-linear processing* eliminates residual echo by introducing progressive loss to the signal over a specific period. The service provider can provision the amount of loss and the time interval for applying loss, thereby eliminating speech clipping, noise bursts, and voice-band signal distortion.

Normalized least mean squares (NLMS)   A method for obtaining a best fit expressed as a linear function between one independent and one or more dependent variables by normalizing (scaling) the data before minimizing the sum of squares.

Operation, administration, maintenance, and provisioning (OAM&P)   The activities routinely performed by operations or maintenance personnel to insure that telecommunications systems function properly. These tasks are often associated with centralized operating systems (OS) that control a large number of different types of telecommunications equipment, but may also be performed locally by technicians working directly on an element.

Operating system (OS)   (1) The software that controls the management and maintenance of equipment. (2) A software-based system used to support OAM&P functions performed on telecommunications equipment.

Optical carrier signal level 3 (OC-3)   A SONET-based optical facility format (155.52 Mbps). Typically used in North American Applications.

Packet control unit (PCU)   An element that converts wireless transmission protocol to packet (IP or ATM) protocol, thereby allowing wireless data signals to be transported over a packet network (e.g., the internet) infrastructure.

Packet media gateway (PMG)   Telecommunications equipment (e.g., a 3G wireless terminal) that is capable of supporting multi-media services, and may have packet facility interfaces.

Pass-through   A provisioning option for a telecommunications element, which allows the values of a signal to be passed from the input to the output without being changed or modified.

Path   A route or connection between any two points in a telecommunications network.

Perceptual-analysis measurement system (PAMS)   A methodology for assessing voice quality MOS. The key enhancement over PSQM focuses on end-to-end measurements utilizing time and level alignment and equalization respectively.

Perceptual evaluation of speech quality (PESQ)   The PESQ tool was designed to compute MOS-LQO values on voice samples consistent with subjective evaluations. To do this, PESQ derives a score from the difference between a reference signal and an output signal emerging from equipment in the signal path. In general, the greater the difference between the reference signal and the output signal, the lower the MOS value.

Perceptual speech-quality measure (PSQM)   This measure was designed to compute MOS-LQO values. Its main purpose was the assessment of speech codecs, and it was the first to evaluate and score voice on a quality scale. When emulating human perceptions, PSQM utilizes key psychophysical processes in the human auditory system by computing an internal representation of loudness in time and frequency.

Permanent virtual circuit (PVC)   A facility connection in which there is a constant (permanent) association that exists between two equipment units and that is identical to the data-transfer phase of a virtual call. No call set-up or clearing procedures are required for PVCs.

Personal communication service (PCS)   A wireless service, which provides secure digital communications in a frequency range around 1900 MHz. This service typically uses small low-powered base stations, and compact (light-weight) personal communicators.

Personal digital cellular (PDC)   A 2G digital wireless telephony service offered primarily in Japan.

Phase   The phase of a wave relates the position of a feature, typically a peak or a trough of the waveform, to that same feature in another part of the waveform (or, which amounts to the same, on a second waveform). The phase may be measured as time, distance, fraction of the wavelength, or angle.

Pitch-synchronous innovation – code-excited linear prediction (PSI-CELP)   The main deviation from CELP is its non-conventional random excitation vectors. It resorts to having the random excitation vectors contain pitch periodicity by repeating stored random vectors as well as by using an adaptive codebook. During silent, unvoiced, and transient frames, the coder stops using the adaptive codebook and switches to fixed random codebooks.

Plain old telephone service (POTS)   Telephone service that is restricted to push-button dialing, local, national, or worldwide direct dialing, and is usually administered with traditional billing practices.

Plesiochronous network   A network that contains multiple sub-networks; each having internal timing that is operating at the same nominal frequency, but whose timing may be slightly different at any particular instant (i.e., timing is not synchronized to or originating from the same identical clock source). For example, a network in which each timing element is traceable to its own stratum-1 clock is considered plesiochronous with respect to other elements.

Port   A physical, electrical, or optical interface on a system (e.g., STS-1 facility connection) that is usually connected to external equipment.

Power dissipation   The power consumed by a device during normal operation and therefore not available in the electrical output of the device.

Public switched-telephone network (PSTN)   (1) A traditional dial-up telephone system used primarily to carry voice communications. (2) The term applied when a path is established (using traditional signaling and circuit switching) to carry voice or data over telephone circuits.

Pulse-code modulation (PCM)   is a modulation technique invented by the British Alec Reeves in 1937. It is a digital representation of an analog signal where the magnitude of the signal is sampled regularly at uniform intervals. Every sample is quantized to a series of symbols in a digital code, which is usually a binary code. Pulse-code modulation (PCM) is used in digital-telephone systems. It is also the standard form for digital audio in computers and various compact-disc formats.

Quality of service (QoS)   A measurement that indicates the quality level (e.g., error rate, stability) of a telecommunications service.

Quantization   In signal processing, quantization is the process of approximating a continuous signal by a set of discrete symbols or integer values.

Radio-network controller (RNC)   3G wireless network equipment (similar to a base station controller) that is capable of handling both packet (e.g., data) and circuit switched services.

Regular pulse-excited (RPE)   One of the first voice-encoding procedures to be introduced in GSM (phase I). The RPE uses uniform spacing between pulses. The uniformity eliminates the need for the encoder to locate the position of any pulse beyond the first one. The RPE distinguishes between voiced and unvoiced signals. When the signal is classified as unvoiced, the RPE ceases to generate periodic pulses. Its pulsing becomes random, and it corresponds to the functioning of the unvoiced signal.

Relaxed code-excited linear predictive coding (RCELP)   A codec, which employs a relaxed CELP procedure. The relaxation is delivered by replacing the original residual with a time-warped version of it, allowing only one pitch parameter per frame, thus using a lesser amount of bandwidth on the pitch information. The relaxed algorithm provides equivalent voice quality to the QCELP-13 version at a much reduced bit rate.

Remote-access system (RAS)   A dedicated operating system that supports monitoring (alarm collection and limited maintenance) from a remote location.

Remote-alarm indication (RAI)   A facility alarm that is enabled when a one is detected in position 3 of the eight-bit word in time slot 0 (TS0). An RAI informs downstream equipment of the error condition and prevents proliferation of spurious alarms, commonly called a **yellow alarm**.

Remote-management system (RMS)   A centralized operating system that supports OAM&P activities from a remote location.

Remote terminal (RT)   Usually refers to equipment installed at an unstaffed site with controlled or uncontrolled environmental conditions (e.g., air conditioning).

Routing   The process of determining the path to be used for forwarding messages through a network.

Selectable-mode vocoder (SMV)   The SMV codec is based on 4 codecs: full rate at 8.5 kbps, half rate at 4 kbps, quarter rate at 2 kbps, and eighth rate at 800 bps. The codec is used in cdma2000.

Self test   The capability of equipment to run software routines with a minimum of human intervention, at regular intervals, to test and verify operation.

Service-affecting   Conditions that affect a network's operability or cause service degradation that impacts quality.

Service GPRS support node (SGSN)   3G wireless equipment used to isolate and transport packet (e.g., data) traffic.

Signal   Information transmitted over a communications system using electrical or optical methods.

Signal-processing control (SPC)   The ability for an echo canceler to enable or disable the echo-cancelation function for an individual 64 kbps channel (or group of channels) as controlled by external commands that are part of the call set-up process (e.g., commands sent by a 5ESS switching system).

Sine wave   A waveform with deviation that can be graphically expressed as the sine curve.

Speech signal   A signal that represents the sound waves (acoustic signals) generated by a speaker's larynx, which have been converted into an electrical or optical signal format.

Superframe (SF)   The format originally used to synchronize DS1 signals containing 24 channels per frame. A superframe requires twelve 125 µs DS-1 frames to establish frame synchronization.

Synchronization messaging   SONET synchronization messaging is used to communicate the quality of network timing, internal timing status, and timing states throughout a sub-network.

Synchronous digital hierarchy (SDH)   An international standard defining physical layer protocols, which is generally used for high-speed backbone networks. Typically referred to as SONET/SDH.

Synchronous optical network (SONET)   A Telcordia (formerly Bellcore) specification used primarily in North American telecommunications networks. It defines a synchronous optical network hierarchy for either public or private applications, and is compatible with SDH standards (see synchronous digital hierarchy).

Synchronous transport-mode level 1 (STM-1)   A composite (155.52 Mbps bandwidth) signal containing 63 E1 facility signals, typically deployed in global applications that utilize optical signal transmission infrastructures.

Synchronous transport signal level 1 (STS-1)   A composite (51.84 Mbps bandwidth) signal containing 28 DS1 signals (672 64 kbps channels), typically used in North American applications that utilize electrical (or optical) signal transmission infrastructures (see EC-1).

T1   The basic 24 channel 1.544 Mbps pulse-code modulation (PCM) telephony signal used in North American applications, also referred to as DS1.

Tail delay   See tail length.

Tail length, a.k.a. echo-path delay   The time it takes a signal to traverse the distance from the R_{in} port through the reflection point and back to the S_{in} port of the echo canceler. It is also defined as the length of the filter that cancels echoes (measured in ms).

Talker   The subscriber who originates a speech signal in a telephone network, or the subscriber who is presently speaking (see listener).

Tandem-free operation (TFO)   An architecture that passes coded signals from one mobile to another where the encoding and decoding functions are performed within the mobile terminals respectively.

Terminal   A point of data input or output in a system where voice or data communication takes place.

Time-division multiple access (TDMA)   A digital multiplexing method primarily used in wireless applications.

Time-division multiplexing (TDM)   A multiplexing process that combines digital signals using time-sharing techniques.

Time slot   An individual and unique time interval in a transmission path (commonly called a 64 kbps channel).

Toll switch   An element of a transmission network that is used to connect one or two end offices to a toll center (i.e., the first stage of concentration for inter-toll traffic).

Traffic   A general term for a variety of revenue-bearing customer messages (e.g., voice or data signals) carried by a telecommunications network.

Transmission frame   Refer to frame, definition (2).

Transmission system   A collection of switches, facilities, and other equipment used to connect two or more points in a telecommunications network.

Two-wire   A connection provided by two metallic wires, in which both directions of transmission use the same pair of wires. Typically, the network interface interconnecting voice band analog signals (via a hybrid) to a switched message network such as the PSTN.

Universal mobile-telephone service (UMTS)   A 2G–3G wireless network standard.

Upstream   See downstream/upstream.

Vector-sum-excited linear prediction (VSELP)   The VSELP codec enhances the codebook search procedure by organizing frequently occurring speech combinations closer together. It employs three excitation sources, one of which is adaptive, and it generates a combined excitation sequence to drive the synthesis filter.

Virtual channel (VC)   A unidirectional communications link used to transport packets (e.g., ATM cells). Usually an "end-to-end connection" that is used to carry packet services for a specific time interval.

Virtual circuit   A communication arrangement, in which data is passed from one point to another point in a network over "real circuit equipment" during a specific time interval. See virtual channel and virtual path.

Virtual path (VP)   A collection of virtual channels that are bundled together to form a unidirectional communication path between two end points. Note that all the virtual channels in a virtual path must have the same origin and destination points in the network.

Visiting location register (VLR)   The temporary registration point used to identify a visiting mobile subscriber in a wireless network. See home-location register.

Voice over ATM (VoATM)   An ATM packet-based telecommunications system used to carry voice signals.

Voice-over-internet protocol (VoIP)   An IP packet-based telecommunications system used to carry voice signals.

Voice-over-packet (VoP)   A packet-based telecommunications system (either ATM or IP) used to carry voice signals.

Voice-path assurance (VPA)   The VPA feature allows service providers to provision an echo canceler to detect a continuous 2010 Hz tone that is used to disable the echo cancelation function automatically. Echo cancelers are usually disabled (turned off) during VPA continuity testing. The VPA feature is typically disabled (turned off) as a default state, and is selectively enabled (turned on) when performing DS-1 testing (i.e., groups of 24 DS0 channels) during equipment provisioning or VPA testing activities.

Voice-quality enhancement (VQE)   Generally used to describe enhanced echo canceler equipment that provides more capabilities than traditional electrical (hybrid) echo cancelation. Enhancement features include AEC, NR, NC, ALC, TFO, etc.

Voice-quality system (VQS)   A product or system designed to cancel echo, reduce noise, and adjust speech levels in telecommunications networks. The system may be either a stand-alone or a built-in application inside a larger system such as a MSC or a BSC.

Waveform   The mathematical representation of a wave, especially a graph obtained by plotting a characteristic of the wave against time.

Wavelength   The horizontal distance between successive wave crests measured perpendicular to the crest, usually expressed in meters.

White noise   Random noise with a frequency spectrum that is continuous and uniform over a wide frequency range.

Wide-area network (WAN)   A communication network that serves a large geographic area containing several separate networks that are linked together.

Wideband CDMA (W-CDMA)   An enhanced wireless service based on CDMA standards and having a channel bandwidth of 5 MHz.

Zero Code Suppression (ZCS)   The technique of inserting a "one" bit value in a transmission stream to prevent consecutive "zero" bits from being generated, thereby facilitating signal recovery. See B8ZS.

Brief summary of echo cancelation and VQS major standards

(The testing sequence in G.165 specifies the use of a "white-noise" signal source.)

ITU-T G.160   This recommendation outlines procedures for assessing the performance of a VQS. It also specifies (in the appendix) a detailed methodology for measuring and assessing the quality of noise-reduction algorithms.

ITU-T G.164   G.164 was originally an echo-suppressor specification that specifies the requirement for tone disabling. Compliance with G.164 is commonly referred to as "straight-tone disable." This implies that receiving a 2100 Hz tone with or without phase reversal shall disable echo suppression or echo cancelation functions. A typical example of this application is the early signal format used to transmit fax (facsimile) messages.

ITU-T G.165   This is the digital echo-cancelation specification that was in effect until 1997.

Echo-cancelation performance   G.165 defines the standards for echo cancelation, and associated tests for evaluating the echo-cancelation function.

The standard includes the following parameters:

(1) echo return loss (ERL),

(2) convergence speed,

(3) double talk,

(4) leak rate (of the echo signal),

(5) infinite echo return loss (also known as "open-tail condition").

ITU-T G.167   This document specifies the performance characteristics for acoustic-echo cancelation (AEC) equipment, and the methods used to verify performance. Techniques are provided for guidance, but are not compulsory.

ITU-T G.168   This is the most recent echo-cancelation standard. It incorporates the G.164 and G.165 tone-disable specifications for testing. The major difference between G.165 and G.168 is that a composite-source signal (CSS) is used in G.168 during the echo-cancelation testing sequence (rather than a "white-noise" signal source, as in G.165). The CSS signal is a more accurate representation of actual field conditions, and the criteria for tests 1 through 5 (see list in G.165) are more stringent.

A revision to G.168 was completed in 2000 and in 2002. The modifications include improved echo-path models to represent the network more realistically, and many of the existing tests have been expanded or improved. The revised standard is commonly called "G.168-200X," where X represents the year that the standard was revised.

ITU-T G.169   This recommendation specifies the performance characteristics for automatic level control (ALC) devices, and the tests used to verify performance. This recommendation does *not* define a standard algorithm, planning rules, deployment, or target levels for ALC devices.

Tone-disable standard   Compliance with this part of G.165 is commonly referred to as "phase-reversal tone disable". Receiving a 2100 Hz tone having a phase reversal every 450 ms shall disable the echo-cancelation function. A typical example of this application is the signal format used by high-speed (i.e., ≥ 9600 baud) modems.

Brief summary of key voice-quality assessment standards

ITU-T P.800   This recommendation provides a general overview of the methods used to determine subjective transmission quality in telecommunications networks, and describes various echo conditions.

ITU-T P.831   This recommendation provides a description of subjective performance evaluation methods used for network echo cancelers.

ITU-T P.861   This recommendation outlines the PSQM methodology for assessing voice-quality performance.

ITU-T P.862   This recommendation provides a comprehensive description of the PESQ methodology for assessing voice quality.

Bibliography

3GPP2 C.S 0030–0, *Selectable Mode Vocoder Service Option 56 for Wideband Spread Spectrum Communication Systems*, version 2.0 (Dec., 2001), see www.3gpp.org/.

3GPP TS 06.77 V8.1.1, *Minimum Performance Requirements for Noise Suppresser Application to the AMR Speech Encoder* (1999), see www.3gpp.org/.

3GPP TS 26.190, *AMR Wideband Speech Codec: Transcoding Functions*, see www.3gpp.org/.

3GPP TS 32.111–1, *3rd Generation Partnership Project; 3G Fault Management, Part 1: Fault Management Requirements*, see www.3gpp.org/.

3GPP TS 32.111–2, *3rd Generation Partnership Project; 3G Fault Management, Part 2: Alarm Integration Reference Point: Information Service*, see www.3gpp.org/.

3GPP TS 32.111–3, *3rd Generation Partnership Project; 3G Fault Management Alarm Integration Reference Point: CORBA Solution Set*, see www.3gpp.org/.

3GPP TS32.300, *3rd Generation Partnership Project; NRM Naming Convention*, see www.3gpp.org/.

3GPP TS 46.008, *Performance Characterization of the GSM Half Rate Speech Codec*, version 4.0.0, 2001–03, see www.3gpp.org/.

A.S0004-B v2.0, *CDMA Tandem Free Operation Specification – Release B*, 2002/08; CCSA: CCSA-MC-A.S0004-B v2.0; TIA: TIA/EIA-895-A pub; TTC: TS-3GB-A.S0004-Bv2.0.

Adrat, M., Brauers, J., Clevorn, T., and Vary, P., The EXIT-characteristic of softbit-source decoders, *IEEE Communications Letters*, **9**: 6 (2005), 540–542.

Ahmadi, S., *Tutorial on the Variable-Rate Multimode Wideband Speech Codec*, Nokia Inc., CommsDesign.com (Sep. 02, 2003).

Anderson, B. W. and Kalb, J. P., English verification of the STI method for estimating speech intelligibility of a communications channel, *Journal of the Acoustical Society of America*, **81** (1987), 1982–1985.

Andrej, B., *Nonlinear Aspects of Telecommunications: Discrete Volterra Series and Nonlinear Echo Cancellation* (CRC Press, 2000).

ANSI/IEEE 743–1984, *Standard Methods and Equipment for Measuring the Transmission Characteristics of Analogue Voice Frequency Circuits*.

Arslan, L., McCree, A., and Viswanathan, V., New methods for adaptive noise suppression, *IEEE International Conference on Acoustics, Speech, and Signal Processing*, **1** (May 1995), 812–815.

Atal, B. S., Rabiner, R., A pattern recognition approach to voiced-unvoiced-silence classification with applications to speech recognition, *IEEE Transactions on Acoustics, Speech, and Signal Processing*, **24**: 3 (1976), 201–212.

Barnwell, T. P. III, Nayebi, K., and Richardson, C. H., *Speech Coding, a Computer Laboratory Textbook* (John Wiley & Sons, Inc., 1996).

Beaugeant, C., Turbin, V., Scalart, P., and Gilloire, A., New optimal filtering approaches for hands-free telecommunication terminals, *Signal Processing*, **64**: 1 (1998), 33–47.

Beerends, J. G., Modelling cognitive effects that play a role in the perception of speech quality, in *Proceedings of the International Workshop on Speech Quality Assessment* (Bochum, Nov. 1994), 1–9.

Beerends, J. G., *Measuring the Quality of Speech and Music Codecs, an Integrated Psychoacoustic Approach*, 98th AES convention, pre-print no. 3945 (1995).

Beerends, J. G. and Stemerdink, J. A., Measuring the quality of audio devices, 90th AES convention, pre-print no. 3070 (1991).

Beerends, J. G. and Stemerdink, J. A., A perceptual audio quality measure based on a psychoacoustic sound representation, *Journal of the Audio Engineering Society*, **40**: 12 (1992), 963–974.

Beerends, J. G. and Stemerdink, J. A., The optimal time-frequency smearing and amplitude compression in measuring the quality of audio devices, 94th AES convention, pre-print no. 3604 (1993).

Bellamy, John, *Digital Telephony*, 2nd Edition (John Wiley & Sons, 1991).

Bellcore, LSSGR: signaling for analog interfaces, *GR-506-CORE*, issue 1 (Jun. 1996).

Bellcore, Call processing, *GR-505-CORE*, issue 1 (Dec. 1997).

Benesty, J., Gansler T., Morgan, D. R., Sondhi. M. M., and Gay, S. L., *Advances in Network and Acoustic Echo Cancellation* (Springer, 2001).

Beranek, L. L., *Noise Reduction* (McGraw-Hill, 1960).

Beranek, L. L., *Noise and Vibration Control* (McGraw-Hill, 1971), 564–566.

Beritelli, F., Casale, S., Ruggeri, G., and Serrano, S., Performance evaluation and comparison of G. 729/AMR/fuzzy voice activity detectors, *IEEE Signal Processing Letters*, **9**: 3 (2002), 85–88, ieeexplore.ieee.org.

Bessette, B., Salami, R., Lefebvre, R., and Jelinek, M. J., The adaptive multi-rate wideband speech codec (AMR-WB), *IEEE Transactions on Speech and Audio Processing*, **10**: 8 (2002), 620–636, ieeexplore.ieee.org.

Boll, S. F., Suppression of acoustic noise in speech using spectral subtraction, *IEEE Transactions on Acoustics, Speech, and Signal Processing*, **27**: 2 (Apr. 1979).

Boyd, I. and Southcott, C. B., A speech codec for the skyphone service, *British Telecom Technology Journal*, **6**: 2 (Apr. 1988).

Brady, P. T., A technique for investigating on–off patterns of speech, *The Bell System Technical Journal*, **44** (1965), 1–22.

Brady, P. T., Equivalent peak level: a threshold-independent speech level measure, *Journal of the Acoustical Society of America*, **44** (1968), 695–699.

Brady, P. T., A statistical analysis of on–off patterns in 16 conversations, *The Bell System Technical Journal*, **47** (1968), 73–91.

Brandenburg, K., Evaluation of quality for audio encoding at low bit rates, 82nd AES convention, pre-print no. 2433 (1987).

British Standard, IEC 1260, *Electro-Acoustics – Octave-Band and Fractional-Octave-Band Filters* (1996).

Butler, L. W. and Kiddle, L., *The Rating of Delta Sigma Modulating Systems with Constant Errors, Burst Errors, and Tandem Links in a Free Conversation Test Using the Reference Speech Link*, Signals Research and Development Establishment, Ministry of Technology, Christchurch, Hants., report no. 69014 (Feb. 1969).

C11–20030714-003-WBC_Stage2_Requirement_Document, *Stage Two Requirements (Test Plan) for CDMA2000 Wideband Speech Codec*, 3GPP2 technical specification (Jul. 2003).

Cappe, O., Elimination of the musical noise phenomenon with the Ephraim and Malah noise suppressor, *IEEE Transactions on Speech and Audio Processing*, **2** (Apr. 1994), 345–349.

Carson, R., A digital speech voltmeter – the SV6, *British Telecommunications Engineering*, **3**: 1 (Apr. 1984), 23–30.

Cattermole, K. W., *Principles of Pulse Code Modulation*, Elsevier Publishing Co. (1969).

Chapanis, R. N., Parrish, R. B., Ochsman, and Weeks, G. D., Studies in interactive communication II. The effects of four communication modes on the linguistic performance of teams during cooperative problem solving, *Human Factors*, **19** (1977), 101–126.

Chapple, E. D., The interaction chronograph: its evolution and present application, *Personnel*, **25** (1949), 295–307.

Chow, Ming-Chwan, *Understanding Wireless, Digital Mobile, Cellular and PCS*, Adnan Publisher.

Claringbold, P. J., The within-animal bioassay with quantal responses, *Journal of the Royal Statistical Society*, Series B, **18**: 1 (1956), 133–137.

Cohen, I. and Berdugo, B. H., Speech enhancement for non-stationary noise environments, *Signal Processing*, **81** (2001), 2403–2418.

Cohen, I., Noise estimation by minima controlled recursive averaging for robust speech enhancement, *IEEE Signal Processing Letters*, **9**: 1 (Jan. 2002).

Coleman A., Gleiss N., and Usai P., A subjective testing methodology for evaluating medium rate codecs for digital mobile applications, *Speech Communications*, **7** (Jun. 1988), 151–166.

COM-D007-E, *Proposed Modification to Draft P.862 to Allow PESQ to be Used for Quality Assessment of Wideband Speech (BT, United Kingdom and KPN, The Netherlands)* (Feb. 2001).

COM12-D187-E, *Performance Evaluation of the Wideband PESQ Algorithm (NTT)* (Mar. 2004).

Cox, R. V., Speech coding, in Madisetti, V. K., and Williams, D. B., ed., *The Digital Signal Processing Handbook* (CRC Press, 1998), 45-1–45-4.

Cox, R. V., Kleijn, W. B., and Kroon, P., Robust CELP coders for noisy backgrounds and noisy channels, *Proceedings of the International Conference on Acoustics, Speech, and Signal Processing* (1989), 739–742.

Crowe, D. P., Selection of voice codec for the aeronautical satellite service, *European Conference on Speech Communication and Technology*, **2**: S37 (Sep. 1989), 320–323.

Crystal, T. H. and House, A. S., Segmental durations in connected speech signals: preliminary results, *Journal of the Acoustical Society of America*, **72**: 3 (Sep. 82), 705.

Daumer, W. R. and Cavanaugh, J. R., A subjective comparison of selected digital codecs for speech, *Bell Systems Technical Journal*, **57**: 9 (Nov. 1978).

Davis, G. M. (ed.), *Noise Reduction in Speech Applications* (CRC Press, 2002).

Dimolitsas, S., Corcoran, F., and Ravishankar, C., Correlation between headphone and telephone handset listener opinion scores for single-stimulus voice coder performance assessments, *IEEE Signal Processing Letters*, **2**: 3 (1995).

Doblinger, G., Computationally efficient speech enhancement by spectral minimum tracking in sub-bands, *EUSPICO* (Sep. 1995), 1513–1516.

Doronan, A., *The Essential Guide to Wireless Communications Applications* (Prentice Hall, 2001).

Douglas, B., *Voice Encoding Methods for Digital Wireless Communications Systems*, Southern Methodist University, EE6302, Section 324 (Fall 1997).

Drucker, H., Speech processing in a high ambient noise environment, *IEEE Transactions on Audio Electroacoustics*, **AU-16** (Jun. 1968), 165–168.

DSPG, *DSPGenie Echo Canceller Tester*, www.dspg.co.uk/.

Duttweiler, D. L., Proportionate normalized least-mean-squares adaptation in echo cancellers, *IEEE Transactions on Speech and Audio Processing*, **8**: 5 (Sep. 2000), 508–518.

Ealey, D., Kellher, H., and Pearce, D., Harmonic tunneling: tracking non-stationary noises during speech, *Eurospeech*, **2001** (2001) 437–440.

EIA/TIA-PN2398 Recommendation IS-54, *The 8 kbits/S VSELP Algorithm*, 1989.

EIA/TIA Recommendation IS-127, *Enhanced Variable Rate Codec (EVRC)* (1998).

EIA/TIA Recommendation IS-853 *TDMA Third Generation Wireless; Noise Suppression Minimum Performance for APR*.

EIA/TIA Recommendation IS-895-A, CDMA *Tandem Free Operation*.

Eisler, F. G., *Psycholinguistics: Experiments in Spontaneous Speech* (New York: Academic Press, 1968).

Eng, M. K. C. B., Adaptive signal processing algorithms for non-gaussian signals, Ph.D. dissertation, Faculty of Engineering, Queen's University Belfast (Sep. 2002).

Enzner, G., Martin, R., and Vary, P., Partitioned residual echo power estimation for frequency-domain acoustic echo cancellation and postfiltering, *European Transactions on Telecommunications*, **13**: 2 (2002), 103–114 and www.ind.rwth-aachen.de/publications/ge_rm_pv_ett02.html.

Ephraim, Y. and Malah, D., Speech enhancement using a minimum mean-square error log-spectral amplitude estimator, *IEEE Transactions on Acoustics, Speech, and Signal Processing*, **33**: 2 (Apr. 1985), 443–445.

Ephraim, Y. and Malah, D., Speech enhancement using a minimum mean-square error short-time spectral amplitude estimator, *IEEE Transactions on Acoustics, Speech, and Signal processing*, **32**: 6 (Dec. 1994), 1109–1121.

Ephraim, Y. and Merhav, N., Hidden Markov processes, *IEEE Transactions on Information Theory*, **48** (Jun. 2002), 1518–1569.

Erdmann, C. and Vary, P., Performance of multistage vector quantization in hierarchical Coding, *European Transactions on Telecommunications (ETT)*, **15**: 4 (July/Aug. 2004), 363–372.

ETSI/SMG11, Speech aspects, Presentation of SMG11 activities to Tiphon, http://portal. etsi.org/stq/presentations/smg11.pdf.

ETSI TR 126 976, *Digital Cellular Telecommunications System (Phase 2+); Universal Mobile Telecommunications System (UMTS); Performance Characterization of the Adaptive Multi-Rate Wideband (AMR-WB) Speech Codec* (3GPP TR 26.976, version 6.0.0, release 6), 2004–12.

ETSI TS 23.057, *UMTS Mobile Station Application Execution Environment (MExE); Functional Description; Stage 2*, version 3.0.0 (2000).

ETSI TS 128 062, *Inband Tandem Free Operation (TFO) of Speech Codecs, Stage 3 – Service Description*, 2002–03.

Etter, W. and Moschytz, G. S., Noise reduction by noise adaptive spectral magnitude expansion, *Journal of the Audio Engineering Society*, **42**: 5 (May 1994), 341–349.

Etter, W., Moschytz, G. S., and Graupe, D., Adaptive noise reduction using discrimination functions, *Proceedings of ICASSP 91* (1991) 965–968.

Farhang-Boroujeny, B., *Adaptive Filters, Theory, and Applications* (Wiley, 1999).

Fingscheidt, T. and Vary, P., Softbit speech decoding: a new approach to error concealment, *IEEE Transactions on Speech and Audio Processing*, **9**: 3 (2001), 240–251.

Forgie, J. W., Feehrer, C. E., and Weene, P. L., *Voice Conferencing Technology Program: Final Report*, Lincoln Laboratory, Massachusetts Institute of Technology, Lexington, MA (1979).

Furuta, S. and Takahashi, S., A noise suppresser for the AMR speech codec and evaluation test results based on 3GPP specifications, *Speech Coding* (2002), *IEEE Workshop Proceedings* (2002), ieeexplore.ieee.org.

Gabrielsson, A., *Statistical Treatment of Data from Listening Tests on Sound-Reproducing Systems*, Report TA No. 92, KTH Karolinska Institutet, Department of Technical Audiology, S-10044 Stockholm, Sweden, IEC publication 268–13, annex 3, subclause 3.3 (a condensed version, Nov. 1979).

Gannot, S., Burshtein, D., and Weinstein, E., Iterative and sequential Kalman filter-based speech enhancement algorithms, *IEEE Transactions on Speech and Audio Processing*, **6** (Jul. 1998), 373–385.

Garg, V. K. and Wilkes, J. E., *Principles and Applications of GSM* (Prentice Hall, 1999).

Gay, S. L. and Tavathia, S., The fast affine projection algorithm, *International Conference on Acoustics, Speech, and Signal Processing*, ICASSP-95, **5**: 9–12 (May 1995), 3023–3026.

Gerson, I. and Jasiuk, M., Vector sum excited linear prediction (VSELP) speech coding at 8 kbits/s, *International Conference on Acoustics, Speech, and Signal Processing*, ICASSP-90 (New Mexico, Apr. 1990), 461–464.

GL-Communications, GL-Comm echo canceller testing solutions, www.gl.com/echocan.html.

Goodman, D. J. and Nash, R. D., Subjective quality of the same speech transmission conditions in seven different countries, *International Conference on Acoustics, Speech and Signal Processing*, ICASSP-82, **2** (Paris, May 1982).

Goodman, D. J. and Wei, S. X., Efficiency of packet reservation multiple access, *IEEE Transactions on Vehicular Technology*, **40**: 1 (1991), 170–176.

Greer, S. C., *Speech Coding for Wireless Communications* (Nokia, Nov. 27, 2001).

GSM 01.04, *Digital Cellular Telecommunications System (Phase 2+); Abbreviations and Acronyms*, see www.etsi.org/.

GSM 03.10, *Digital Cellular Telecommunications System (Phase 2+); GSM Public Land Mobile Network (PLMN) Connection Types*; version 7.0.1 (Release 1998), see www.etsi.org/.

GSM 03.34, *Digital Cellular Telecommunications System (Phase 2+): High Speed Circuit Switched Data (HSCSD) – Stage 2*; version 7.0.0 (Release 1998), see www.etsi.org/.

GSM 03.60, *GPRS Service Description Stage 2*, version 7.0.0 (Phase 2+) (1998), see www.etsi.org/.

GSM 04.21, *Digital Cellular Telecommunications System (Phase 2+); Rate Adaptation on the Mobile Station – Base Station System (MS – BSS) Interface*, version 8.2.0 (Release 1999), see www.etsi.org/.

GSM 05.02, *Digital Cellular Telecommunications System (Phase 2+); Multiplexing and Multiple Access on the Radio Path*, version 8.5.1 (Release 1999), see www.etsi.org/.

GSM 06.01, *Digital Cellular Telecommunications System (Phase 2+); Full Rate Speech; Processing Functions*, see www.etsi.org/.

GSM 06.02, *Digital Cellular Telecommunications System (Phase 2+); Half Rate Speech; Half Rate Speech Processing Functions*, see www.etsi.org/.

GSM 06.06, *Digital Cellular Telecommunications System (Phase 2+); Half Rate Speech; ANSI-C Code for the GSM Half Rate Speech Codec*, see www.etsi.org/.

GSM 06.07, *Digital Cellular Telecommunications System (Phase 2+); Half Rate Speech; Test Sequences for the GSM Half Rate Speech Codec*, see www.etsi.org/.

GSM 06.08, *Digital Cellular Telecommunications System (Phase 2+); Half Rate Speech; Performance Characterization of the GSM Half Rate Speech Codec*, see www.etsi.org/.

GSM 06.10, *Digital Cellular Telecommunications System (Phase 2+); Full Rate Speech; Transcoding*, see www.etsi.org/.

GSM 06.11, *Digital Cellular Telecommunications System (Phase 2+); Full Rate Speech; Substitution and Muting of Lost Frames for Full Rate Speech Channels*, see www.etsi.org/.

GSM 06.12, *Digital Cellular Telecommunications System (Phase 2+); Full Rate Speech; Comfort Noise Aspect for Full Rate Speech Traffic Channels*, see www.etsi.org/.

GSM 06.20, *Digital Cellular Telecommunications System (Phase 2+); Half Rate Speech; Half Rate Speech Transcoding*, see www.etsi.org/.

GSM 06.21, *Digital Cellular Telecommunications System (Phase 2+); Half Rate Speech; Substitution and Muting of Lost Frame for Half Rate Speech Traffic Channels*, see www.etsi.org/.

GSM 06.22, *Digital Cellular Telecommunications System (Phase 2+); Half Rate Speech; Comfort Noise Aspects for Half Rate Speech Traffic Channels*, see www.etsi.org/.

GSM 06.31, *Digital Cellular Telecommunications System (Phase 2+); Full Rate Speech; Discontinuous Transmission (DTX) for Full Rate Speech Traffic Channels*, see www.etsi.org/.

GSM 06.32, *Digital Cellular Telecommunications System (Phase 2+); Voice Activity Detector (VAD)*, see www.etsi.org/.

GSM 06.41, *Digital Cellular Telecommunications System (Phase 2+); Half Rate Speech; Discontinuous Transmission (DTX) for Half Rate Speech Traffic Channels*, see www.etsi.org/.

GSM 06.42, *Digital Cellular Telecommunications System (Phase 2+); Half Rate Speech; Voice Activity Detector (VAD) for Half Rate Speech Traffic Channels*, see www.etsi.org/.

GSM 06.51, *Digital Cellular Telecommunications System (Phase 2+); Enhanced Full Rate (EFR) Speech Coding Functions; General Description*, see www.etsi.org/.

GSM 06.53, *Digital Cellular Telecommunications System (Phase 2+); ANSI-C Code for the GSM Enhanced Full Rate (EFR) Speech Codec*, see www.etsi.org/.

GSM 06.54, *Digital Cellular Telecommunications System (Phase 2+); Test Sequences for the GSM Enhanced Full Rate (EFR) Speech Codec*, see www.etsi.org/.

GSM 06.55, *Digital Cellular Telecommunications System (Phase 2+); Performance Characterization of the GSM Enhanced Full Rate (EFR) Speech Codec*, see www.etsi.org/.

GSM 06.60, *Digital Cellular Telecommunications System (Phase 2+); Enhanced Full Rate (EFR) Speech Transcoding*, see www.etsi.org/.

GSM 06.61, *Digital Cellular Telecommunications System (Phase 2+); Substitution and Muting of Lost Frames for Enhanced Full Rate (EFR) Speech Traffic Channels*, see www.etsi.org/.

GSM 06.62, *Digital Cellular Telecommunications System (Phase 2+); Comfort Noise Aspects for Enhanced Full Rate (EFR) Speech Traffic Channels*, see www.etsi.org/.

GSM 06.81, *Digital Cellular Telecommunications System (Phase 2+); Discontinuous Transmission (DTX) for Enhanced Full Rate (EFR) Speech Traffic Channels*, see www.etsi.org/.

GSM 06.82, *Digital Cellular Telecommunications System (Phase 2+); Voice Activity Detection (VAD) for Enhanced Full Rate (EFR) Speech Traffic Channels*, see www.etsi.org/.

GSM 06.85, *Digital Cellular Telecommunications System (Phase 2+); Subjective Tests on the Interoperability of the Half Rate/Full Rate/Enhanced Full Rate (HR/FR/EFR) Speech Codecs, Single, Tandem and Tandem Free Operation*, version 8.0.0 (Release 1999), ETSI TR 101 294 V8.0.0 (2000–06), see www.etsi.org/.

GSM 08.20, *Digital Cellular Telecommunications System (Phase 2+); Rate Adaptation on the Base Station System – Mobile-Services Switching Center (BSS – MSC) Interface*, version 8.1.0 (Release 1999), see www.etsi.org/.

GSM 09.18, *Visitors Location Register (VLR), Gs Interface, Layer 3 Specification*, version 1.2.0 (1997), see www.etsi.org/.

GSM World, *What is General Packet Radio Service?* www.gsmworld.com/technology/gprs/intro.shtml.

Gustafsson, S., *Enhancement of Audio Signals by Combined Acoustic Echo Cancellation and Noise Reduction*, Ph.D. dissertation, ABDN Band 11, P. Vary (Aachen: Hrsg. Verlag der Augustinus Buchhandlung, 1999), www.ind.rwth-aachen.de/publications/sg_diss99.html.

Gustafsson, S., Martin, R., and Vary, P., Combined acoustic echo control and noise reduction for hands-free telephony, *Signal Processing*, **64**: 1 (1998), S 21–32.

Gustafsson, S., Martin, R., Jax, P., and Vary, P., A psychoacoustic approach to combined acoustic echo cancellation and noise reduction, *IEEE Transactions on Speech and Audio Processing*, **10**: 5 (2002), 245–256.

Han, Y., Bahk, H. G., and Yang, S., CDMA mobile system overview: introduction, background, and system concepts, *Electronics and Telecommunications Research Institute Journal*, **19**: 3 (1997), 83–97.

Hansen, J. J., Speech enhancement using a constrained iterative sinusoidal model, *IEEE Transactions on Speech and Audio Processing*, **9**: 7 (2001) 731–740.

Hansler, E. and Schmidt, G., *Acoustic Echo and Noise Control, a Practical Approach* (John Wiley & Sons, 2004).

Harte, L., *Introduction to WCDMA, Physical Channels, Logical Channels, Network, and Operation* (Fuquay Varina: Althos, Aug. 2004).

Haug, T., Overview of GSM: philosophy and results, *International Journal of Wireless Information Networks*, **1**: 1 (1994), 7–16.

Haykin, S., *Adaptive Filter Theory*, 3rd edn. (Prentice Hall, 1997).

Haykin, S. and Widrow, B., *Least-Mean-Square Adaptive Filters* (John Wiley & Sons, 2003).

He, C. and Zweig, G., Adaptive two-band spectral subtraction with multi-window spectral estimation, *IEEE International Conference on Acoustics, Speech, and Signal Processing*, **2** (1999), 793–796.

He, P. P., Dyba, R. A., and Pessoa, F. C., *Network Echo Cancellers; Requirements, Applications and Solutions* (Motorola Inc., 2002).

Heinen, S. and Vary, P., Source Optimized Channel Coding for Digital Transmission Channels, *IEEE Transactions on Communications*, **53**: 4 (2005), 592–600.

Heinle, F., Rabenstein, R., and Stenger, A., A measurement method for the linear and nonlinear properties of electro-acoustic transmission systems, *Signal Processing*, **64**: 1 (1998), 49–60.

Hennbert, J., Ris, C., Bourlard, H., Renals, S., and Morgan, N., Estimation of global posteriors and forward backward training of hybrid HMM/ANN systems, *Enrospeech* (Sep. 1997), 1951–1954.

Hera, C., Echo cancellation in cellular networks, private communication (Aug. 2003).

Hess, W., *Pitch Determination of Speech Signals* (Berlin: Springer-Verlag, 1983).

Hindelang, T., Heinen, S., Vary, P., and Hagenauer, J., Two approaches to combined source-channel coding: a scientific competition in estimating correlated parameters, *International Journal of Electronics and Communications (AEÜ)*, **54**: 6 (2000), 364–378.

Hirsch, H. G. and Ehrlicher, C., Noise estimation techniques for robust speech recognition, *IEEE International Conference on Acoustics, Speech, and Signal Processing* (1995), 153–156.

Hollier, M. P., Hawksford, M. O., and Guard, D. R., Characterisation of communications systems using a speech-like test stimulus, *Journal of the Audio Engineering Society*, **41**: 12 (1993), 1008–1021.

Hollier, M. P., Hawksford, M. O., and Guard, D. R., Error activity and error entropy as a measure of psychoacoustic significance in the perceptual domain, *IEEE Proceedings – Vision, Image and Signal Processing*, **141**: 3 (1994), 203–208.

Holma, H. and Toskala, A., *WCDMA for UMTS* (John Wiley & Sons, 2004).

Hoth, D. F., Room noise spectra at subscribers' telephone locations, *Journal of the Acoustic Society of America*, **12** (Apr. 1941), 99–504.

IEC Publication 581.5, *High Fidelity Audio Equipment and Systems; Minimum Performance Requirements – Part 5: Microphones* (1981).

IEC Publication 581.7, *High Fidelity Audio Equipment and Systems; Minimum Performance Requirements – Part 7: Loudspeakers* (1986).

IEC Publication 651, *Sound Level Meters*, (1979, Amendment 1–1993, Corrigendum March 1994).

IETF RFC 791 (STD-5), *Internet Protocol, DARPA Internet Program Protocol Specification* (1981).

ISO 1996–1, *Acoustics – Description and Measurement of Environmental Noise – Part 1: Basic Quantities and Procedures* (1982).

ISO 1996–2, *Acoustics – Description and Measurement of Environmental Noise – Part 2: Acquisition of Data Pertinent to Land Use* (1984).

ISO 1996–3, *Acoustics – Description and Measurement of Environmental Noise – Part 3: Application to Noise Limits* (1987).

ISO 266, *Acoustics – Preferred Frequencies for Measurements* (1975).

ISO/IEC, International Standard 11172–3:1993, *Information Technology – Coding of Moving Pictures and Associated Audio for Digital Storage Media at up to about 1.5 Mbit/s – Part 3, Audio* (1993).

ITU-R Recommendation BS1116, *Subjective Assessment of Audio Systems with Small Impairments Including Multi-Channel Sound Systems* (Geneva, Oct. 1997).

ITU-R Recommendation BS1116, *Methods for the Subjective Assessment of Small Impairments in Audio Systems Including Multichannel Sound Systems* (Jul. 1998).

ITU-R Recommendation BS1116–1, *Methods for the Subjective Assessment of Small Impairments in Audio Systems Including Multi-Channel Sound Systems* (1997).

ITU-R Recommendation BS1387, *Method for Objective Measurements of Perceived Audio Quality* (Jan. 1999).

ITU-T, Absolute category rating (ACT) method for subjective testing of digital processors, *Red Book*, **V** (1984), 111–114.

ITU-T, *Additions to the Handbook on Telephonometry* (1999).

ITU-T, *Handbook on Telephonometry* (1993).

ITU-T P-series Supplement 23, *ITU-T Coded-Speech Database* (1998).

ITU-T Recommendation E.164, *The International Public Telecommunication Numbering Plan* (1997).

ITU-T Recommendation E.651, *Reference Connections for Traffic Engineering of IP Access Networks* (2000).

ITU-T Recommendation G.100, *Definitions Used in Recommendations on General Characteristics of International Telephone Connections and Circuits* (2001).

ITU-T Recommendation G.107, *The E-Model, a Computational Model for Use in Transmission Planning* (1998).

ITU-T Recommendation G.108, *Application of the E-Model – a Planning Guide* (1999).

ITU-T Recommendation G.109, *Definition of Categories of Speech Transmission Quality* (1999).

ITU-T Recommendation G.113, *Transmission Impairments* (1996).

ITU-T Recommendation G.114, *One-Way Transmission Time* (1996).

ITU-T Recommendation G.120, *Transmission Characteristics of National Networks* (1998).

ITU-T Recommendation G.121, *Loudness Ratings (LRS) of National Systems* (1993).

ITU-T Recommendation G.122, *Influence of National Systems on Stability and Talker Echo in International Connections* (1993).

ITU-T Recommendation G.131, *Control of Talker Echo* (1996).

ITU-T Recommendation G.160, *Voice Enhancement Devices*, draft 14 (2004).

ITU-T Recommendation G.161, *Interaction Aspects of Signal Processing Network Equipment* (2002).

ITU-T Recommendation G.164, *Echo Suppressors* (1988).

ITU-T Recommendation G.165, *Echo Cancellers* (1993).

ITU-T Recommendation G.167, *Acoustic Echo Controllers* (1993).

ITU-T Recommendation G.168, *Digital Network Echo Cancellers* (1997, 2000, 2002).

ITU-T Recommendation G.169, *Automatic Level Control Devices* (1999).

ITU-T Recommendation G.191, *Software Tools for Speech and Audio Coding Standardization* (2000).

ITU-T Recommendation G.192, *A Common Digital Parallel Interface for Speech Standardization Activities* (1996).

ITU-T Recommendation G.212, *Hypothetical Reference Circuits for Analogue Systems* (1988).

ITU-T Recommendation G.223, *Assumptions for the Calculation of Noise on Hypothetical Reference Circuits for Telephony* (1988).

ITU-T Recommendation G.703, *Physical/Electrical Characteristics of Hierarchical Digital Interfaces* (2001).

ITU-T Recommendation G.711, *Pulse Code Modulation (PCM) of Voice Frequencies* (1988).

ITU-T Recommendation G.712, *Transmission Performance Characteristics of Pulse Code Modulation Channels* (2001).

ITU-T Recommendation G.722, *7 kHz Audio-Coding within 64 kbit/s* (1988).

ITU-T Recommendation G.722.2, *Wideband Coding of Speech at Around 16 kbit/s Using Adaptive Multi-Rate Wideband (AMR-WB)* (Geneva, January 2002).

ITU-T Recommendation G.726, *40, 32, 24 and 16 kbit/s Adaptive Differential Pulse Code Modulation (ADPCM)* (1990).

ITU-T Recommendation G.727, *5-, 4-, 3- and 2-Bit/Sample Embedded Adaptive Differential Pulse Code Modulation (ADPCM)* (1990).

ITU-T Recommendation G.728, *Coding of Speech at 16 kbit/s Using Low-Delay Code Excited Linear Prediction* (1992).

ITU-T Recommendation G.729, *Coding of Speech at 8 kbit/s Using Conjugate-Structure Algebraic-Code-Excited Linear-Prediction (CS-ACELP)* (1996).

ITU-T Recommendation G.763, *Digital Circuit Multiplication Equipment Using (Recommendation G.726) ADPCM and Digital Speech Interpolation* (1998).

ITU-T Recommendation G.772, *Protected Monitoring Points Provided on Digital Transmission Systems* (1993).

ITU-T Recommendation G.826, *Error Performance Parameters and Objectives for International, Constant Bit Rate Digital Paths at or Above the Primary Rate* (1999).

ITU-T Recommendation G.1000, *Communications Quality of Service: a Framework and Definitions* (2001).

ITU-T Recommendation G.1010, *End-User Multimedia QOS Categories* (2001).

ITU-T Recommendation I.350, *General Aspects of Quality of Service and Network Performance in Digital Networks, Including ISDN* (1993).

ITU-T Recommendation I.460, *Multiplexing, Rate Adaptation and Support of Existing Interfaces*.

ITU-T Recommendation M.3010, *Principles for a Telecommunications Management Network* (2000).

ITU-T Recommendation O.41, *Psophometer for Use on Telephone-Type Circuits* (1994).

ITU-T Recommendation O.131, *Quantizing Distortion Measuring Equipment Using a Pseudo-Random Noise Test Signal* (1988).

ITU-T Recommendation O.132, *Quantizing Distortion Measuring Equipment Using a Sinusoidal Test Signal* (1988).

ITU-T Recommendation O.133, *Equipment for Measuring the Performance of PCM Encoders and Decoders* (1993).

ITU-T Recommendation Series P, *Telephone Transmission Quality, Methods for Objective and Subjective Assessment of Quality*.

ITU-T Recommendation P.10, *Vocabulary of Terms on Telephone Transmission Quality and Telephone Sets* (1998).

ITU-T Recommendation P.11, *Effect of Transmission Impairments* (1993).

ITU-T Recommendation P.48, *Specification for an Intermediate Reference System* (1998).

ITU-T Recommendation P.50, *Artificial Voices* (1999).

ITU-T Recommendation P.51, *Artificial Mouth* (1996).

ITU-T Recommendation P.56, *Objective Measurement of Active Speech Level* (1993).

ITU-T Recommendation P.57, *Artificial Ears* (1996).

ITU-T Recommendation P.58, *Head and Torso Simulator for Telephonometry* (1996).

ITU-T Recommendation P.76, *Determination of Loudness Ratings; Fundamental Principles* (1988).

ITU-T Recommendation P.78, *Subjective Testing Method for Determination of Loudness Ratings in Accordance With Recommendation P.76* (1993).

ITU-T Recommendation P.79, *Calculation of Loudness Ratings for Telephone Sets* (1993).

ITU-T Recommendation P.79, *Wideband Loudness Rating Algorithm*, annex G (Nov. 2001).

ITU-T Recommendation P.80, *Methods for Subjective Determination of Transmission Quality* (1993).

ITU-T Recommendation P.81, *Modulated Noise Reference Unit (MNRU)*, vol. of blue book (Geneva: ITU, 1989), 198–203.

ITU-T Recommendation P.82, *Method for Evaluation of Service from the Standpoint of Speech Transmission Quality* (1984).

ITU-T Recommendation P.84, *Subjective Listening Test Method for Evaluating Digital Circuit Multiplication and Packetized Voice Systems* (1993).

ITU-T Recommendation P.85, *A Method for Subjective Performance Assessment of the Quality of Speech Voice Output Devices* (1994).

ITU-T Recommendation P.310, *Transmission Characteristics for Telephone Band (300–3400 Hz) Digital Telephones* (2000).

ITU-T Recommendation P.311, *Transmission Characteristics for Wideband (150–7000 Hz) Digital Handset Telephones* (Feb. 1998).

ITU-T Recommendation P.340, *Transmission Characteristics and Speech Quality Parameters of Hands-Free Terminals* (2000).

ITU-T Recommendation P.341, *Transmission Characteristics for Wideband (150–7000 Hz) Digital Hands-Free Telephony Terminals* (Feb. 1998).

ITU-T Recommendation P.501, *Test Signals for Use in Telephonometry* (1996).

ITU-T Recommendation P.502, *Objective Test Methods for Speech Communication Systems Using Complex Test Signals* (2000).

ITU-T Recommendation P.561, *In-Service Non-Intrusive Measurement Device – Voice Service Measurements* (1996).

ITU-T Recommendation P.563, *Single Ended Method for Objective Speech Quality Assessment in Narrow-Band Telephony Applications* (2004).

ITU-T Recommendation P.581, *Use of Head and Torso Simulator (HATS) for Hands-Free Terminal Testing* (2000).

ITU-T Recommendation P.800, *Methods for Subjective Determination of Transmission Quality* (Aug. 1996).

ITU-T Recommendation P.810, *Modulated Noise Reference Unit (MNRU)* (1996).

ITU-T Recommendation P.830, *Subjective Performance Assessment of Telephone-Band and Wideband Digital Codecs* (Aug. 1996).

ITU-T Recommendation P.831, *Subjective Performance Evaluation of Network Echo Cancellers* (1998).

ITU-T Recommendation P.832, *Subjective Performance Evaluation of Hands-Free Terminals* (2000).

ITU-T Recommendation P.861, *Objective Quality Measurement of Telephone-Band (300–3400 Hz) Speech Codecs* (Feb. 1998).

ITU-T Recommendation P.862, *Perceptual Evaluation of Speech Quality (PESQ): an Objective Method for End-to-End Speech Quality Assessment of Narrow-Band Telephone Networks and Speech Codecs* (2001).

ITU-T Recommendation Q.23, *Technical Features of Push-Button Telephone Sets* (1988).

ITU-T Recommendation Q.24, *Multi-Frequency Push-Button Signal Reception* (1988).

ITU-T Recommendation Q.35, *Technical Characteristics of Tones for the Telephone Service* (1988).

ITU-T Recommendation Q.50, *Q.50 Signalling Between Circuit Multiplication Equipments (CME) and International Switching Centres (ISC)* (1993).

ITU-T Recommendation T.4, *Standardization of Group 3 Facsimile Terminals for Document Transmission* (1996).

ITU-T Recommendation T.30, *Procedures for Document Facsimile Transmission in the General Switched Telephone Network* (1996).

ITU-T Recommendation V.8, *Procedures for Starting Sessions of Data Transmission over the Public Switched Telephone Network* (1998).

ITU-T Recommendation V.18, *Operational and Interworking Requirements for DCEs Operating in the Text Telephone Mode* (2000).

ITU-T Recommendation V.25, *Automatic Answering Equipment and General Procedures for Automatic Calling Equipment on the General Switched Telephone Network Including Procedures for Disabling of Echo Control Devices for Both Manually and Automatically Established Calls* (1996).

ITU-T Recommendation V.110, *Support of Data Terminal Equipments (DTEs) with V-Series Interfaces by an Integrated Services Digital Network.*

ITU-T Recommendation Y.1231, *IP Access Network Architecture* (2000).

ITU-T Recommendation Y.1540, *Internet Protocol Data Communication Service – IP Packet Transfer and Availability Performance Parameters* (2000).

ITU-T Recommendation Y.1541, *Network Performance Objectives for IP-Based Services* (2001).

ITU-T SG-12 question 9, contribution 32, *Cause Analysis of Objective Speech Quality Degradations Used in SQuad* (Swissqual, Aug. 2001).

ITU-T SG-15 question 6, delayed contribution, *On Objective Method for Evaluation of Noise-Reduction Systems*, Swissqual, SQuad, (Geneva, Apr. 2002).

ITU-T, Subjective performance assessment of digital encoders using the degraded category rating (DCR) procedure, *Red Book*, **V**: annex B to supplement 14 (1984).

ITU-T Supplement no. 5 to Recommendation P. 74, The SIBYL method of subjective testing, *Red Book*, **V**.

Jaffe, J., Cassotta, L., and Feldstein, S., Markovian model of time patterns in speech, *Science*, **144** (1964), 884–886.

Jaffe, J. and Feldstein, S., *Rhythms of Dialogue* (New York: Academic Press, 1970).

Jax, P. and Vary, P., On artificial bandwidth extension of speech signals, *Signal Processing*, **83**: 8 (2003), 1707–1719.

Johnson, S., Interactive teaching: breaking television viewing habits, *The Distance Education Network Report*, **21** (1988), 4–6.

Kao, Y. H., *Low Complexity CELP Speech Coding at 4.8 kbps*, University of Maryland, Electrical Engineering Department (1990).

Karjalainen, J., A new auditory model for the evaluation of sound quality of audio systems, *IEEE International Conference on Acoustics, Speech, and Signal Processing* (1985), 608–611.

Kataoka, A., *et al.*, An 8 kbps speech coder based on conjugate structure CELP, *IEEE International Conference on Acoustics, Speech, and Signal Processing* (1993), 595–598.

Kitawaki, N. and Itoh, K., Pure delay effects on speech quality in telecommunications, *IEEE Journal on Selected Areas in Communications*, **9**: 4 (1991), 586–593.

Klatt, D. H., Linguistic uses of segmental duration in English: acoustic and perceptual evidence, *Journal of the Acoustical Society of America*, **59**: 5 (1976), 1208.

Kondoz, A. M., *Digital Speech* (John Wiley & sons, 1994).

Krauss, R. M. and Bricker, P. D., Effects of transmission delay and access delay on the efficiency of verbal communication, *Journal of the Acoustical Society of America*, **41** (1966), 286–292.

Lev-Ari, H. and Ephraim, Y., Extension of the signal subspace speech enhancement approach to colored noise, *IEEE Signal Processing Letters*, **10** (Apr. 2003), 104–106.

Lim, J. S., ed., *Speech Enhancement* (New Jersey: Prentice-Hall, Inc., 1983).

Lim, J. S. and Oppenheim, A. V., All-pole modelling of degraded speech, *IEEE Transactions on Acoustics, Speech, and Signal Processing*, **ASSP-26** (Jun. 1978), 197–210.

Lin, Yi-Bing and Chlamtac, I., *Wireless and Mobile Network Architectures* (John Wiley & Sons, 2001).

Lockwood, P. and Boudy, J., Experiments with a nonlinear spectral subtractor (NSS), hidden Markov models and the projection, for robust speech recognition in cars, *Speech Communication*, **11** (1992) 215–228.

Lotter, T. and Vary, P., Speech enhancement by MAP spectral amplitude estimation using a super-Gaussian speech model, *EURASIP Journal on Applied Signal Processing*, **2005**: 7 (2005), 1110–1126.

Mack, M. A. and Gold, B., *The Intelligibility of Non-Vocoded and Vocoded Semantically Anomalous Sentences*, technical report 703 (Boston, MA: Lincoln Laboratories, 26 Jul. 1985).

Makinen, J., Ojala, P., and Toukamaa, H., *Performance Comparison of Source Controlled AMR and SMV Vocoders*, Nokia Research Center, www.nokia.com/library/files/docs/Makinen2.pdf.

Martin, R. and Gustafsson, S., The echo shaping approach to acoustic echo control, *Speech Communication*, **20**: 3–4 (1996), S.181–190.

Martin, R., Noise power spectral density estimation based on optimal smoothing and minimum statistics, *IEEE Transactions on Speech and Audio Processing*, **9**: 5 (2001), 504–512.

McAulay, R. J. and Malpass, M. L., Speech enhancement using a soft-decision noise suppression LTER, *IEEE Transactions on Acoustic, Speech, and Signal Processing*, **ASSP-28** (Apr. 1980), 137–145.

McElroy, Ciarán., Speech coding, www.dsp.ucd.ie/speech/tutorial/speech_coding/speech_tut.html (Oct. 1997).

Mehrotra, A., *GSM System Engineering* (Boston: Artech House, 1997).

Miller, G. A., Speaking in general, review of J. H. Greenberg (ed.), *Universals of Language, Contemporary Psychology*, **8** (1963), 417–418.

Moore, B. C. J., *An Introduction to the Psychology of Hearing*, 4th edn. (Academic Press, 1997).

Mouly, Michel and Pautet, Marie-Bernadette, *The GSM System for Mobile Communications* (Telecom Publishing, 1992).

Musicus, B. R., An iterative technique for maximum likelihood parameter estimation on noisy data, S.M. thesis, M.I.T., Cambridge, MA (1979).

National Communications System, FS 1016 CELP speech coding @ 4800 bps, NCS Technical Information Bulletin.

Norwine, C. and Murphy, O. J., Characteristic time intervals in telephonic conversation, *Bell System Telephone Journal*, **17** (1938), 281–291.

Omologo, M., Svaizer, P., and Matassoni, M., Environmental conditions and acoustic transduction in hands-free speech recognition, *Speech Communication*, **25** (1998), 75–95.

Osgood, C. E., Suci, G. J., and Tannenbaum, P. H., *The Measurement of Meaning* (University Of Illinois Press, 1957).

Padgett, Jay E., Gunter, G., and Hattori, Takeshi, Overview of wireless personal communications, *IEEE Communications Magazine* (Jan. 1995), www.cs.bilkent.edu.tr/~korpe/courses/cs515-fall2002/papers/overview-wireless-pcs.pdf.

Proakis, John G. and Manolakis, Dimitris G., *Digital Signal Processing Principles, Algorithms, and Applications* (Prentice Hall, 1996).

Psytechnics Limited, PAMS, measuring speech quality over networks, white paper (May 2001), www.psytechnics.com.

Quackenbush, S. R., Barnwell III, T. P., and Clements, M. A., *Objective Measures of Speech Quality* (New Jersey: Prentice Hall Advanced Reference Series, 1988).

Rabiner, L. and Schafer, R., *Digital Processing of Speech Signals* (Prentice Hall, 1978).

Ramabadran, T., Meunier, J., Jasiuk, M., and Kusher, B., Enhancing distributed speech recognition with backend speech reconstruction, *EUROSPEECH* (Sep. 2001) 1859–1862.

Rappaport, T. S., *Wireless Communications* (Prentice Hall, Inc., 1996).

Rasalingham, Gobu, and Vaseghi, Saeed, Subband acoustic echo cancellation for mobile and hands-free phones in car, http://dea.brunel.ac.uk/cmsp/Home_Gobu_Rasalingam/Gobu_new_way.htm.

Reed, C. M. and Bilger, R. C., A comparative study of S/N_0 and E/N_0, *Journal of the Acoustical Society of America*, **53** (1973), 1039–1044.

Richards, D. L., *Telecommunication by Speech* (London: Butterworth, 1973).

Ris, Christophe, and Dupont, Stephane, Assessing local noise level estimation methods: application to noise robust ASR, *Speech Communication*, **34**: 1–2 (2001), 141–158.

Rix, A. W. and Hollier, M. P., Perceptual speech quality assessment from narrowband telephony to wideband audio, 107th AES Convention, pre-print no. 5018 (Sep. 1999).

Rix, A. W. and Hollier, M. P., The perceptual analysis measurement system for robust end-to-end speech quality assessment, *IEEE International Conference on Acoustics, Speech, and Signal Processing* (Jun. 2000).

Rix, A. W., Reynolds, R., and Hollier, M. P., Perceptual measurement of end-to-end speech quality over audio and packet-based networks, 106th AES convention, pre-print no. 4873 (May 1999).

Rysavy, P., Voice capacity enhancements for GSM evolution to UMTS, white paper (Jul. 18, 2002).

Salami, R. A., Hanzo, L., *et. al.*, Speech coding, in R. Steele (ed.), *Mobile Radio Communications* (Pentech, 1992), 186–346.

Schmidt-Neilson and Everett, S. S., *A Conversational Test for Comparing Voice Systems Using Working Two-way Communication Links*, Naval Research Laboratory, Washington D.C., NRL report no. 8583 (Jun. 1982).

Schmidt-Neilson, *Voice Communication Testing Using Naive and Experienced Communicators*, Naval Research Laboratory, Washington, D.C., NRL report no. 8655 (3 Feb. 1983).

Schmidt-Neilson and Kallman. H. J., *Evaluating the Performance of the LPC 2.4 kbps Processor with Bit Errors Using a Sentence Verification Task*, Naval Research Laboratory, Washington, D.C., NRL report no. 9089 (30 Nov. 1987).

Schnitzler, J. and Vary, P., Trends and perspectives in wideband speech coding, *Signal Processing* **80** (2000), 2267–2281.

Schobben, D. W. E., *Real-Time Adaptive Concepts in Acoustics: Blind Signal Separation and Multi-Channel Echo Cancellation* (Kluwer Academic Pub., 2001).

Schroeder, M. R., Atal, B. S., and Hall, J. L., Optimizing digital speech coders by exploiting masking properties of the human ear, *Journal of the Acoustical Society of America*, **66**: 6 (1979), 1647–1652.

Schroeder, M. R. and Atal, B., Code-excited linear prediction (CELP): high quality speech at very low bit rates, *IEEE International Conference on Acoustics, Speech, and Signal Processing* (Tampa, Apr. 1985), 937.

Shankar, H., EDGE in wireless data (Aug. 2005), www.commsdesign.com/main/2000/01/0001feat2.htm.

Sharpley, A. D., 3G AMR NB characterization experiment 1A – Dynastat results, 3GPP TSG-S4#14 meeting (Bath, UK, Nov. 27–Dec. 1, 2000).

Skyworks Solutions, Selectable mode vocoder (SMV), enhanced voice quality and capacity for CDMA networks (May 2004), www.skyworksinc.com/docs/smv/smv_bd.pdf.

Sondhi, M. M., An adaptive echo canceller, *Bell Systems Technical Journal*, **46** (Mar. 1967), 497–510.

Spanias, A., Speech coding: a tutorial review, *Proceedings of the IEEE*, **82**: 10 (1994), 1541–1582.

Steeneken, H. J. M. and Houtgast, T., A physical method for measuring speech-transmition quality, *Journal of the Acoustical Society of America*, **67** (1980), 318–326.

Stevens, S. S., *Psychophysics – Introduction to its Perceptual, Neural and Social Prospects* (John Wiley & sons, 1975).

Su, Huan-yu, Shlomot, Eyal, and Nakamura, Herman K., *Selectable Mode Vocoder Emerges for cdma2000 Designs*, Mindspeed Technologies, Inc., CommsDesign.com (Jan. 07, 2003).

Tanner, R. and Woodard JP. (ed.), *WCDMA: Requirements and Practical Design* (John Wiley & Sons, 2004).

Tardelli, Kreamer, La Follette, and Gatewood, *A Systematic Investigation of the Mean Opinion Score (MOS) and the Diagnostic Acceptability Measure (DAM) for Use in the Selection of Digital Speech Compression Algorithms* (ARCON Corp., Sep. 1994).

Tardelli, J. D., Sims, C. M., LaFollette, P. A., and Gatewood, P. D., *Research and Development for Digital Voice Processors*, R86-01 W, ARCON corporation (30 May 1986); and final report (Jan. 1984–Feb. 1986).

Thiede, T., Treurniet, W. C., Bitto, R., *et al.*, PEAQ–the ITU standard for objective measurement of perceived audio quality, *Journal of the Audio Engineering Society*, **48**: 1/2 (2000), 3–29.

TIA-728, *Inter System Link Protocol (ISLP)* (Jan. 2002).

TIA TR-45.5 subcommittee, *TDMA IS-136 (Time Division Multiple Access) Mobile Telephone Technology* (1994).

TR 25.953, *TrFO/TFO R3*, release 4 version 4.0.0.

Treurniet, W. C. and Soulodre, G. A., Evaluation of the ITU-R Objective Audio Quality Measurement Method, *Journal of the Audio Engineering Society*, **48**: 3 (2000), 164–173.

TS 02.53, *Tandem Free Operation (TFO); Service Description; Stage 1*, release 99 version 8.0.1.

TS 03.53, *Tandem Free Operation (TFO); Service Description; Stage 2*, release 99 version 8.0.0.

TS 08.62, *Inband Tandem Free Operation (TFO) of Speech Codecs; Service Description; Stage 3*, release 99 version 8.0.1.

TS 22.053, *Tandem Free Operation (TFO); Service Description; Stage 1*, release 4 version 4.0.1 or release 5 version 5.0.0.

TS 23.053, *Tandem Free Operation (TFO); Service Description; Stage 2*, release 4 version 4.0.1 or release 5 version 5.0.0.

TS 28.062, *Inband Tandem Free Operation (TFO) of Speech Codecs; Service Description; Stage 3*, release 4 version 4.5.0 or release 5 version 5.3.0.

Tuffy, M., The removal of environmental noise in cellular communications by perceptual techniques, Ph.D. thesis, Department of Electronics and Electrical Engineering, University of Edinburgh (Dec. 1999).

Vary, P., Noise suppression by spectral magnitude estimation – mechanism and theoretical limits, *Signal Processing*, **8**: 4 (1985).

Vary, P., An adaptive filterbank equalizer for speech enhancement, *Signal Processing*, special issue on applied speech and audio processing, 2005.

Vary, P. and Martin, R., *Digital Speech Transmission – Enhancement, Coding and Error Concealment* (Wiley, 2005).

Vaseghi, S. V., *Advanced Signal Processing and Digital Noise Reduction* (John Wiley and Teubner, 1996).

Virag, N., Single channel speech enhancement based on masking properties of the human auditory system, *IEEE Transactions on Speech and Audio Processing*, **7** (1999), 126–137.

Voiers, W. D., *Methods of Predicting User Acceptance of Voice Communication Systems*, Dynastat Inc., final report, 10 Jun. 1974–30 Jun. 1976 (15 Jul. 1976).

Voiers, W. D., Diagnostic acceptability measure for speech communication systems, *Proceedings of the IEEE International Conference on Acoustics, Speech, and Signal Processing* (May 1977), 204–207.

Voiers, W. D., *Exploratory Research on the Feasibility of a Practical and Realistic Test of Speech Communicability*, Dynastat inc., final report (Apr. 1978).

Voiers, W. D., Evaluating processed speech using the diagnostic rhyme test, *Speech Technology*, **1**: 4 (1983), 30–39.

Voiers, W. D., Cohen, M. F., and Mickunas, J., *Evaluation of Speech Processing Devices, I. Intelligibility, Quality, Speaker Recognizability*, final report AF19-628-4195, AFCRL (1965).

Voiers, W. D., Sharpley, A. D., and Hehmsoth, C. J., *Research on Diagnostic Evaluation of Speech Intelligibility*, final report AFCRL-72-0694, AFCRL (24 Jan. 1973).

Volker, S., Fischer, A., and Bippus, R., Quantile based noise estimation for spectral subtraction and Wiener filtering, *IEEE International Conference on Acoustics, Speech, and Signal Processing* (2000), 1875–1878.

Wan, E. A. and Nelson, A. T., Removal of noise from speech using the dual EKF algorithm, *IEEE International Conference on Acoustics, Speech, and Signal Processing* (1998), 381–384.

Wang, S., Sekey, A., and Gersho, A., An objective measure for predicting subjective quality of speech coders, *IEEE Journal on Selected Areas in Communications*, **10**: 5 (1992), 819–829.

Ward, W. C. G., Elko, W., Kubli, R. A., and McDougald, W. C., Residual echo signal in critically sampled sub-band acoustic echo cancellers, *IEEE Transactions on Signal Processing*, **45**: 4 (1997).

Webb, W. T. and Hanzo, L., *Modern Quadrature Amplitude Modulation: Principles and Applications for Fixed and Wireless Communications* (IEEE Press–Pentech Press, 1994).

Webster, J. C., *A Compendium of Speech Testing Material and Typical Noise Spectra for Use in Evaluating Communications Equipment*, Naval Electronics Laboratory Center, San Diego, CA. (Sep. 1972).

Wheddon, C. and Linggard, R., *Speech and Language Processing* (Chapman and Hall, 1990).

Wisniewski, J. S., *The Colors of Noise*, www.ptpart.co.uk/colors.htm (Oct. 1996).

Wong, K. H. H. and Hanzo, L., Channel coding, in R. Steele (ed.) *Mobile Radio Communications* (London: Pentech Press, 1992), 347–488.

Wu, K. and Chen, P., Efficient speech enhancement using spectral subtraction for car hands-free application, *International Conference on Consumer Electronics*, **2** (2001), 220–221.

Xu, W., Heinen, S., Adrat, M., *et al.*, An adaptive multirate (AMR) codec proposed for GSM speech transmission, *International Journal of Electronics and Communications (AEÜ)*, **54**: 6 (2000).

Yost, W. A., *Fundamentals of Hearing – an Introduction* (Elesvier Academic Press, 2000).

Zwicker, E. and Fastl, H., *Psychoacoustics: Facts and Models* (New York: Springer-Verlag, 1990).

Index

2100 Hz 11, 63, 130, 80, 136, 154, 155, 156, 211, 212, 213
3G 11, 17, 63, 130
A Law 79, 151
ACELP 16–18, 24
ACR 32, 39, 40, 41, 42
Adaptive excitation signal 13
ADPCM 22
Alcatel 3
AMPS 23
AMR 17–18, 23, 24, 35, 48, 255, 307
Analog Communications 11
Analysis by synthesis (AbS) 13
ATM voice platform 3, 166, 273, 279
audio transparency 268
average rate per user (ARPU) 256

Background music 78, 86, 112, 128, 129–130, 185, 189, 207, 217, 267–269, 283, 284
bi-directional NR 130, 132, 196–197
black noise 99

C5 279–281
CDMA codecs 36–37
    QCELP 15, 17, 23, 24, 36
    EVRC 17, 24, 36, 126, 255, 282, 307
    SMV 18–24, 36, 282
CDMA, cdmaOne 11, 18–24, 112, 114, 161, 163, 165
cdma2000 11, 17–18, 24, 165, 168
cdmaTwo 167
center clipper 268
center clipper transfer function 78
Chinese service providers 3
Code excited linear prediction (CELP) 15, 22, 44
comfort-noise injection
    colored noise matching 64
Comfort-noise injection, Comfort-noise matching 4, 21, 43, 51, 63, 64, 96, 105, 107, 124, 176, 179, 180, 183, 189, 271, 272, 279–281, 305, 308, 309
CSS 176–183

D-AMPS 23, 24
DCME 43
DCR 32
Dispersion 56, 85, 187, 285
Ditech Communications 2, 3, 250
DMOS 31, 32–37, 255
double talk 65, 71, 72, 73, 80, 94, 179–181, 184, 185–187, 190, 203, 217, 266, 278, 283, 284
DSN 48, 191, 192–202, 199, 200
DTMF 128, 129, 130, 136, 207–209, 211, 219, 272, 278, 291–293, 294, 296, 298, 299, 305, 311, 312
DTX 20–22, 43, 113, 124–125, 272
    VAD 21
    VOX 21

echo-path delay, tail delay 69, 72–73, 78, 169, 180, 284, 285
EDGE 168
EFR 18, 24, 28, 36, 89, 195, 255, 307
E-Model, G.107 43, 45–46
Ericsson 2
ERLE 65, 70, 71, 87, 89, 139, 178, 179, 263

FDMA 23
FFT 95, 106, 117, 120, 123, 176, 212
Flat delay 76
forward-backward procedure 71
FR 28, 89, 195, 307

G.160 47–48, 191, 197, 198, 199, 200
G.164 68, 80, 130, 211, 273
G.165 68, 80, 130, 211
G.167 185–187
G.168 46–47, 68, 88, 176–183, 184, 273, 285
G.169 48, 136, 203
G.711 89, 159, 161
global noise 43
GSM 11, 14, 15–26, 28, 89, 112, 155, 161, 164, 165, 166
GSM codecs 4, 16–20, 24, 29, 32–37, 89

h-register 72, 73, 81, 180, 181, 183, 266, 267
hangover time 60
hot signal 144, 268, 278
Hoth noise source 179
HR 28, 89, 159, 195, 255, 307
HSCSD 213, 273
Hybrid 52, 56, 58, 60, 63, 66, 68, 177, 180,
    181, 182, 185, 187, 255, 285
Hybrid coding 22

IFFT 95, 120, 212
InterWorking Function (IWF) 153–156, 213
intrusive measures 42, 195
Isurus Market Research 256

KDDI 3
Korean service providers 3

leak rate 71, 181, 267
Least squares method 13, 73
Linear prediction (LP) 13
Linear predictive coding (LPC) 13, 22
Literature survey xiii
    Books on Noise Reduction xiii
    Books on Echo Cancelation xiii–xiv
    Books on Acoustic\ Echo Control xiv
local noise 42–43
Long-term prediction (LTP) 13
loudness loss 42
Lucent 2, 249, 251

μ law 79, 151
MNB 41
MOS 29, 30, 32–37, 39, 41, 194–201, 220
MOS interpretation of R factor 46, 194
MOS-LQO 29, 30, 32, 37, 42–43, 194
MOS-LQS 29, 42
musical noise 282
My Sound 7, 294

Near-end speech
    detector 71
Nextel 3
NLMS 72, 73
NMS Communications 256
noise clipping 64, 83, 86, 106, 111, 199, 255
Nokia 2
non-intrusive procedure, P.563 42–45, 195, 218
NPLR 31, 48, 191, 199
NTT DoCoMo 3, 16

One-way communications 42, 198, 199, 200
orange noise 99

P.563 43–45, 195
PAMS 39, 40

Per call control, PCC 81, 155, 159
PESQ, P.862 30–31, 40–42, 43, 44, 45–46,
    195, 196, 197–199, 300
phase reversal 81, 211–212
pink noise 99
Plain old telephone service (POTS) 11
PNLMS 72, 73
pooled configuration 250, 251
PSI-CELP 16
PSQM, P.861 38–39, 41
Pulse-code modulation (PCM) 12, 22, 89, 108
    Toll Quality 12
    Waveform Coding 12, 22

Q.50 81
QCELP-13 15, 17, 24
QCELP-8 15, 24
Qualcomm CELP 15
quantization 123, 159

R factor 44, 45–46
RCELP 17
    EVRC 17, 21, 24
red noise 99
Regular pulse excited (RPE) 14
residual error 62, 180–182, 203
Ring back tone 86, 128–129, 209, 311
RPE-LTP 14, 15–16, 23, 24

saturation, saturated speech 136, 142, 144, 310
shadow echo canceler 71
Siemens 3, 250
singing side-tone 53, 54
SMV 18–20, 21, 23, 24, 126, 307
    eX-CELP 19
    VAD 19, 20
    Noise suppression 20, 21–22
SNR, S/N 31, 34, 37, 43, 113–114, 120, 123,
    126, 130, 141
SNRI 31, 48, 191, 192–202, 199, 200
Source or parametric codecs 22
speech clipping 43, 60, 63, 65, 113, 144, 151,
    179, 184, 185–187, 189, 190, 277
SPLR 31, 199
SQuad 32, 195, 196–197, 199
straight-tone disable 80
sub-band NLP 94, 95
SwissQual 32, 191, 199
switch integrated 250–252, 254

talk-off 278
tandem configurations of echo cancelers 70
Tandem treatment of a single impairment xi
TDMA 124, 161, 163, 165, 166, 168
Tellabs 2, 250
TFO 25, 159–162, 213–215

Theory of sleep 7, 301
T-Mobile 3, 249
TNLR 31, 48, 191, 192–202
Tracer Probe 6, 289
TRAU 25, 108, 110, 148, 151, 153, 159, 161,
    261–265
tripper's trap 169, 170
TrueVoice 2
TrueVoice2 2

UMTS 35

V.110 273
VAD 42, 43, 113, 116, 124, 125, 139, 278
Verizon Wireless 3
via net loss 58
VoATM 55, 169, 274, 279
Voice signal classifier (VSC) 212, 273

voicemail 129
VoIP 41, 55, 82, 169, 171, 274, 279, 289,
    293, 296
VoIP PBX xi, 274, 275
VoIP voice platform 3
VOP 55, 72, 165, 169, 172, 252, 284, 285
VPA 72, 75, 81, 279
VQS evolution to 3G 5, 163
VSELP 15–16, 24
    D-AMPS 15, 16
    PDC 16

Walsh code 167, 168
W-CDMA 11, 17, 164, 165, 168
Wireless network architectures 5, 163–172,
    283, 274, 294, 296

Zero Mile Policy 1